NANOMATERIALS IN ENERGY DEVICES

Energy Storage Derivatives and Emerging Solar Cells

NANOMATERIALS IN ENERGY DEVICES

Energy Storage Derivatives and Emerging Solar Cells

Editor

Hieng Kiat Jun

Department of Mechanical and Material Engineering
Universiti Tunku Abdul Rahman
Sungai Long Campus, Jalan Sungai Long
Bandar Sungai Long
43000 Kajang, Selangor D.E., Malaysia

CRC Press is an imprint of the
Taylor & Francis Group, an informa business
A SCIENCE PUBLISHERS BOOK

Cover Credit: Cover images have been taken from Chapter 7 and are reproduced by kind courtesy of Dr. Pei Yi Chan (author of Chapter 7).

CRC Press
Taylor & Francis Group
6000 Broken Sound Parkway NW, Suite 300
Boca Raton, FL 33487-2742

First issued in paperback 2021

Version Date: 20171012

ISBN-13: 978-0-367-78168-2 (pbk)
ISBN-13: 978-1-4987-6351-6 (hbk)

Library of Congress Cataloging-in-Publication Data

Names: Kiat, Jun Hieng, editor.
Title: Nanomaterials in energy devices / editor Jun Hieng Kiat, Department of Mechanical and Material Engineering, Universiti Tunku Abdul Rahman, Sungai Long Campus, Jalan Sungai Long, Bandar Sungai Long, Kajang, Malaysia.
Description: Boca Raton, FL : CRC Press, [2017] | Includes bibliographical references and index.
Identifiers: LCCN 2017027220| ISBN 9781498763516 (hardback : acid-free paper) | ISBN 9781498763523 (e-book)
Subjects: LCSH: Energy storage--Equipment and supplies. | Nanostructured materials.
Classification: LCC TK2896 .N355 2017 | DDC 621.31/24240284--dc23
LC record available at https://lccn.loc.gov/2017027220

Visit the Taylor & Francis Web site at
http://www.taylorandfrancis.com

and the CRC Press Web site at
http://www.crcpress.com

Preface

We live in a technological era. Without doubt, our lives are bound up with computer science, telecommunication, manufacturing and energy, whether directly or indirectly. The energy field is one of the most discussed and researched area nowadays. When nanotechnology and nanoscience are applied in the field of energy, we have a very promising and high potential for application in a wide range of energy devices. Such potential is seen in production or conversion and storage of energy by utilizing the fundamental of nanotechnology.

This book is intended to bring forth some important developments in the research of nanoscaled materials for the application in energy devices. It aims to summarize the fundamentals and novel techniques of nanoscale materials applied in energy devices. Since the field is developing and accelerating at a fast pace, it is impossible for this book to cover all types of energy devices (production or storage); and it is not the intention of the editors to cram everything into a single volume as well.

In this volume, the central theme is energy storage and conversion. This narrows down to the application of batteries and their derivatives as well as light-to-electric conversion devices. Each chapter presents not just different applications but also different methods of fabrication as well as the material's characteristics. All the chapters are a unique combination of these approaches. This is because each application itself can have a diverse fabrication method and nanomaterials used. Nevertheless, the focus is on the application of nanoscale materials which are responsible for the process and functionality of the energy devices. More importantly, all the chapters are contributed by prominent researchers from academia.

It is hoped that this book will serve as a general introduction to anyone who is just entering the field of energy devices utilizing nanoscaled materials. It is also for experts who are seeking up-to-date information in this field. This book is not a handbook or a comprehensive review of a particular subject. In brief, Chapter 1, which is an introduction, gives an overview of the field of nanomaterials in energy devices. The following chapter focuses on the application of biopolymer materials as solid electrolytes in dye-sensitized solar cell. Chapter 3 gives a short review on the current status of nanomaterials in lithium ion batteries while the subsequent chapter brings us the specific application in electric double layer capacitors. Chapter 5 introduces the various nanomaterials used in emerging solar cells. This is followed by the introduction of metal-organic frameworks for the application of hydrogen storage. Finally, the last chapter delves into general application of nanomaterials in electrochemical capacitors.

The editor would like to express his sincere gratitude to all the contributors of this book, whose excellent support resulted in the successful realization of this collaboration. Their passion and support in sharing of knowledge are commendable. The editor would also like to thank the publisher, Science Publishers, an imprint of CRC Press, for recognizing the demand for such niche theme for the book.

August 2017 **Hieng Kiat Jun**

Content

1

Introduction to Nanomaterials in Energy Devices

Hieng Kiat Jun

1. A Brief Introduction to Current Scenario

According to the report released by the World Energy Council, global energy demand will increase from year to year. In fact, the demand for energy will double by the year 2060. However, the primary energy demand will peak until year 2030. This primary energy source includes energy derived from coal, oil and gas. The slow growth of the primary energy is forecast due to the disruptive trends of the emerging energy sources. Coupled with environmental concern like global warming, the adoption of emerging energy sources is beginning to encroach into the dominance of the primary energy source. Generally, these emerging energy sources are referred to as alternative energy resources which are not derived from fuel and coal. Some examples of such alternative energy sources include (but not limited to) solar energy, hydro energy and wind energy. The ultimate purpose of the energy demand is to be able to power up various equipments, be it portable or stationary, in order to achieve certain tasks. Therefore, it is the goal of each inventor and researcher to design an efficient energy device where maximum useful energy can be extracted.

Department of Mechanical and Material Engineering, Universiti Tunku Abdul Rahman Sungai Long Campus, Jalan Sungai Long, Bandar Sungai Long, 43000 Kajang, Selangor D.E., Malaysia.
E-mail: junhk1@gmail.com

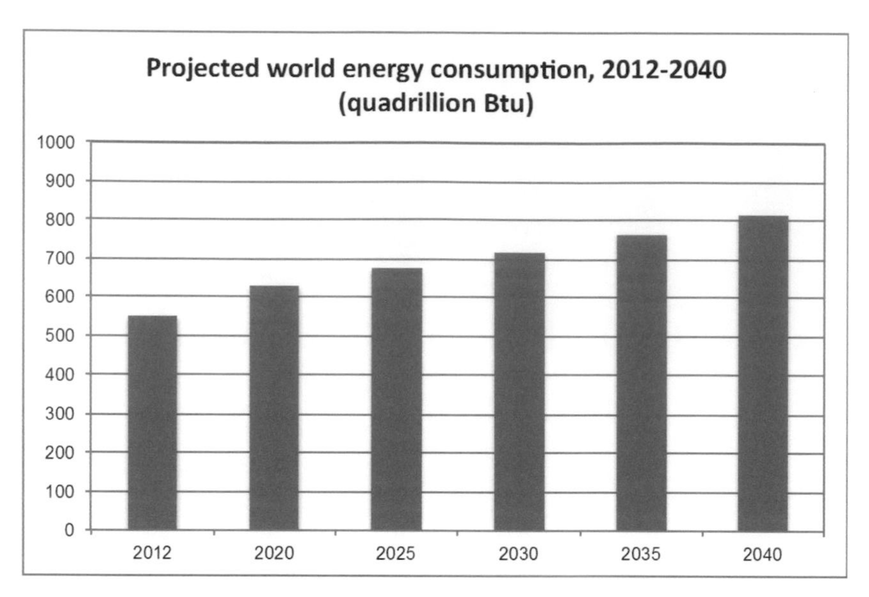

Figure 1. Projected world energy consumption (source: EIA, International Energy Outlook 2016).

2. Energy Devices

The ability of a system to produce useful energy is tantamount to an efficient energy converter system. Such energy converter system usually produces ready-to-use energy. Energy can be in various forms such as chemical energy, thermal energy, kinetic energy and electric energy. These types of energies are useful for performing work. In some cases, there will be surplus energy. In such an event, we may need to store the energy for future use. The system that allows us to convert, produce and store a specific energy type is termed as an energy device. Generally, the energy device will be connected externally for the supply of electric energy. In a reverse process, external circuit maybe connected to the device as well for converting and storing the energy into the device.

Due to a rising need for low cost, reliable and accessible energy, low-priced and efficient energy devices are the key for the mass adoption. Such avenues could reduce our dependency on fossil fuels (i.e., coal, petroleum and natural gas). It could also transform the current way of production, delivery and utilization of energy. In the long run, it will bring benefits to the environment due to low or near zero green house gases emission. Therefore, the development of new or improved energy devices plays an important role for the energy revolution.

A desirable energy system should be able to convert a raw energy source into useful energy. For example, a solar cell is used to convert

Figure 2. Types of energy devices used to convert and supply energy. From left to right: solar panel, wind turbine and hydro dam.

sunlight into electric energy. Generally, the working mechanism of an energy device is based on fundamental physical principles. From the definition of thermodynamics, energy cannot be created or destroyed in a system. It can only be converted from one form to another form. The law also states that the flow within the system is unidirectional until the achievement of equilibrium at the final state. These approaches are applied in every energy device system.

The requirement for large effective area of an energy device prompts the researchers to look for a system that is cost effective yet efficient. This is particularly important when portable energy device is in demand. One of the methods is to scale down the size of the active layer in the energy device where it is responsible for energy production and conversion processes. Such technique involves the application of nanotechnology.

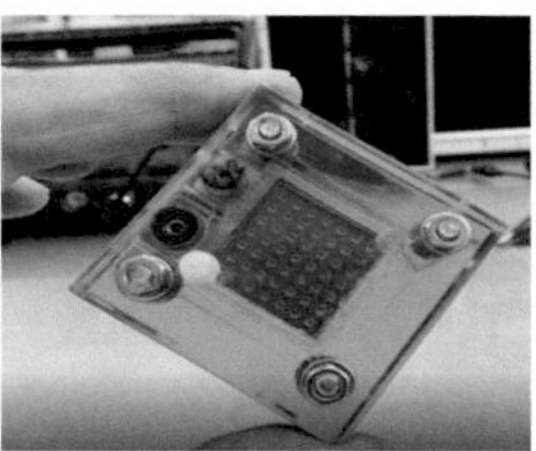
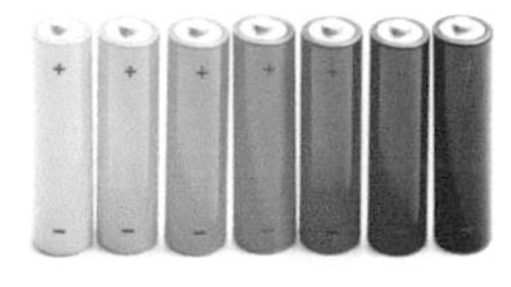

Figure 3. Examples of small energy devices which utilized nano-sized materials as their active materials: third generation solar cells, fuel cells and batteries.

3. Nanomaterials

Nanotechnology is a general term used to define the technology of design, fabrication and application of nano-sized materials (also known as nanomaterials). Typically, nanomaterials involve the manipulation of the materials properties at extremely small scale, ranging from 1–100 nm. As the size of the materials reduced, the surface area is increased which could lead to better volume exchange for the chemical reaction or process. Large surface area is also beneficial for creating effective pathway for charge transport through the interfaces. When nano-sized materials are applied into energy devices, especially for devices that require energy conversion through porous membrane or layer, the nano-size of the active materials will have an impact on the performance of the energy conversion. This is evidenced from the varying performance result in nanostructured solar cell (Jun et al. 2014). For instance, in dye-sensitized solar cell, the porosity of the titania on the working electrode is crucial for the penetration and the attachment of dye molecules onto the titania. The porosity in this case can be manipulated by reducing the particle size of the titania (Fig. 4).

In another example of semiconductor quantum dot-sensitized solar cell, the size of the semiconductor quantum dot will determine the band gap energy of the semiconductor itself due to quantum confinement effect (Jun et al. 2013). This will indirectly influence the light absorption range of the semiconductor materials as well. Thus, solar cells which have different sizes of semiconductor quantum dots will produce different range of performance. Apart from altering the particle size, the surface properties of the particles can be engineered where different functional groups could be attached onto the particles. The addition of functional groups could elevate or improve the physico-chemical properties of the materials. Essentially, the overall performance of that energy device will change.

The effect of the nano-size materials can be demonstrated from the following example where different commercial brands of titania were used to prepare the same type of dye-sensitized solar cell (DSSC). DSSC is an emerging solar cell technology where molecular structure of dye is used as light harvester. During the dye attachment process, the dye molecules will penetrate the porous titania and attach onto the titania particles. The challenge in this device structure is to have a good level of porosity to enable

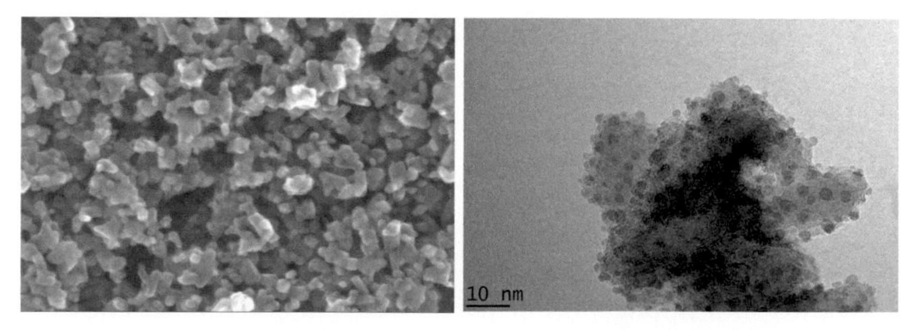

Figure 4. Left: FESEM image of titania structure at 200 k magnification. Right: The same titania structure as viewed from TEM. The "dots" attached on the titania particles are quantum dot semiconductor.

large amount of dye molecules attached onto the titania structure. Different commercial brands of titania exhibited different average titania particle size and its porosity level (Fig. 5). Thus, when all the titania electrodes are further processed for dye attachment under the same condition, it was assumed that the same amount of dye molecules would be attached on the titania surface. However, upon verification of the assembled DSSC, all the samples had different cell performance (Fig. 6 and Table 1). Such difference

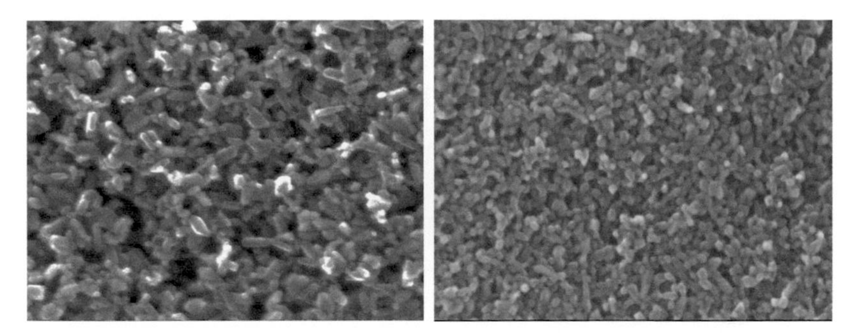

Figure 5. FESEM images (200 k magnification) of titania structure of two different commercial brands.

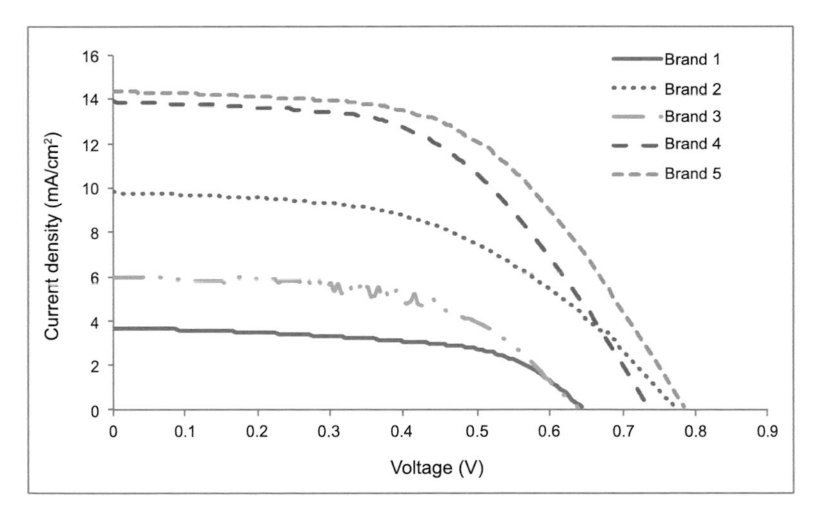

Figure 6. Current density vs. potential for dye-sensitized solar cells with various brands of titania used.

Table 1. Efficiency of dye-sensitized solar cells with various brands of titania used.

	Brand 1	Brand 2	Brand 3	Brand 4	Brand 5
Efficiency (%)	1.4	3.7	2.0	5.4	6.1

in terms of cell performance could be attributed to distinct nanostructure properties of the titania used.

In this book, five chapters are dedicated to the application of nanomaterials in different types of energy devices. It is not the intention of this book to cover every type of energy devices but rather on the selected few based on the impact in recent years. The content of this book is mainly focused on energy conversion devices such as third generation solar cells, lithium ion batteries, and electric double layer capacitors. One chapter is dedicated to fuel storage. In this case, hydrogen fuel is referred. All the devices discussed in the following chapters utilized the intrinsic properties of the nanomaterials. In most cases, polymer materials have a substantial role in energy devices as they can be applied as electrolyte and electrode materials due to their versatility and ease of manipulating the properties *via* alteration of the nanostructure. The engineering of polymer electrolyte promotes the performance enhancement of the related energy devices.

4. Acknowledgments

Part of the work has benefited from the grant awarded by the Malaysia's Ministry of Higher Education (Grant no. FRGS/1/2015/SG06/UTAR/02/1).

References

Jun, H.K., M.A. Careem and A.K. Arof. 2013. Quantum dot-sensitized solar cells—perspective and recent developments: a review of Cd chalcogenide quantum dots as sensitizers. Renew. Sustainable Energy Rev. 22: 148–167.

Jun, H.K., M.A. Careem and A.K. Arof. 2014. Fabrication, characterization and optimization of CdS and CdSe quantum dot-sensitized solar cells with quantum dots prepared by successive ionic layer adsorption and reaction. Int. J. Photoenergy. Volume 2014, Article ID 939423, 14 pages.

https://www.worldenergy.org/work-programme/strategic-insight/global-energy-scenarios/ (Accessed: 1 Feb 2017).

2

Recent Scenario of Solid Biopolymer Electrolytes Based Dye-Sensitized Solar Cell

Rahul Singh, Pramod K. Singh** and *B. Bhattacharya*

1. Introduction

Polymer electrolytes are considered as one of the most important materials used in fabricating many electrochemical devices (MacCallum et al. 1987, 1989, Gray et al. 1991, Singh et al. 2011, Singh et al. 2012, Steinbuchel et al. 2003, Chandra et al. 1998, Özdemir et al. 2007, Amelia et al. 2014, Lenz et al. 1993, Stephen et al. 2006). Polymers are broadly classified into two main categories: (1) synthetic and (2) natural. Most of the electrochemical devices available in market are based on liquid electrolyte; however, solid polymer electrolytes (SPEs), which come under the category of synthetic polymers, are offering more advantages over liquid electrolytes, such as high energy density, flexible geometry, higher temperatures of operation and safety, no-leakage of electrolyte and ease of application. Due to wide variety of synthetic polymers during the last three decades, different SPEs have been studied extensively. The most commonly studied polymer electrolytes are the complexes of metal salts with high molecular weight polymer polyethylene oxide (PEO) (MacCallum et al. 1987, 1989, Gray et al. 1991). PEO qualifies as a host polymer for electrolytes because of its ability to

Material Research Laboratory, Department of Physics, School of Basic Sciences and Research, Sharda University, G. Noida 201 310, India.
E-mail: b.bhattacharya@sharda.ac.in
* Corresponding authors: singhrs1195@gmail.com; pramodkumar.singh@sharda.ac.in

form complex with various metal cations and compatibility with electrodes (Singh et al. 2011). However, one of the major drawbacks of PEO-based solid polymer electrolytes is their low ionic conductivity (10^{-7} S/cm) at ambient temperature, which limits their practical applications (Singh et al. 2012, Steinbuchel et al. 2003). To this date, a large number of other synthetic polymers, such as Polymethyl methacrylate (PMMA), Poly Acrylonitrile (PAN), Poly(vinylidene fluoride-co-hexafluoropropylene) (PVdF-HFP), Polyvinyl alcohol (PVA), and Polyvinylpyrrolidone (PVP), etc., have been studied for electrolyte applications (Chandra et al. 1998, Özdemir et al. 2007, Amelia et al. 2014, Lenz et al. 1993).

Biopolymer electrolyte (BPE) falls into synthetic polymers category. In developing countries, environmental pollution by synthetic polymers is becoming a serious threat. Petroleum-derived plastics are not biodegradable. They do not undergo microbial degradation and hence accumulated in the environment. At the same time, the price of fossil fuel has increased tremendously. These are the major issues why the research focus is now being shifted to biodegradable polymers. Initially introduced in 1980s, biodegradable plastics and polymers from renewable resources have attracted an increasing amount of attention over the last two decades. In recent times, a vast number of biodegradable polymers have been synthesized (Stephen et al. 2006, Finkenstadt et al. 2005, Chamy et al. 2013, Dumitriu et al. 1998, Rees et al. 1977). Biopolymer electrolytes (BPE) are solid ion conductors formed by dissolving salts in polymers having high molecular weight. They can be prepared in semi-solid or solid form, which is a cheap and reliable process (Habibi et al. 2012, Finkenstadt et al. 2004, 2005, Nijenhuis et al. 1997, Campo et al. 2009, Augst and Rinaudo 2006, Kalia et al. 2011, Le and Sequeira 2010, Ohno et al. 2005). BPE materials possess high ionic conductivity which is essential for any electrochemical device, like fuel cell (Wang et al. 2011), supercapacitors, batteries (Stephan and Fonseca 2006, Ozer et al. 2002) dye-sensitized solar cell (Gratzel et al. 2003, 2000, 2001, 2009, Bhattacharya et al. 1996, Gerischer et al. 1976, Skotheim et al. 1985, 1981, Ellis et al. 1976, Meyer et al. 2010, Robertson et al. 2006, Yella et al. 2011, Hagfeldt et al. 2000, Toyoda et al. 2004, Li et al. 2006, ORegan et al. 1991, Hardin et al. 2012, Narsaisah et al. 1995, Yohannes et al. 1998, Jung et al. 2010, Mathew et al. 2014, Bai et al. 2008, Longo and Haque 2003, Kim et al. 2004a, 2005, Kang et al. 2003, 2005, 2007, 2006, 2008, Ren and Katsaros 2002, Nogueira et al. 1999, 2001a, 2004, Kalaignan et al. 2006, Chatzivasiloglou and Han 2005, Stergiopoulos et al. 2002, Chen et al. 2007, Anandan et al. 2006, 2008, Freitas et al. 2008b, 2009). For any efficient electrochemical device, the prime condition is to achieve high conductivity, i.e., liquid like conductivity (10^{-2} to 10^{-4} S/cm) which can easily be obtained in these BPE (Ohno et al. 2005, Ileperuma et al. 2004, Kumar et al. 2002, Croce et al. 1998, Preechatiwong et al. 1996, Wang et al. 2005, Yang and Gorlov

2008, Iwakia et al. 2012, Chandra et al. 1981, Agarwal et al. 2008, 1999). One of the main objectives in polymer research is the development of stable bio-polymeric systems with excellent electrical and mechanical properties.

Biopolymer based devices have the advantages of high ionic conductivity, high energy density, solvent-free condition, leak proof, wide electrochemical stability windows, easy processability and lightweight. Good quality with mechanically strong semi-crystalline films could be obtained but unfortunately, in most cases the obtainable conductivity is between 10^{-4} and 10^{-5} S/cm while the device requirements for conductivity values are at least 2–4 orders of magnitude higher than this value. To obtain high ionic conductivity, various approaches have been adopted as reported in literature, i.e., copolymerization, plasticization, dispersion of organic/inorganic and nanoparticles, dispersion of low viscosity ionic liquid (MacCallum et al. 1987, 1989, Gray et al. 1991, Singh et al. 2012). Due to the wide spread application of BPE, we confine this review towards the application of BPE in dye-sensitized solar cell and battery area.

The first generation of solar cells was based on crystalline silicon. With the next generation, thin film solar cells were developed while the third generation solar cell came after the development of their previous counterparts. Normally, 3rd generation solar cells are based on nanoparticles and polymers. The two main members of this group are dye-sensitized solar cells (DSSC) and organic photovoltaic cells (OPV). As we have mentioned above, due to broad scope of BPE in electrochemical devices, we will focus our attention only to DSSC and batteries area. Dye-sensitized solar cells have been under extensive research for the last 24 years, which were introduced by Prof. Grätzel in year 1991 (Gratzel et al. 1991, 2001). The concept of DSSC is believed to reduce the production costs and energy payback time significantly as compared to standard silicon. The conversion efficiency varies between 6–13% depending on the module size and technology. The highest conversion efficiency till date for DSSC was achieved by using liquid electrolyte but due to the leakage and packaging problem, researchers focused their attention towards solid electrolytes. The quasi gel electrolyte is proposed as a novel alternative but liquid entrapped in pores creates a problem and hence affects the life time of DSSC (Bai et al. 2008, Gratzel et al. 2009, Longo et al. 2003, Kim 2004). The next proposed alternative is solid polymer electrolyte in which polyether has played a dominant role (Haque et al. 2003, Anandan et al. 2006, 2008, Ganesan et al. 2008b, Freitas et al. 2008, 2009, Flores et al. 2007, Han et al. 2005, Zhang et al. 2007a,b). Biopolymer-salt complex is also presenting itself as a novel alternative since most of the biopolymers have high conductivity as well as easy complex formation with salts (Singh et al. 2013, Hsu et al. 2013, Guo et al. 2013, Siti et al. 2013, Weijia et al. 2011). Apart from high conductivity, good thermal stability, biodegradability, thin/thick film formation, etc. make them to be suitable candidates for electrolyte in DSSC.

2. Dye Sensitized Solar Cell Using Solid Biopolymer Electrolyte

There are huge numbers of reviews, books and papers on the solar cells with different types of materials, different varieties of junctions and their combinations (Gratzel et al. 2000, 2001, Meyer et al. 2010, Robertson et al. 2006, Hagfeldt et al. 2000, Toyoda et al. 2004). DSSC is considered as the most promising third generation photovoltaic technology. It is a totally new type of solar cell which offers cost effective solar cells. It also has great potential for commercialization as its materials are comparably cheaper and is easy to fabricate. It also has a number of attractive features such as, it is semi-flexible and semi-transparent that can be used for various application, it can easily be fabricated from conventional roll to roll printing techniques. But, in practice it is observed that, it is very difficult to replace a number of costly materials, such as ruthenium dye and platinum. Liquid electrolyte is also a serious challenge for fabricating a stable and reliable solar cell which can be use in all weather. Through this chapter we are trying to contribute in paving a path for choosing suitable solid biopolymer electrolyte for making leakage proof cost-effective device (Singh et al. 2012, 2013, 2014, Finkenstadt et al. 2005, Nemoto et al. 2007, Khanmirzaei and Park 2013, Buraidah et al. 2011).

2.1 Background of DSSC

Dye-sensitized solar cells (DSSC) have been under extensive research for the last 25 years, which were introduced by Prof. Grätzel in year 1991 (Gratzel et al. 2001, Hardin et al. 2012). Briefly, DSSC consists of nanoporous wide band gap semiconductors films infiltrated with dye molecules. The liquid electrolyte provides high conductivity path and makes an impressive contact between electrode and electrolyte which overall gives high efficiency DSSC (Narsaiah et al. 1995, Yohannes et al. 1998, Jung et al. 2010, Mathews et al. 2014, Bai et al. 2008, Gratzel et al. 2009, Longo et al. 2003). The dye-sensitized solar cell is comprised of a working electrode made up of a transparent conducting glass sheet coated with porous nanocrystalline TiO_2 and dye molecules attached to the surface of TiO_2, an electrolyte containing a reduction-oxidation couple such as I^-/ I^{3-} and a Pt catalyst coated counter electrode (Kim et al. 2004, Kang et al. 2005, 2006, 2007, 2008, Ren et al. 2002, Akhtar et al. 2007, Nogueira et al. 1999, 2001b). Under light illumination, the cell produces voltage over and current through an external load connected to the electrodes (Fig. 1). The absorption of light in the DSSC occurs in dye molecules and results in charge separation by electron injection from the dye to the TiO_2 at the semiconductor electrolyte interface. The color of the device can be easily varied by choosing different dyes (Nogueira et al. 2001c, 2004, Kim and Chatzivasiloglou 2005, Kalaignan et al. 2006, Katsaros and Stergiopoulos 2002, Chen et al. 2007). The concept of DSSC is believed to reduce the production costs and energy payback time significantly as compared to standard silicon photovoltaic. The conversion efficiency varies between 6–13% depending on the module size and technology. The highest conversion efficiency till date obtained from DSSC by using liquid electrolyte whereas, liquid electrolytes have known problem as explained earlier which force us to find alternative electrolytes. The quasi gel electrolyte proposed as novel alternative but liquid entrapped in pores creates a problem and hence affects the life time of DSSC (Bai et al. 2008, Gratzel et al. 2009, Longo et al. 2003, Kim et al. 2004). The next proposed alternative is solid polymer electrolyte in which polyether has played a dominant role (Haque et al. 2003, Anandan et al. 2006, 2008, Ganesan et al. 2008a,b, Benedetti et al. 2008, Freitas et al. 2008, 2009, Flores and Zhang 2007, Kang et al. 2003, Han et al. 2005). Biopolymer-salt complex is also presenting itself as a novel alternative since most of the biopolymers can make complex with well known salts producing high conductivity (Hsu et al. 2013, 2014, 2011, Guo et al. 2013a,b). Biopolymers are suitable candidates for preparing electrolyte in DSSC because they offer several interesting properties such as biodegradable in nature, very high ionic conductivity, excellent thermal stability, etc.

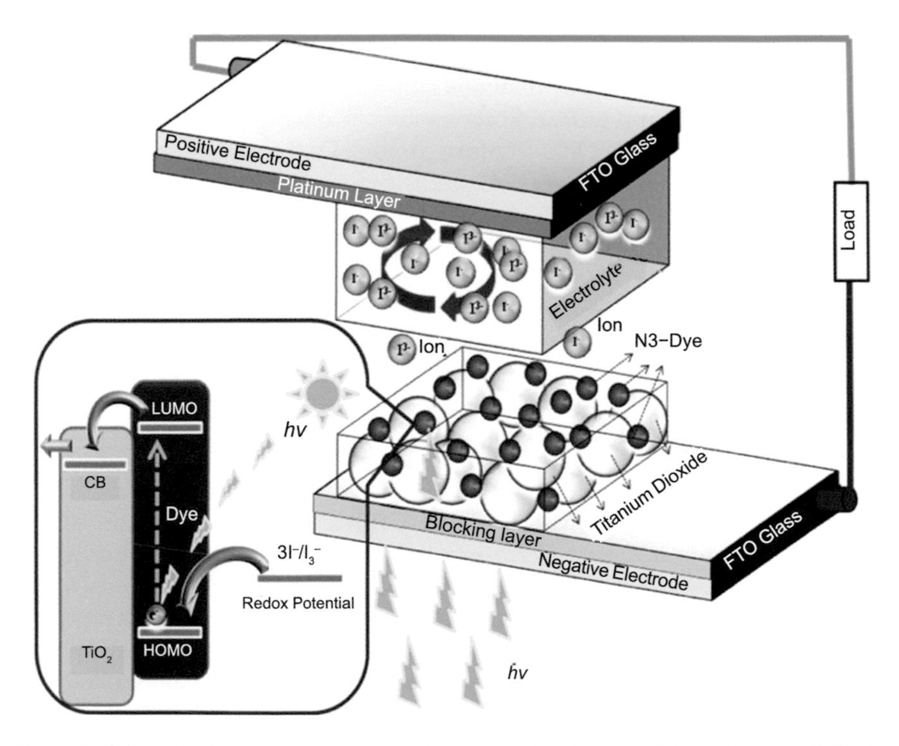

Figure 1. Schematic diagram presenting working principle of dye-sensitized solar cells (DSSC).

3. Solid Biopolymer Electrolyte Used in Electrochemical Devices

3.1 Solid biopolymer electrolytes for electrochemical devices

Polysaccharides are a class of natural biopolymers which are already treated as dominant player in developing good and efficient electrochemical devices. They are used as composite matrices due to several characteristics of relevance for biological and electrochemical device applications. Polysaccharides are well known natural polymers playing a major role in developing efficient electrochemical devices, i.e., dye-sensitized solar cells, supercapacitor, actuators, batteries and fuel cells (Ohno et al. 2005, Tarascon et al. 2001, Ozer et al. 2002, Zhang et al., Singh and Hsu 2013). A large number of BPE are available in literature and some of them are briefly described below and tabulated in Table 1.

Agarose/Agar: Agarose is a linear polymer consisting of alternating beta-D-galactose and 1, 4-linked 3, 6-anhydro-alpha-L-galactose units (Stephen et al. 2006). It has very few sulfate groups. The gelling temperature ranges from 32–45°C, and the melting temperature range is normally 80–95°C, depending on the type of Agarose preparation used. Methylation, alkylation and hydroxyalkylation of the polymer chain can change the melting and gelling temperatures (Finkenstadt et al. 2005). Generally, agarose/agar is insoluble in cold water, but it swells considerably. However, it can easily dissolve in hot water (H_2O) and other solvents at temperatures between 95° and 100°C such as DMF (Dimethylformamide), DMSO (Dimethyl sulfoxide), dimethylacetamide (DMAc), Glycol, orthophosphoric acid, (NMP) N-Methyl-2-pyrrolidone, etc. Mechanically strong gel with a small amount of agarose can be produced, while maintaining the ionic conductivity of the liquid electrolyte. Agarose shows good ionic conductivity in the order of 10^{-2}–10^{-4} S/cm. Complex impedance spectroscopic analysis revealed that doping of KI provided the additional charge carries (cations/anions) enhancing the overall conductivity, while the decrease in conductivity is due to charge pair formation phenomena. Maximum conductivity achieved till now is of the order of 10^{-2} S/cm (MacCallum et al. 1987, 1989, Ozdemir et al. 2007).

Carrageenan: Carrageenan is obtained from the red seaweeds of the class Rhodophyceae. It is a group of linear galactan with ester sulfate content of 15–40% (w/w) and containing alternating (1→3)-α-D and (1→4)-β-Dgalactopyranosyl (or 3,6-anhydro-α-D-galactopyranosyl) linkages. Three types of commercially available carrageenans are κ, ι, and λ. Anionic polysaccharides with molecular weight: 100,000–1,000,000, form gels with potassium or calcium ions (Chandra et al. 1998, Stephen et al. 2006, Chamy et al. 2013). Carrageenan can also easily dissolve in hot water (H_2O) even at room temperature. The solubility temperature is between

Table 1. List of biopolymer and its structure and film type commonly used in developing electrochemical devices.

Source	Biopolymer	Functional Groups	Film Type	Structure	Reference
Marine red Algae	Agarose/Agars	OH	Gel	Linear	Nijenhuis et al. 1997
Red seaweeds	Carrageenans	OH, OSO_3^-	Gel	Linear	Campo et al. 2009
Brown algae	Alginate	OH, COO^-	Gel	Linear	Augst et al. 2006
Plant cell walls insoluble	Cellulose	OH	Gel	Linear	Kalia et al. 2011
Plant cell wall soluble	Pectin	OH, COO^-	Gel	Linear	Kalia et al. 2011
Plant seeds/cereal starch	Sago Starch	OH	Thickening/viscosity agents	Linear/Branch-on-branch	Le et al. 2010
	Corn Starch	OH	Thickening/viscosity agents	Linear/Branch-on-branch	Le et al. 2010
	Pea Starch	OH	Thickening/viscosity agents	Linear/Branch-on-branch	Le et al. 2010
	Rice Starch	OH	Thickening/viscosity agents	Linear/Branch-on-branch	Le et al. 2010
	Wheat Starch	OH	Thickening/viscosity agents	Linear/Branch-on-branch	Le et al. 2010
	Oats Starch	OH	Thickening/viscosity agents	Linear/Branch-on-branch	Le et al. 2010
Plant tuber & root	Potato Starch	OH	Thickening/viscosity agents	Linear/Branch-on-branch	Le et al. 2010
	Tapioca/Arrowroot Starch	OH	Thickening/viscosity agents	Linear/Branch-on-branch	Le et al. 2010
Plant exudates/ Acacia trees	Gum Arabic	OH, COO^-	Thickening/viscosity agents	Branch-on-branch	Ali et al. 2009
Plant exudates	Gum Tragacanth	OH, COO^-	Thickening/viscosity agents	Short branched	

Table 1 contd. ...

...Table 1 contd.

Source	Biopolymer	Functional Groups	Film Type	Structure	Reference
Microorganism	Xanthan Gum	OH, COO^-	Gel	Short branched	
	Gellan Gum	OH, COO^-	Gel	Linear	
Shelfish & fungi cell wall/Insects	Chitin	OH, NH_3^+	Gel	Linear	Rinaudo et al. 2006
	Chitosan	OH, NH_3^+	Gel	Linear	Rinaudo et al. 2006
Derived	Carboxymethyl Cellulose	OH, COO^-	Thickening/viscosity agents	Linear	Kalia et al. 2011
	Methylcellulose (MC)	OH, COO^-	Gel/Thickening/viscosity agents	Linear	Kalia et al. 2011

40° and 70°C in solvents such as DMSO (Dimethyl sulfoxide) but it is insoluble in ethanol, acetone and organic solvents (Finkenstadt et al. 2005). Ionic conductivity of carrageenan was found in the range of 10^{-7} to 10^{-3} in water and DMSO. The highest conductivity achieved for κ-carrageenan-AN/MOZ-TBP-I_2/LiI-Pr_4NI+I^{3-} was 2.98×10^{-3} S/cm at room temperature.

Alginate: Alginate is obtained from the brown seaweeds of the class Phaeophyceae, as a structural material. Linear polysaccharide is composed of β-D-mannuronopyranosyl and α-L-guluronopyranosyl units. The units occur in M blocks (containing solely mannuronopyranose residues), G blocks (containing solely guluronopyranose residues), or MG blocks. The ratio of G-, M-, and MG-blocks affects the gel strength, calcium reactivity, and other properties (Stephen et al. 2006). Alginate forms gel with calcium ions. Alginate with high G-blocks results in greater gel strength. Alginate with high M-blocks is more calcium tolerant and less likely to have problem with syneresis. It dissolves slowly in water, forming a viscous solution, but is insoluble in ethanol and ether (Chamy et al. 2013). Gel polymer electrolytes (GPEs) are based on sodium alginate plasticized with glycerol containing either CH_3COOH or $LiClO_4$. The membranes showed ionic conductivity results of 3.1×10^{-4} S/cm for the samples with $LiClO_4$ and 8.7×10^{-5} S/cm for the samples with CH_3COOH at room temperature. Results obtained indicate that alginate-based GPEs can be used as electrolytes in electrochemical devices (Agarwal et al. 1999).

Pectin: Pectin is found in virtually all land-based plant as a structural material. Commercial pectin is extracted from citrus peel, apple pomace, sugar beet, or sunflower heads. A linear chain of galacturonic acid units has molecular weight about 110,000–150,000. Pectins are soluble in pure water. Monovalent cation (alkali metal) salts of pectinic and pectic acids are usually soluble in water; di- and trivalent cations salts are weakly soluble or insoluble (Chamy et al. 2013). The plasticized pectin and $LiClO_4$-based gel electrolyte were prepared and analyzed by spectroscopic, thermal, structural, and microscopic analyses. The best ionic conductivity values of 2.536×10^{-2} were obtained at room temperature for the composition based on Diethanolamine modified pectin (DAP) and glutaraldehyde (GA) dissolved in pectin based biopolymer electrolyte (Mishra et al. 2012).

Cellulose: Cellulose is the most abundant polymer available worldwide. Cellulose is composed of polymer chains consisting of unbranched β (1→4) linked D-glucopyranosyl units (anhydroglucose unit). Nowadays, there are various procedures for extraction of cellulose microfibrils like pulping methods, acid hydrolysis, steam explosion, etc. Cellulose is attacked by a wide variety of microorganisms (Finkenstadt et al. 2005, Chamy et al. 2013, Dumitriu et al. 1998, Ree et al. 1977). When the samples of HEC (hydroxyethylcellulose) plasticized with glycerol and addition of lithium trifluoromethane sulfonate ($LiCF_3SO_3$) salt, the ionic conductivity were obtained in the range of 10^{-4}–10^{-5} S/cm. The best ionic conductivity obtained was 4.68×10^{-2} S cm^{-1} at room temperature by using lithium bis(trifluoromethanesulfonyl)imide (LiTFSI) doped biopolymer cellulose acetate (CA) matrix in 1-allyl-3-methylimidazolium chloride ([Amim] Cl) (Ramesh et al. 2013).

Plant seeds, plant tuber & root, cereal starch: The principal crops used for starch production include potatoes, corn and rice. In all of these plants, starch is produced in the form of granules, which varies in size and somewhat in composition from plant to plant (Chandra et al. 1998). The starch granule is essentially composed of two main polysaccharides, amylose and amylopectin with some minor components such as lipids and proteins. Amylose is a linear molecule of (1→4)-linked α-D-glucopyranosyl units and molecular weights ranging from 10^5 to 10^6 gmol^{-1} (Chamy et al. 2013) (Fig. 2). Amylopectin is a highly branched molecule composed of chains of α-D-glucopyranosyl residues linked together mainly by (1→4)-linkages but with (1→6) linkages at the branch points and having molecular weights ranging from 10^6 to 10^8 g mol^{-1}. Amylose is water soluble but amylopectin is insoluble in cold water and swells in it thereby giving rise to a thick paste upon boiling with water. It is a biopolymer that contains about 23% starch (20 to 25% amylose and 75 to 80% amylopectin). It is clear that adding KI in arrowroot matrix enhances the ionic conductivity and conductivity maxima were obtained by doping NaI and KI concentration where conductivity value approached 6.7×10^{-4} and 1.04×10^{-4} S/cm respectively (Singh et al. 2014a, Tiwari et al. 2011a). Adding (Glycerol+LiCl) and KI in Sago Palm matrix enhances the ionic conductivity and conductivity maxima were obtained

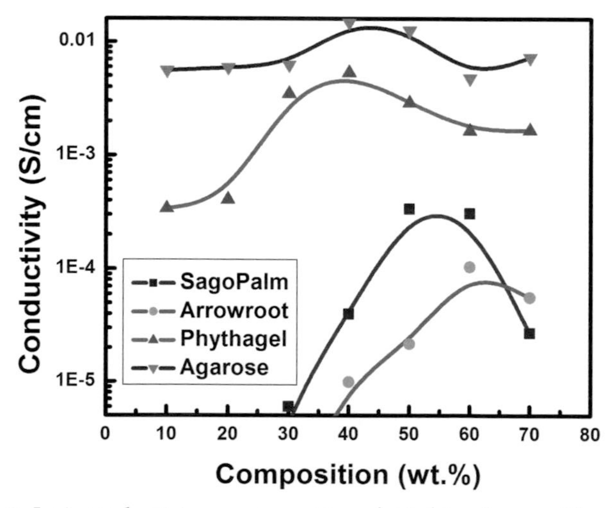

Figure 2. Ionic conductivity vs. composition plot in biopolymers: KI polymer electrolyte system.

by adding LiCl concentrations where conductivity value approached 10^{-3} for LiCl (Skotheim et al. 1981) and 3.4×10^{-4} S/cm for KI (Singh et al. 2014b, 2016b). Potato starch with NH_4I based biopolymer electrolyte prepared by solution casting technique gives the best ionic conductivity of $\sim 2.4 \times 10^{-4}$ S/cm (Kumar et al. 2012).

Corn starch-based biopolymer electrolytes have been prepared by solution casting technique. Lithium hexafluorophosphate (LiPF6) and 1-butyl-3-methylimidazolium trifluoromethanesulfonate (BmImTf) were used as lithium salt and ionic liquid, respectively (Shukur et al. 2015). In other system with ionic liquid, 1-butyl-3-methylimidazolium hexafluorophosphate ($BmImPF_6$) was doped into the corn based biopolymer matrix and maximum ionic conductivity of 1.47×10^{-4} S/cm was reported for this system. Whereas, the blending of corn starch with chitosan was used for making biopolymer electrolyte, which was doped with ammonium iodide (NH_4I). A polymer blend at 80 wt% starch and 20 wt% chitosan was found to be the most amorphous blend. The highest achieved ionic conductivity at room temperature was 3.04×10^{-4} S/cm at NH_4I (Yusof et al. 2014). The overall best ionic conductivity $\sim 10^{-0.5}$ was reported by using N, N-dimethylacetamide (DMAc) along with lithium chloride (LiCl) doped corn based biopolymer (Ahmed et al. 2012).

Rice Starch doped with Lithium Iodide bio-polymer electrolyte was prepared using solution casting method, at room temperature and the highest ionic conductivity achieved $\sim 4.68 \times 10^{-5}$ S/cm (Khanmirzaei et al. 2014a). Tapioca is a polymer containing heteroatoms in their structure. Hence, they can interact with protons or lithium ions leading to ionic conduction. Among the different natural polymers, starch-based biopolymer electrolyte shows good opto-electrochemical characteristics and can be applied to electrochemical devices. The ionic conductivity results obtained for these bio polymer electrolytes varied from 10^{-6} S/cm to 10^{-4} S/cm at room temperature. Conductivity reaches 8.1×10^{-3} S/cm for cassava doped with lithium perchlorate. The amount of acetic acid and NH_4NO_3 was found to influence the proton conduction. Wheat can easily dissolve in acetic acid and room temperature conductivity was found in the order of 10^{-5} S/cm to 10^{-4} S/cm. The common starch based biopolymer electrolytes used in electrochemical devices are shown in Table 2.

Chitin and Chitosan: Chitosan is a linear polysaccharide consisting of β (1-4) linked D-glucosamine with randomly located *N*-acetylglucosamine groups depending upon the degree of deacetylation of the polymer. Chitin is basically found in the shells of crabs, lobsters, shrimps and insects. Chitosan is the deacylated derivative of chitin. Chitin is insoluble in its native form but chitosan, is water soluble. Chitosan is soluble in weakly acidic solutions resulting in the formation of a cationic polymer with a high charge density and can, therefore, form polyelectrolyte complexes with wide range of

Table 2. Conductivity chart of starch based biopolymer electrolyte.

Starch	Amylose (%)	Starch (%)	Type	Gelatinization Temperature Range [°C]	Dispersoids	σ (S/cm)	Ref.
Arrowroot	20.5	84	Root	70–75	KI	1.04×10^{-4}	Singh et al. 2014a, 2016a
Corn	28	31–50	Cereal	62–72	LiI	1.83×10^{-4}	Shukur et al. 2013b
Pea	27	40	Legume	60–75	-	-	-
Potato	20	65	Tuber	59–68	NaI	1.3×10^{-4}	Tiwari et al. 2014, 2011
Rice	18.5	87	Grain	-	LiI	4.68×10^{-5}	Khanmirzaei et al. 2013, 2014
Sago	25.8	84	Root	70–72	KI	3.4×10^{-4}	Singh et al. 2014b
Tapioca/ Cassava	16.7	84	Root	58–70	Li salt	10^{-6}–10^{-4}, 8.1×10^{-3}	-
Wheat	26	25	Cereal	58–64	CH_3COOH	10^{-5} to 10^{-4}	-

anionic polymers. Chemical modification of chitosan can significantly affect its solubility and degradation rate (Stephen et al. 2006, Finkenstadt et al. 2005, Chamy et al. 2013, Rees et al. 1977).

The electrical properties of polymer electrolytes based on chitosan complexed with lithium and ammonium salts have been reported (Mohamed et al. 1995, Buraidah et al. 2010, 2011, Idris et al. 2007). Conductivities of the order of 10^{-6} S/cm at room temperature were reported for chitosan with poly (ethylene oxide) (PEO) blends and doped with LiTFSI salt (Agrawal et al. 1999) and also for the complex formed by chitosan, poly (aminopropylsiloxane) (pAPS) and $LiClO_4$ (Hsu et al. 2013). Conductivity ranging from 10^{-5} and 10^{-4} S/cm were reported for proton-conducting polymer electrolytes, based on chitosan and ammonium salts (NH_4NO_3 and $NH_4CF_3SO_3$) (Hamdam et al. 2014, Majid et al. 2005) and conductivity in the range of 10^{-6}–10^{-4} S/cm for chitosan and κ-carrageenan containing ammonium nitrate-based film (Nijenhuis et al. 1997, Campo et al. 2009). $LiMn_2O_4$ doped biopolymer based chitosan with carbon has been reported where the biopolymer-in-salt based electrolyte has achieved the best ionic conductivity of 3.9×10^{-3} at room temperature (Kamarulzaman et al. 2001).

Gum Arabic: A gummy exudate obtained from Acacia trees, with molecular weight of about 250,000, is highly soluble with low viscosity even at 40% concentration. Gum Arabic, e.g., Acacia arabica, Acacia babul, exhibits a conductivity of approximately 1.5×10^{-6} S/cm after drying. Gum Arabica produces salt complexes with inorganic materials like $FeSO_4$, [K_2SO_4, $Al_2(SO_4)_3$, $24H_2O$], $LiClO_4$, iodine, etc. (Mallick et al. 2000). Gum Arabic functions as a proton conductor through hydronium ions H_3O^+ (Mallick et al. 2000).

Gum Tragacanth: Gum Tragacanth is an exudate of Astragalus, a perennial short brush in Asia. It is slightly acidic and occurs as Ca, Mg, or Na salt. It contains neutral highly branched arabinogalactan and tragacanthic acid (linear (1→4)-linked α-D-galacturonopyranosyl units, with some substitutions). It is highly viscous with some emulsification properties. The highest conductivity reported for NaOH based biopolymer was 88.8×10^{-3} S/cm at room temperature (Arora et al. 2014).

Xanthan gum: Xanthan gum is prepared through culturing Xanthomonas campestris, a single-cell organism producing gum as protective coating. A trisaccharide side chain is attached to alternate D-glucosyl units at the O-3 position. The side chain consists of a D-glucuronosyl unit between two D-mannosyl units. Molecular weight is about 2,000,000–3,000,000. Its viscosity is stable at wide temperature and pH. Among Gum Xanthan + PVP, Gum tragacanth + PVP and Gum Acacia + PVP, Gum Acacia + PVP has better compatiblilty because this system has stronger intermolecular interaction compared to other systems. In the same manner, Gum Xanthan +

PEG, Gum Acacia + PEG and Gum tragacanth + PEG, among these systems, Gum Tragacanth + PEG has better compatibility (Park et al. 2013).

Gellan gum: Gellan gum is prepared by culturing Pseudomonas elodea and is composed of a four-sugar repeating sequence containing one D-glucuronopyranosyl, two D-glucopyranosyl, and one L-rhamnopyranosyl unit. Its molecular weight is about 1,000,000–2,000,000. It requires either monovalent or divalent cations to form a gel (Stephen et al. 2006, Finkenstadt et al. 2005, Chamy et al. 2013, Dumitriu et al. 1998, Rees et al. 1977). The ionic conductivity measurements revealed that the ionic conductivity of the Gellan gum doped with 40 wt.% of lithium trifluoromethanesulfonate ($LiCF_3SO_3$) electrolyte varied with the salt concentration reaching the highest conductivity value of 5.4×10^{-4} S/cm at room temperature (Higgins et al. 2011). When doped with lithium iodide (LiI), it exhibited ionic conductivity of 3.8×10^{-4} S/cm order at room temperature (Singh et al. 2010). To achieve good ionic conductivity, plasticizers, such as glycerol, ethylene glycol, ethylene carbonate, propylene carbonate and others are used; lithium salts, such as $LiClO_4$, $LiBF_4$, $LiCF_3SO_3$, LiI/I_2 or acetic acid are added to promote the proton conduction (Halim et al. 2012, Higgins et al. 2011, Singh et al. 2010).

Carboxymethyl cellulose (CMC): Carboxymethyl cellulose (CMC) is prepared by soaking cellulose in aqueous sodium hydroxide and reacting with monochloroacetic acid (Stephen et al. 2006, Finkenstadt et al. 2005). Carboxymethyl cellulose doped with Lithium perchlorate and plasticizer Polycarboxylate based transparent solution of $CMC/LiClO_4/PC$ have been reported and ionic conductivity of the biopolymer electrolyte was found to be 2×10^{-4} S/cm. In another system, oleic acid based biopolymer electrolyte and NH_4Br based electrolyte achieved ionic conductivity at 2.11×10^{-5} and 1.12×10^{-4} S/cm respectively (Samsudimi et al. 2010, Chai et al. 2013). Carboxymethyl cellulose can be doped with different concentration of DTAB/EC *via* solution casting technique. The highest ionic conductivity, σ, was found to be 2.37×10^{-3} S/cm at room temperature (Isa et al. 2013).

3.2 *Dye-sensitized solar cell using solid biopolymer electrolyte*

The prime aim of the development of solid polymer electrolyte is to avoid the disadvantages caused by liquid electrolyte. Biopolymer electrolytes are free from leakage, moderate conductivity, free from corrosion and more stable and hence frequently used in DSSC. Variety of polysaccharides with various salts are reported in literature and tabulated in Table 3.

Table 3. Status of dye-sensitized solar cell (DSSC) and other electrochemical devices using solid biopolymer electrolyte.

Bio-polymer	Additive	Conductivity (S/cm)	Parameters	Device	References
Agarose/Agars	KI	9.02×10^{-3}	0.54%	DSSC	Singh et al. 2013
	LiI	3.98×10^{-3}	-	DSSC	Wejia et al. 2011
	NH_4I	4.89×10^{-3}	0.008%	DSSC	Siti et al. 2013
	NaI	12.41×10^{-4}	-	-	Koh et al. 2012
	1-allyl-3-ethylimidadolium iodide	-	5.89%	DSSC	Hsu et al. 2013
	1-allyl-3-ethylimidadolium iodide	-	7.43%	DSSC	Hsu et al. 2014
	DMSO/PC-(MPII)	14.2×10^{-3}	1.97%	DSSC	Hsu et al. 2011
	DMSO/4EG-(MPII)	4.4×10^{-3}	1.38%	DSSC	Hsu et al. 2011
	DMSO/3EG-(MPII)	4.6×10^{-3}	1.39%	DSSC	Hsu et al. 2011
	DMSO/PG-(MPII)	6.2×10^{-3}	1.06%	DSSC	Hsu et al. 2011
	Pure DMSO-(MPII)	5.0×10^{-3}	1.15%	DSSC	Hsu et al. 2011
	Polysorbate 80/Fe_3O_4 nanoparticles	2.98×10^{-3}	1.83%	DSSC	Guo et al. 2013
	PEG 200/Fe_3O_4 nanoparticles	2.88×10^{-3}	-	DSSC	Guo et al. 2013
	TiO_2-modified	2.66×10^{-3}	1.71%	DSSC	Yang et al. 2014
	Co_3O_4-modified	4.37×10^{-3}	2.11%	DSSC	Yang et al. 2014
	NiO-modified	3.33×10^{-3}	2.02%	DSSC	Yang et al. 2014
	NMP/LiI	3.94×10^{-4}	4.14%	DSSC	Yang et al. 2011b
	NMP/LiI/nanoparticle TiO_2	4.4×10^{-4}	4.74%	DSSC	Yang et al. 2011a
	1-alkyl-3-methyl-imidazolium salts	-	2.93%	DSSC	Suzuki et al. 2006
	Acetic acid/glycerol	1.1×10^{-4}	-	-	Raphael et al. 2010
	1-ethyl-3-methylimidazolium acetate	2.35×10^{-5}	-	-	Leones et al. 2012

Table 3 contd. ...

...Table 3 contd.

Bio-polymer	Additive	Conductivity (S/cm)	Parameters	Device	References
Carrageenans	3-methyl-2-oxazolidinone	-	6.87%	DSSC	Nemoto et al. 2007
	AN/MOZ-TBP-I_2/LiI-Pr_4NI+I^{3-}	2.92×10^{-3}	6.87%	DSSC	Masao et al. 2004
	Chitosan + carrageenan	1.38×10^{-6}	-	-	Shuhaimi et al. 2008
	Chitosan + carrageenan + NH_4NO_3	2.39×10^{-4}	13 to 18.5 Fg^{-1}	EDLCs	Shuhaimi et al. 2008
	Chitosan + carrageenan + H_3PO_4 + PEG	6.29×10^{-4}	35 F g^{-1}	EDLCs	Arof et al. 2010
Cellulose	NH_4NO_3	2.1×10^{-6}	-	-	Shuhaimi et al. 2010a
	NH_4NO_3 PEG	-	31.52 mA cm^{-2}	Fuel cells	Shuhaimi et al. 2010b
	Cellulose acetate (CA) + LiBOB + GBL	5.36×10^{-3}	4.7 Voc	Battery	Abidin et al. 2013
	Cellulose acetate +TiO_2	1.37×10^{-2}	54.1 mA h	Battery	Johari et al. 2012
	Cellulose acetate + $NH_4CF_3SO_3$ + EC	$\sim 10^{-4}$	1·4 Voc	Battery	Saaid et al. 2009
	Cellulose acetate electrolyte	-	22·41 mA h	Battery	Johari et al. 2009
	PEO–HPC + LiTFSI	2.5×10^{-4}	3.65 Voc	Battery	Chelmecki et al. 2007
	$LiCF_3SO_3$	5.3×10^{-7}	-	-	Jafirin et al. 2013
	CA-LiTFSI-[Amim] Cl	4.68×10^{-2}	-	-	Ramesh et al. 2013
	CA-NH_4BF_4	2.18×10^{-7}	-	-	Harun et al. 2012
	CA-NH_4BF_4 + PEG600	1.41×10^{-5}	-	-	Harun et al. 2012
	CA-LiTFSI-DES	2.61×10^{-3}	-	-	Ramesh et al. 2012
	BC–TEA	$1.8\,9 \times 10^{-5}$	-	-	Salvi et al. 2014
Pectin	Amidated pectin + Glutaraldehyde (GA)	1.098×10^{-3}	-	-	Mishra et al. 2009
	DAP + glutaraldehyde (GA)	2.536×10^{-2}	-	-	Mishra et al. 2012
	$LiClO_4$	4.7×10^{-4}	-	-	Andrade et al. 2009
	KCl	1.45×10^{-3}	-	-	Leone et al. 2014

Sago Starch	KI	3.4×10^{-4}	0.57%	DSSC	Singh et al. 2014b, 2016b
	Glycerol + LiCl	10^{-3}	-	-	Pang et al. 2014
Corn Starch	Starch-chitosan-NH4I	3.04×10^{-4}	-	-	Yusof et al. 2014
	NH_4Br	5.57×10^{-5}	-	-	Shukur et al. 2015
	NH_4Br + Glycerol	1.80×10^{-3}	-	-	Liew et al. 2013
	$LiPF_6$ + BmImTf	3.21×10^{-4}	-	-	Shukur et al. 2014
	LiOAc	2.07×10^{-5}	33.31 Fg^{-1}	EDLC	Liew et al. 2012
	Glycerol-LiOAc	1.04×10^{-3}	-	-	Liew et al. 2012
	$LiPF_6$–BmImTf at 80°C	6.00×10^{-4}	-	-	Teoh et al. 2014
	$LiClO_4$	1.28×10^{-4}	-	-	Ramesh et al. 2012
	LiTFSI-DES (Choline chloride & urea)	1.04×10^{-3}	-	-	Shukur et al. 2013
	LiI	1.83×10^{-4}	-	-	Ramesh et al. 2011
	LiI-Glycerol	9.56×10^{-4}	-	-	Ramesh et al. 2011
	LiTFSI-[Amim] Cl	$4{\cdot}18 \times 10^{-2}$	-	-	Ramesh et al. 2011b
	LiTFSI-[Amim] Cl	5.68×10^{-2}	-	-	Ramesh et al. 2012c
	$LiPF_6$-$BmImPF_6$	1.47×10^{-4}	-	-	Ramesh et al. 2011a
	$LiClO_4$-SiO_2	1.23×10^{-4}	-	-	Teoh et al. 2012
	[Amim]Cl	$10^{-1.6}$	-	-	Ahmad and Ning 2009
	DMAc-LiCl	$10^{-0.5}$	-	-	Ning et al. 2009b
	Glycerol	10^{-8}	-	-	Ma et al. 2008
	Glycerol-Carbon black	7.08	-	-	Ma et al. 2008
	[BMIM]Cl	$10^{-4.6}$	-	-	Sankri et al. 2010
	Starch-Chitosan blend-NH_4I-Glycerol	1.28×10^{-3}	1.8–4.0 Fg^{-1}	EDLC	Yusof et al. 2014

Table 3 contd. …

...Table 3 contd.

Bio-polymer	Additive	Conductivity (S/cm)	Parameters	Device	References
Pea Starch	KI	2.28×10^{-4}	-	-	Ummartyotin et al. 2015
Rice Starch	LiI	4.68×10^{-5}	-	-	Khanmirzaei et al. 2014, 2014
	LiI-TiO	2.27×10^{-4}	-	-	Khanmirzaei et al. 2014
	LiI-MPII-TiO2	3.63×10^{-4}	0.17%	DSSC	Khanmirzaei et al. 2014
Wheat Starch	CH3COOH (Acetic acid)	10^{-5} to 10^{-4}	-	-	-
Potato Starch	NaI	1.3×10^{-4}	-	-	Tiwari et al. 2013, 2014
	$NaClO_4$	7.19×10^{-6}	-	-	Tiwari et al. 2014
	NaSCN	1.12×10^{-4}	-	-	Tiwari et al. 2011
	NH_4I	2.4×10^{-4}	-	-	Kumar et al. 2012
	Methanol-GA	2.50×10^{-6}	-	-	Tiwari et al. 2014
	Methanol-GA-NaI	8.40×10^{-6}	-	-	Tiwari et al. 2014
	Methanol-GA-PEG300-NaI	1.80×10^{-4}	-	-	Tiwari et al. 2014
	Acetone-GA	8.80×10^{-6}	-	-	Tiwari et al. 2014
	Acetone-GA-NaI	3.22×10^{-5}	-	-	Tiwari et al. 2014
	Acetone-GA-PEG300-NaI	4.30×10^{-5}	-	-	Tiwari et al. 2014
	-	-	335 Fg^{-1}	EDLC	Zhao et al. 2009
Arrowroot Starch	KI	5.68×10^{-4}	0.63%	DSSC	Singh et al. 2014a
	NaI	6.7×10^{-4}	-	-	Tiwari et al. 2011
	Chitosan blend-NH_4NO_3	3.89×10^{-5}	-	-	Khiar et al. 2011

Gum Arabic	-	1.5×10^{-6}	-	-	Mallick et al. 2000
Gum Tragacanth	NaOH	88.8×10^{-3}	-	-	Arora et al. 2014
Xanthan Gum	Water-based thixotropic	-	4.78%	DSSC	Park et al. 2013
Gellan Gum	$LiCF_3 SO_3$	5.4×10^{-4}	-	-	Noor et al. 2012
	LiI	1.5×10^{-3}	-	-	Halim et al. 2012
	Polypyrrole-Gellan gum	-	–0.8 to + 0.4 V	EDLC	Higgins et al. 2011
Chitin/Chitosan	NaI	-	0.13%	DSSC	Singh et al. 2010a
	EMImSCN-NaI	2.60×10^{-4}	0.73%	DSSC	Singh et al. 2010a
	NH_4SCN–Al2TiO_5	2.10×10^{-4}	-	-	Hassan et al. 2013
	$LiNO_3$	2.7×10^{-4}	1.113 V	Battery	Mohamed et al. 1995
	NH_4I	3.73×10^{-7}	0.29%	DSSC	Buraidah et al. 2010b
	NH_4I–EC	7.34×10^{-6}	0.51%	DSSC	Buraidah et al. 2010b
	NH_4I–BMII	8.47×10^{-4}	1.24%	DSSC	Buraidah et al. 2010
	Tartaric-NH_4I-BMII	3.02×10^{-4}	0.38%	DSSC	Buraidah et al. 2011b
	Tartaric-PEO-NH_4I-BMII	5.52×10^{-4}	0.39%	DSSC	Buraidah et al. 2011b
	Tartaric-Phthaloyl chitosan-NH_4I-BMII	5.86×10^{-4}	0.43%	DSSC	Buraidah et al. 2011b
	Tartaric-Phthaloyl chitosan-PEO-NH_4I-BMII	6.24×10^{-4}	0.46%	DSSC	Buraidah et al. 2011b
	PEO blend-NH_4I	4.32×10^{-6}	0.46%	DSSC	Mohamad et al. 2007
	PEO blend–NH_4I (Dye-Sumac/Rhus)	1.18×10^{-5}	1.5%	DSSC	Al-Bat'hi et al. 2013
	$NH_4CF_3SO_3$	8.91×10^{-7}	-	-	Ahmad et al. 2006
	Methylcellulose blend-$H_4CF_3SO_3$	4.99×10^{-6}	-	-	Hamdan et al. 2014
	EC-LiOAc	7.6×10^{-6}	-	-	Yahya et al. 2004

Table 3 contd. ...

...Table 3 contd.

Bio-polymer	Additive	Conductivity (S/cm)	Parameters	Device	References
Chitin/Chitosan	chitosan–LiOAc-oleic acid	10^{-5}	-	-	Yahya et al. 2003
	chitosan, palmitic acid (PA)-LiOAc	5.5×10^{-6}	-	-	Yahya et al. 2002
	Corn starch-NH_4Cl-Glycerol	5.11×10^{-4}	-	-	Shukur et al. 2014
	NH_4NO_3	2.53×10^{-5}	-	-	Majid et al. 2005
	NH_4NO_3-Acetic acid	1.46×10^{-1}	27.90 mA h^{-1}	Battery	Jamaludin et al. 2010
		-	3.67 mW cm^{-2}	-	-
	NH_4NO_3-EC-Acetic acid	9.93×10^{-3}	8.70 mW cm^{-2}	Battery	Ng et al. 2006
		-	17.0 mAh	-	-
	Polyethylene oxide-NH_4NO_3	1.02×10^{-4}	-	-	Kadir et al. 2011b
	$LiCF_3SO_3$	2.75×10^{-5}	-	-	Winie et al. 2006
	CMC-Chitosan-DTAB	1.85×10^{-6}	-	-	Bakar et al. 2014
	PVA-Chitosan-NH_4NO_3	2.07×10^{-5}	1.6 and 1.7 V	Battery	Kadir et al. 2010
	PVA-chitosan-NH_4NO_3-EC	1.60×10^{-3}	27.1 Fg^{-1}	EDLC	Kadie et al. 2011a
	PVA–NH_4I	1.77×10^{-6}	-	-	Buraidah et al. 2011a
	NH_4I	3.73×10^{-7}	0.22%	DSSC	Buraidah et al. 2011a, 2010
	NH_4I-EC	7.34×10^{-6}	0.18 FF	DSSC	Buraidah et al. 2011
	NH_4I-BMII	3.43×10^{-5}	0.22 FF	DSSC	Buraidah et al. 2011
	$LiCF_3SO_3$-DEC-EC	4.26×10^{-5}	-	-	

	H_3PO_4	5.36×10^{-5}	-	Fuel cell	Winie et al. 2009
	H_3PO_4–Al_2SiO_5	1.12×10^{-4}	-	Fuel cell	Majid et al. 2009
	H_3PO_4–NH_4NO_3	1.16×10^{-4}	-	Fuel cell	Majid et al. 2009
	H_3PO_4–NH_4NO_3–Al_2SiO_5	1.82×10^{-4}	-	Fuel cell	Majid et al. 2009
	PEO-LiTFSI	1.40×10^{-6}	-	-	Idris et al. 2007
	$LiCF_3SO_3$-DEC-EC	4.26×10^{-5}	-	Battery	Winie et al. 2009
	$LiCF_3SO_3$-PC/EC	1.09×10^{-4}	-	Battery	Winie et al. 2009
	LiOAc	2.20×10^{-7}	-	-	Yahya et al. 2006
	NH_4SCN	1.81×10^{-4}	-	-	Shukur et al. 2013
	NH_4SCN-Glycerol	1.51×10^{-3}	-	-	Shukur et al. 2013
	PVA-NH_4Br	7.68×10^{-4}	-	-	Yusof et al. 2014
	PVA-$LiClO_4$	3.0×10^{-6}	-	-	Rathod et al. 2014
	NH4 CH3COO-EC	1.47×10^{-4}	1.83 mW cm^{-2}	Battery	Alias et al. 2014
	NH_4 CH_3COO-EC	1.47×10^{-4}	1.36 mW cm^{-2}	Battery	Alias et al. 2014
	Chitosan–NH_4Br	4.38×10^{-7}	-	-	Ummartyotin et al. 2015
	Chitosan–NH_4Br–glycerol	2.15×10^{-4}	-	-	Ummartyotin et al. 2015
	Starch blend- NH_4Br	9.72×10^{-5}	-	-	Shukur et al. 2014, 2013
	Starch blend- NH_4Br+EC	1.44×10^{-3}	0.1400 $mAhg^{-1}$	EDLC	Shukur et al. 2014, 2013
	$LiCF_3SO$-EC-DMC	3.9×10^{-3}	1.5 V	Battery	Kamarulzaman et al. 2001
	EC–$LiCF3SO_3$	4.0×10^{-5}	-	-	Ozman et al. 2001
	Acetic acid-$NaClO_4$	4.6×10^{-5}	1.114 V	Battery	Subban et al. 1996

Table 3 contd. …

...Table 3 contd.

Bio-polymer	Additive	Conductivity (S/cm)	Parameters	Device	References
Carboxymethyl Cellulose	NH_4Br	1.12×10^{-4}	-	Battery	Samsudin et al. 1996
	NH_4F	2.68×10^{-7}	-	-	Ramlli et al. 2014
	Salicylic acid	9.50×10^{-8}	-	-	Ahmad et al. 2012
	DTAB	7.72×10^{-4}	-	-	Isa et al. 2013
	DTAB-EC	2.37×10^{-3}	-	-	Isa et al. 2013
	CH_3COONH_4	5.77×10^{-4}	-	-	Rani et al. 2014
	Oleic acid	2.11×10^{-5}			Chai et al. 2013
Methylcellulose (MC)	-	3.08×10^{-11}	-	-	Shuhaimi et al. 2010
	NH_4NO_3	2.10×10^{-6}	-	-	Shuhaimi et al. 2010
	NH_4F	6.40×10^{-7}	-	-	Aziz et al. 2010
	Glycolic acid	7.16×10^{-10}	-	-	Harun et al. 2010
	MC-NH_4NO_3	1.17×10^{-4}	-	-	Harun et al. 2010
	MC-NH_4NO_3-PC	4.91×10^{-3}	-	-	Harun et al. 2010
	MC-NH_4NO_3-EC	1.74×10^{-2}	-	-	Harun et al. 2010
	PVDF-MC/PVDF-$LiFePO_4$	1.5×10^{-3}	34 mA g^{-1}	Battery	Xiao et al. 2014
	NH_4NO_3-PEG	1.14×10^{-4}	31.52 mA cm^{-2}	Fuel Cell	Shuhaimi et al. 2010

3.3 *Recent works in author's laboratory*

In our laboratory, we have tested a series of biopolymers (like Agarose/Agars, Sago Starch, Arrowroot Starch, Phythagel) doped with different alkali metal salts based system. We have carried out detail electrical, structural and photovoltaic (DSSC) performances (Singh et al. 2013, 2014, 2016). The experimental observations and discussions are given in the following section.

4. Experimental Details

4.1 *Synthesis of biopolymer electrolyte*

BPE films are prepared by the standard solution cast technique. In a common procedure, biopolymers and the desired amount of salts will be weighed separately and dissolved in distilled methanol or any other suitable organic solvents. The solution will be stirred for a long time for uniform mixing and complexation and the mixed viscous polymer electrolyte will be poured into polypropylene/Teflon moulds for casting at room temperature (Singh et al. 2014 a,b, 2016a,b). The solvent will be slowly evaporated at room temperature and a free standing BPE film will be obtained. The obtained BPE film matrix will be further characterized using various techniques.

4.2 *Fabrication of DSSC*

A dye-sensitized solar cell (DSSC) generally comprises a sandwich structure of two electrodes (anode, cathode) and electrolyte. The anode is coated with a layer of mesoporous wide band gap semiconductors (e.g., TiO_2) and the other, the cathode, is coated with a thin layer of platinum on a glass having a transparent conductive oxide (TCO) coating. The space between the two electrodes is filled with an electrolyte containing a reduction-oxidation couple such as I^-/I^{3-} that ensures charge transportation through a redox couple (Yella et al. 2011). Generally, the nanoporous wide band gap semiconductors film is immersed in dye solution. The gel polymer electrolyte provides high conductivity path and makes an impressive contact between electrode and electrolyte which overall gives high efficiency DSSC (Narsaiah et al. 1995, Yohannes et al. 1998, Jung et al. 2010, Mathew et al. 2014, Bai et al. 2008, Gratzel et al. 2009, Longo et al. 2003). Under light illumination, the cell produces voltage over and current through an external load connected to the electrodes. The absorption of light in the DSSC occurs in dye molecules and results in charge separation by electron injection from the dye to the TiO_2 at the semiconductor electrolyte interface (Fig. 1). The color of the device can be easily varied by choosing different dyes (Nagueira et al. 2001, 2004,

Kim and Chatzivasiloglou 2005, Kalaignan et al. 2006, Katsaros and Stergiopoulos 2002, Chen et al. 2007).

4.3 *Electrical conductivity measurement*

The ionic conductivity measurement of the biopolymers-KI based biopolymer electrolyte films was carried out using CH instrument workstation (model CHI604D, USA) over frequency range of 100 Hz to 1 MHz. To measure ionic conductivity, we sandwiched the free standing biopolymer electrolyte films between the steel electrodes and the electrical conductivity was evaluated using the formula

$$\sigma = \frac{1}{R_b}\left(\frac{l}{A}\right) \quad \text{(Equation 1)}$$

where, σ is ionic conductivity, R_b is the bulk resistance, l is the thickness of sample and A is the area of given sample.

The calculated values of ionic conductivity with different KI % for biopolymers based electrolyte are shown in Fig. 2. It was found that adding KI in biopolymer matrix enhances the ionic conductivity at room temperature with mixed humidity which gives information of interaction between salt and biopolymer matrix. The increase in the ionic conductivity with increasing KI concentration can be further affirmed related to the increase in the number of mobile charge carriers while the possible decrease in the ionic conductivity at a KI concentration greater than maximum conductivity composition can be attributed to the formation of ion multiples (Singh et al. 2014a,b, 2016a,b, Kumar et al. 2012).

The complex permittivity ε* of a sample is obtained from the measured capacitance C as in $C_0 = \varepsilon_0$ A/d, where A and d are the electrode area and electrode separation distance, respectively, and ε_0 is the permittivity of free space (8.85×10^{-12} F/m) and ε′ and ε″ are the real and imaginary parts of the permittivity, respectively. We can understand the effect of salt concentration with the help of dissociated ions model shown in Fig. 3. The impedance measurement data explains clearly the role of charge carriers in the biopolymers electrolyte matrix. It can be seen that both mobile concentration factor and dissociated charge fraction, which is related with mobile ion concentration data+, are following the conductivity trend (discussed in later section). Hence, ion pair formation and re-dissociation theory is applicable in biopolymer–salt complex systems. The minimum conductivity is due to the decreased dissociation. Increased ion association (i.e., decreased dissociation) or triplet ion formation, both contribute to decrease mobility (Singh et al. 2013, 2014, Kumar et al. 2012).

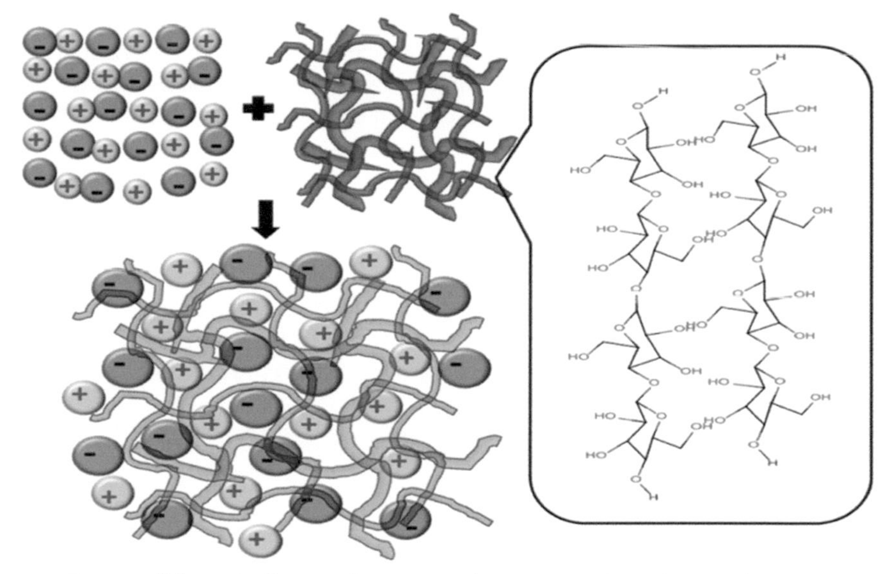

Figure 3. Schematic diagram showing ion dissociation in biopolymer salt matrix.

The ionic conductivity (σ) in the case of electrolyte system is given as

$\sigma = n.\ q.\ \mu$ (Equation 2)

where n is the charge carrier density, q is the charge of the carrier, μ is the mobility of the carriers. Therefore, any increase in either of the parameters, n or q, will certainly affect the value of ionic conductivity.

Here the concentration of free charge carrier's density can be given by:

$$n = \left[\frac{\sigma_{DC}}{\left(\sqrt{\frac{\varepsilon'\omega}{\varepsilon's}} - 1 \right) \varepsilon_0 \varepsilon'_s \omega_x} \right]^4 \varepsilon_0 \varepsilon'_s \frac{kT}{e^2 d^2} \qquad \text{(Equation 3)}$$

where σ_{DC} = Conductivity at high frequency, ε_0 = Vacuum permittivity, k = Boltzmann constant, ε'_s & ε'_ω= Real permittivity at high frequency and at the frequency ω respectively, d = thickness of the sample. Here, ε'_s is the real part of the dielectric permittivity in the high frequency region and ω_x is the angular frequency for which $\varepsilon'(\omega_x) = 10\ \varepsilon'_s$.

Adding KI in biopolymer starch matrix provides additional mobile charge species, i.e., in present case, K^+ cation and I^- anion, and hence, ionic conductivity is enhanced while decrease in ionic conductivity could be explained on the basis of charge pair model. Variation in the number of charge carriers and mobility with KI concentration can be easily understood by using Eq. 2. Basically, the number of charge carriers remains almost the same. Beyond certain fixed wt% of dispersoid, the increase in mobility dominates by the decrease in the carrier concentration. This is due to the fact that as the salt concentration increased, the amorphicity of the biopolymer matrix increased, thereby allowing more pathways to ions for migration. As a result, the conductivity increased with the addition of the salt concentration and attains maxima in agarose, arrowroot, sago starch and phythagel respectively. This means that the conductivity is not affected by the value of n, and depends on the movement and distribution of charge carriers. At high salt concentration, ion moves in the sample within very short distance and the mobility does not affect the conductivity.

4.4 Temperature dependence conductivity

Temperature dependent ionic conductivity of solid biopolymers electrolyte is shown in Fig. 4. It was clear that with increasing temperature, the ionic conductivity values increased and followed the Arrhenius type behavior. Linear relations are observed in all biopolymer electrolytes, which mean that there is no phase transition in the biopolymer matrix or no domain formed by addition of KI. It is known that above T_g (glass transition) temperature, biopolymer starts to burn, hence it restricted our experiment. Variation of conductivity by considering Arrhenius type behavior is given by the Eq. 4.

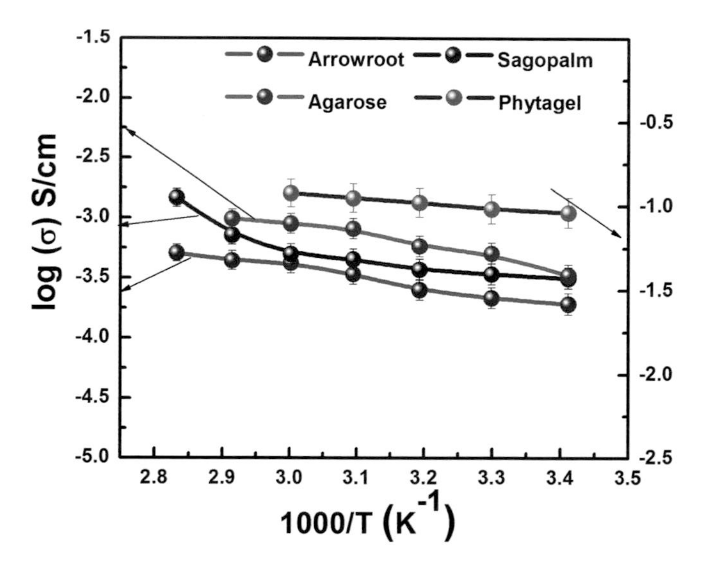

Figure 4. Variation of conductivity with temperature plot in biopolymers: KI systems.

$$\sigma = \sigma_0 \exp\left(\frac{-E_a}{kT}\right) \quad \text{(Equation 4)}$$

where, σ_0 is the pre-exponential factor, E_a is the activation energy and k is the Boltzmann constant. From this relationship, the activation energy E_a for maximum conductivity of biopolymer electrolyte was calculated, where the activation energy varies such that E_a = 0.33, 0.40, 0.39, 0.24e V are the energy values required for an ion to initiate movement. When the ion has acquired sufficient energy, it will move away from the donor ion.

4.5 *Ion dissociation factor*

To further clarify the role of charge carriers in biopolymer electrolyte matrix, we have evaluated the room temperature dissociation factor (n/n_0) for all the samples. The change in the relative number of charge carriers (n/n_0) with increasing amount of salt (KI) concentration in the samples are shown in the Fig. 5. It is observed that at the maximum doping limit, the n/n_0 is maximum and then shows a decreasing trend. It follows the similar trend as we have observed in conductivity as well as dielectric measurements. The conductivity, therefore, in biopolymers: KI matrix is predominantly governed by the number of charge carriers and follow the electrolyte dissociation theory given by Barker (Baker et al. 1964):

$$n = n_0 \exp\{-U/2\varepsilon kT\} \quad \text{(Equation 5)}$$

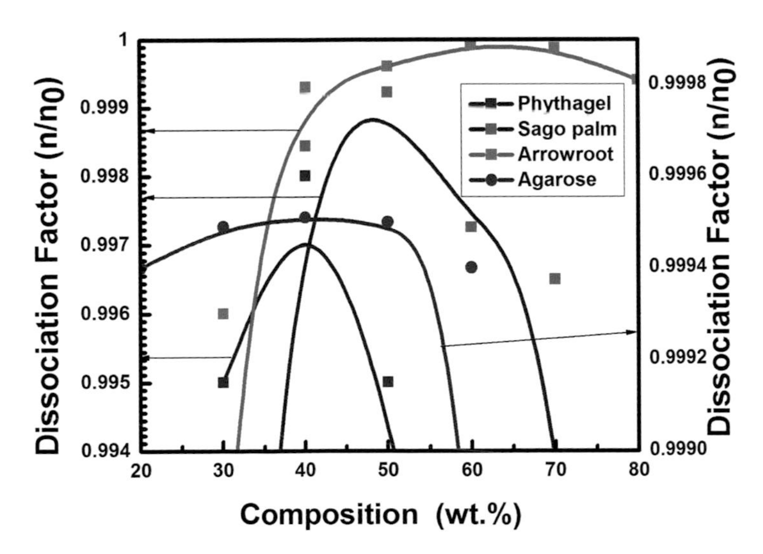

Figure 5. Change in relative number of charge carriers with increasing amount of KI in biopolymers: KI polymer electrolyte matrix.

where, U = Dissociation energy of the salt, k = Boltzmann constant, ε = Dielectric constant of the system, T = Temperature of the sample.

4.6 Dielectric studies

The information about different molecular motion and relaxation processes are well correlated with dielectric studies. Dielectric measurements are extremely sensitive to small changes in material properties (molecular relaxation of the order of only a few nanometer involves dipole changes that can be observed by dielectric study). To justify the overall contribution of the number of free charge carriers in the total conductivity, the dielectric constants (ε) of the biopolymer electrolyte films were calculated. Figure 6 shows the change in dielectric constant with increasing amount of KI in the biopolymers electrolyte at frequency of 82520 Hz. It is clear that conductivity data matches well with dielectric data. The addition of KI results in the change in dielectric constant of the matrix, which results in the change in the number of free charge carriers and thereby the conductivity.

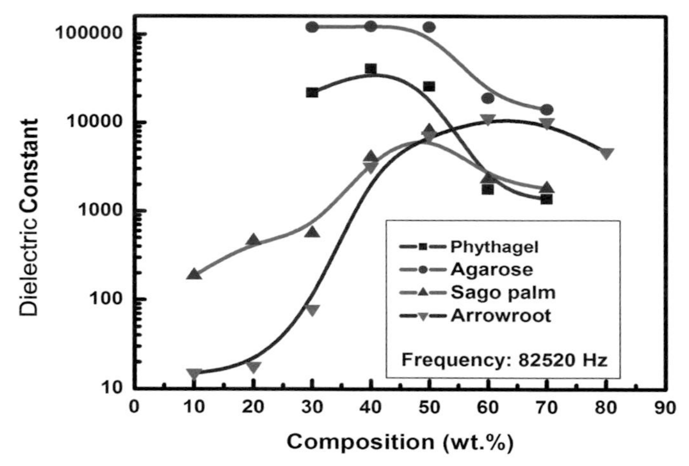

Figure 6. Dielectric constant vs. composition plot of biopolymers: KI polymer electrolyte system.

4.7 Ionic transference number measurement using DC polarization method

Using this simple method we have evaluated the percentage of ionic or electronic nature in biopolymer: KI polymer electrolyte films. In this method, we have applied a fixed DC potential of 0.25 V for ~ 6 hours to the steel plate/biopolymer electrolyte/steel plate system. The DC current was monitored with respect to time by using Keithley source measure unit

2400. By monitoring the initial current and final current and using Eq. 6, we have calculated the ionic transference number. The ionic transference numbers of biopolymer electrolytes, i.e., Sago Palm, Arrowroot, Agarose and Phythagel systems were calculated as 0.94, 0.93, 0.82, 92 respectively, which confirm the ionic nature of the biopolymers electrolyte system.

$$t_{\text{ion}} = \frac{I_{\text{initial}} - I_{\text{final}}}{I_{\text{initial}}} \quad \text{(Equation 6)}$$

4.8 *FTIR studies*

Fourier Transform Infrared Spectroscopy (FTIR) spectra of pure biopolymer and biopolymers doped with KI were recorded using Perkin Elmer 883 IR spectrophotometer between 4000 and 450 cm^{-1} (Khanmirzaei et al. 2013, Ahmad et al. 2006, Sing et al. 2013, 2014). The different bands assigned with various functional groups are listed in Table 4. It is clear from the table and spectra (Fig. 7) that almost all the peaks related to host materials (Biopolymers and KI) are present in biopolymers doped KI sample. It is also clear from the table that there is some shifting in peak portion in biopolymers-salt complex which indicates complex formation as we observed in common polyethers-salt complex polymers (Singh et al. 2012, Sequeira et al. 2010).

4.9 *XRD studies*

Figure 8 shows the X-ray diffraction patterns (XRD) of different samples of biopolymers-salt electrolyte using Rigaku D/max-2500 with scan rate at 2°/min for Agarose, Arrowroot, Sago palm and phythagel. It is evident that all salts are well dissolved in biopolymer matrix which is affirmed by the disapperance of XRD peaks related to salts in biopolymer-salt complex XRD. It is also clear that doping of salt broadens the peak which is clearly an indication of increase in the amorphous behavior of biopolymer matrix/KI system (decreses in crystallinity). Both decrease in crystallinity or increase in amorphicity by salt doping enhances the overall ionic conductivity of the system.

4.10 *Dye-sensitized solar cell performance (I–V curve)*

The photovoltaic performance (*I–V* curve) of the DSSCs were measured with Keithley 2400 source meter under 1 sun light condition. For developing redox couple in Biopolymers-KI matrix, we have added iodine

Table 4. Infrared spectra data of pure biopolymer and KI doped bio-polymer electrolyte system.

Wavelength in cm^{-1}								Freq. ranges	Groups (bonds)	Functional groups
Pure Arrowroot	**Arrowroot + KI**	**Pure Sago Palm**	**Sago Palm + KI**	**Pure Agarose**	**Agarose + KI**	**Pure Phythagel**	**Phythagel + KI**			
3584	3372	3313	3252	3445.86	3434.15	3358	3626	3500–3200 (s,b) 3640–3610 (s,sh)	O–H stretch, free hydroxyl, H–bonded	Alcohols, phenols
-	2929	2929	2931	-	-	2926	2921	3000–2850 (m)	C-H (stretch)	Alkane
2149		2149		2124.66	2105.69	2126	2142	2260–2100 (w)	–CC– stretch	Alkynes
1645	-	1642	-	-	-	-	1667, 1663, 1660, 1651	1680–1640 (m)	-C=C- stretch	Alkenes
-	1626	-	1633	1636.32	1637.57	1619	1644, 1614	1650–1580 (m)	N–H bend	1° amines
1462	1408	1412	1454, 1416	1400.34	1400.22	1409	1445, 1427, 1416	1500–1400 (m)	C–C stretch (in–ring)	Aromatics
	-	1364	1372	-	-	-	-	1370–1350 (m)	C–H rock	Alkanes
1169	1234, 1203, 1075	1241, 1204, 1052, 1151	1239, 1203, 1165, 1047	1121.00	1109.07	1337, 1297, 1235, 1194	1237,1196, 1171	1300–1150 (m) 1300–1150 (m)	C–O stretch C–H wag (–CH_2X)	Alcohols, carboxylic acids, esters, ethers alkyl halides
936	935	932	935	-	-	942, 924	943, 923	950–910 (m)	O-H bend	Carboxylic acid
863, 787	860, 762, 709	859, 762, 708	860, 762, 715	687.28	-	890, 836, 812, 749, 699	891, 836, 815, 755	900–675 (s)	C-H "oop"	Aromatics
482	607, 588	-	-	666.77	666.72, 545.30	666, 644, 608, 565, 546, 539	666, 462, 421, 407	690–515 (m)	C–Br stretch	Alkyl halides

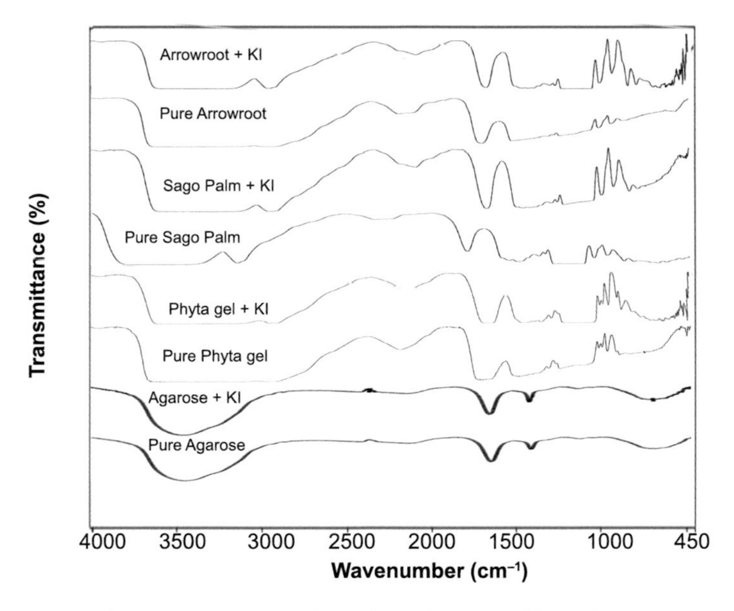

Figure 7. Infrared spectroscopy of pure biopolymers and biopolymers-KI samples.

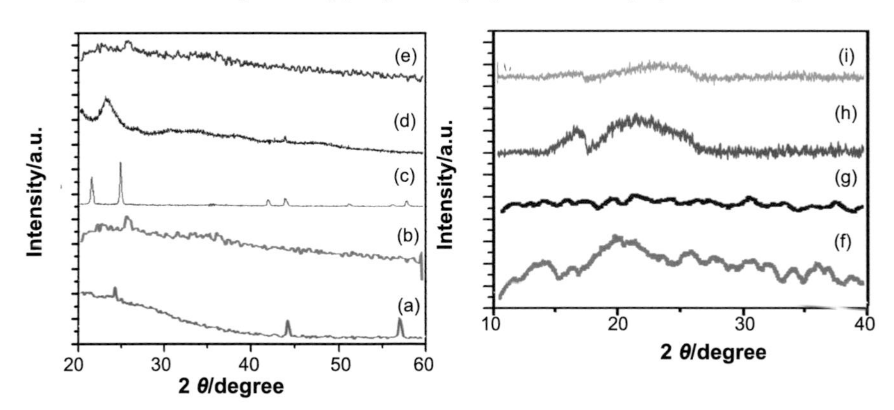

Figure 8. The XRD pattern of pure biopolymers and KI doped biopolymers systems. (a) Pure Sago Palm, (b) Sago palm + KI, (c) Potassium Iodide, (d) Pure Arrowroot, (e) Arrowroot + KI, (f) Pure Phytagel, (g) Phytagel + KI, (h) Pure Agarose and (i) Agarose + KI.

in biopolymers: KI matrix. DSSCs have been fabricated using the highest ionic conductivity composition of biopolymers: KI matrix. The different photovoltaic parameters such as J_{sc} (Acm^{-2}), V_{oc} (V), FF (%), Area (cm^2) and light intensity in biopolymers: salt matrix developed in our laboratory as well as the ones existing in other literature are listed in Table 5.

Table 5. The recorded photovoltaic parameters of DSSC observed in biopolymers-salt matrix.

Composition for Agarose/Agar based Biopolymer	Dye	Working Electrode	σ (Scm^{-1})	*Jsc* $mAcm^{-2}$	Voc (*V*)	FF (%)	Area (cm^2)	η (%)	Intensity m Wcm^{-2}	References
$KI\text{-}I_2$	N719	$FTO\text{-}TiO_2$	9.02×10^{-3}	3.27	0.670	0.24	0.6	0.54	100	Singh et al. 2013, 2016, 2017
$NH_4I\text{-}I_2$-Glycerol	N719	ITO-ZnO	4.89×10^{-3}	0.007	0.29	-	1	-	100	Siti et al. 2013
MPII-PC/DMSO-GuSCN-NMBI-I_2	N719	$FTO\text{-}TiO_2$	-	11.73	0.70	0.64	0.283	5.25	100	Hsu et al. 2013
AEII-PC/DMSO-GuSCN-NMBI-I_2	N719	$FTO\text{-}TiO_2$	-	11.71	0.72	0.65	0.283	5.45	100	Hsu et al. 2013
APII-PC/DMSO-GuSCN-NMBI-I_2	N719	$FTO\text{-}TiO_2$	-	11.53	0.70	0.62	0.283	4.97	100	Hsu et al. 2013
DAII-PC/DMSO-GuSCN-NMBI-I_2	N719	$FTO\text{-}TiO_2$	-	11.84	0.70	0.60	0.283	4.96	100	Hsu et al. 2013
DMSO/PC-MPII-GuSCN-NMBI-I_2	N719	ITO/TiO_2	14.2×10^{-3}	4.65	0.73	0.58	0.25	1.97	100	Hsu et al. 2011
DMSO/4EG-MPII-GuSCN-NMBI-I_2	N719	ITO/TiO_2	4.4×10^{-3}	3.69	0.65	0.57	0.25	1.38	100	Hsu et al. 2011
DMSO/3EG-MPII-GuSCN-NMBI-I_2	N719	ITO/TiO_2	4.6×10^{-3}	4.01	0.61	0.57	0.25	1.39	100	Hsu et al. 2011
DMSO/PG-MPII-GuSCN-NMBI-I_2	N719	ITO/TiO_2	6.2×10^{-3}	3.21	0.58	0.57	0.25	1.06	100	Hsu et al. 2011
DMSO-MPII-GuSCN-NMBI-I_2	N719	ITO/TiO_2	5.0×10^{-3}	3.69	0.58	0.54	0.25	1.15	100	Hsu et al. 2011

TiO_2-NMP-LiI/I_2	N719	FTO-TiO_2	2.66×10^{-3}	5.28	0.605	0.55	0.25	1.71	100	Yang et al. 2014
Co_3O_4-NMP-LiI/I_2	N719	FTO-TiO_2	$4 37 \times 10^{-3}$	7.24	0.635	0.46	0.25	2.11	100	Yang et al. 2014
NiO-NMP-LiI/I_2	N719	FTO-TiO_2	3.33×10^{-3}	6.20	0.625	0.52	0.25	2.02	100	Yang et al. 2014
LiI/I_2-NMP	-	TiO_2	3.94×10^{-4}	8.24	0.597	0.625	0.25	4.14	73	Yang et al. 2011
LiI/I_2-NMP-nanoparticle TiO_2	-	TiO2	4.4×10^{-4}	10.96	0.545	0.570	0.25	4.74	73	Yang et al. 2011
AN/MOZ-TBP-I_2/LiI-Pr_4NI+I^{3-}	N3	FTO/ITO/TiO_2	4.36×10^{-3}	14.00	0.76	0.66	0.20	7.06	100	Masao et al. 2004
1-alkyl–3-methylimidazolium Iodides-LiI/I_2-NMP	N719	FTO/TiO_2	-	6.77	0.650	0.666	0.25	2.93	100	Suzuki et al. 2006
LiI/I_2-NMP-Fe_3O_4-polysorbate 80	N719	FTO/TiO_2	2.98×10^{-3}	5.00	0.7	0.531	-	1.83	100	Guo et al. 2013
LiI/I_2-NMP-Fe_3O_4-PEG200	N719	FTO/TiO_2	-	3.70	0.67	0.612	-	1.50	100	Guo et al. 2013
LiI/I_2-NMP-Fe_3O_4-PVP	N719	FTO/TiO_2	-	3.00	0.67	0.598	-	1.19	100	Guo et al. 2013
LiI/I_2-NMP-Fe_3O_4-SDS	N719	FTO/TiO_2	-	3.18	0.66	0.626	-	1.29	100	Guo et al. 2013

Table 5 contd. ...

... Table 5 contd.

Composition for Agarose/Agar based Biopolymer	Dye	Working Electrode	σ (Scm^{-1})	*Jsc* $mAcm^{-2}$	*Voc* (*V*)	FF (%)	Area (cm^2)	η (%)	Intensity m Wcm^{-2}	References
Carrageenan-AN/ MOZ-TBP-I_2/LiI-Pr_4NI+I^{3-}	N3	FTO/ITO/ TiO_2	2.92×10^{-3}	14.67	0.73	0.67	0.20	6.87	100	Masao et al. 2004
Carrageenan-AN/ MOZ-TBP-I_2/LiI-I^-/ I^{3-}	N3	FTO/ITO/ TiO_2	-	14.67	0.73	0.67	0.20	6.87	100	Masao et al. 2004
Carrageenan-AN/ MOZ-$(Pr)_4$NI-I_2-LiI –TBP	N3	PEN/ITO/ TiO_2	-	2.68	0.80	0.72	0.25	1.54	100	Nemoto et al. 2007
Carrageenan-AN/ MOZ-$(Pr)_4$NI-I_2-LiI-TBP	N3	FTO-TiO_2	-	8.67	0.75	0.65	0.25	4.23	100	Nemoto et al. 2007
Gellan gum + KI	N719	FTO-TiO_2	0.00534	0.0666	0.57	0.57	0.45	0.56	100	-
Sago Starch + KI	N719	FTO-TiO_2	3.4×10^{-4}	0.0291	0.582	0.60	0.45	0.578	100	Singh et al. 2016b
Amylopectin-DMSO-I_2-LiI-DMHIm-(4-tertbutylpyridine)	N719	FTO-TiO_2	-	6.83	0.74	0.66	0.16	3.34	100	Ramesh et al. 2012
Arrowroot starch + KI	N719	FTO-TiO_2	1.04×10^{-4}	0.0568	0.56	-	0.45	0.63	100	Zhao et al. 2009

Rice-LiI: MPII: TiO_2	N3	ITO/TiO_2	3.63×10^{-4}	0.49	0.45	0.75	0.25	0.17	100	Khanmirzaei et al. 2013
Chitosan-NaI-I_2	-	FTO-TiO_2	-	1.05	0.349	0.34	0.16	0.13	100	Singh et al. 2010a
Chitosan-NaI-I_2-EMImSCN	-	FTO-TiO_2	2.60×10^{-4}	2.62	0.53	0.52	0.16	0.73	100	Singh et al. 2010a
Chitosan-NH_4I-I_2	-	ITO-TiO_2	3.73×10^{-7}	0.0049	0.15	0.22	0.09	0.29	56.4	Buraidah et al. 2010
Chitosan-NH_4I-I_2-EC	-	ITO-TiO_2	7.34×10^{-6}	0.0072	0.22	0.18	0.09	0.51	56.4	Buraidah et al. 2010
Chitosan–NH_4I– I_2-BMII	-	ITO-TiO_2	8.47×10^{-4}	0.0192	0.26	0.14	0.09	1.24	56.4	Buraidah et al. 2010
Chitosan-NH_4I-BMII	Black rice anthocyanin	ITO/TiO_2	3.02×10^{-4}	0.9	3.65	0.45	0.16	0.15	100	Buraidah et al. 2011
Chitosan-PEO-NH_4I-I_2-BMII	Black rice anthocyanin	ITO-TiO_2	5.52×10^{-4}	1.213	4.00	0.47	0.16	0.23	100	Buraidah et al. 2011
Tartaric-Phthaloyl Chitosan-NH_4I-BMII	Red cabbage anthocyanin	ITO-TiO_2	5.86×10^{-4}	3.472	3.65	0.34	0.16	0.43	100	Buraidah et al. 2011
Tartaric-Phthaloyl Chitosan-PEO-NH_4I-BMII	Red cabbage anthocyanin	ITO-TiO_2	6.24×10^{-4}	3.503	3.40	0.39	0.16	0.46	100	Buraidah et al. 2011
Chitosan-PEO-NH_4I-I_2	Lawsonia inermis	ITO/TiO_2	1.18×10^{-5}	0.38	336	0.57	0.05	0.7	100	Al-Bat'hi et al. 2013
Chitosan-PEO-NH_4I-I_2	Sumac/Rhus	ITO/TiO_2	1.18×10^{-5}	0.93	394	0.41	0.05	1.5	100	Al-Bat'hi et al. 2013
Chitosan-PEO-NH_4I-I_2	Curcuma longa	ITO/TiO_2	1.18×10^{-5}	0.20	280	0.65	0.05	0.36	100	Al-Bat'hi et al. 2013

5. Conclusions

This chapter deals with the overview of the structure, occurrence, properties (chemical, physical, electrical) and with technical applications of the most biopolymers generated by different sources. We have successfully presented collective data on the additives used by the researchers to enhance the ionic conductivity of biopolymer based electrolytes for device performance particularly in DSSC. It was observed that the biopolymer electrolytes in general followed Arrhenius type of behavior and its ionic conductivity is predominantly governed by the ionic charge carriers. Structural studies revealed that the complexion of biopolymer with dispersoids (salts) decreases the crystallinity of biopolymer matrix and consequently enhances the amorphous region and hence, conductivity increase drastically. These biopolymer-salt complexed systems possess high conductivity values (10^{-3}–10^{-4} S/cm) which further affirm that biopolymer electrolyte could be a novel alternative in developing highly efficient DSSC.

6. Abbreviations

DMSO	:	Dimethyl sulfoxide
PC	:	Propylene carbonates
PG	:	Propylene glycol
3EG	:	Triethylene glycol
4EG	:	Tetraethylene glycol
MPII	:	1-methyl-3-proplylimidazolium iodide
AEII	:	1-allyl-3-ethylimidadolium iodide
APII	:	1-allyl-3-propylimidazolium iodide
DAII	:	1-3-diallylimidazolium iodide
SDS	:	Sodium dodecyl sulfate
PVP	:	Polyvinylpyrrolidone
PEG200	:	Polyethylene glycol
TW-80	:	Polysorbate 80
NMP	:	1-methyl-2-pyrrolidinone
GBL	:	γ-butyrolactone
PEO–HPC	:	Poly(ethylene oxide)–2-hydroxypropylcellulose
BmImTf	:	1-butyl-3-methylimidazolium trifluoromethanesulfonate
LITFSI	:	Lithium bis(trifluoromethanesulfonyl)imide
DES	:	Deep eutectic solvent
[Amim] Cl	:	1-allyl-3-methylimidazolium chloride
BmImPF6	:	1-butyl-3-methylimidazolium hexafluorophosphate
DMAc	:	N, N-dimethylacetamide
LiCl	:	Lithium chloride
[BMIM]Cl	:	1-butyl-3-methylimidazolium chloride

EC	:	Ethylene carbonates
PEG	:	Poly(ethylene glycol)
LiTFSI	:	Lithium trifluoromethanesulfonimide
SPEEK-CS	:	Sulfonated poly(ether ether ketone)-chitosan
DAP	:	Diethanolamine modified pectin
BC	:	Bacterial cellulose
TEA	:	Triethanolamine
GA	:	Glutaraldehyde
DTAB	:	Dodecyltrimethyl ammonium bromide
EMImSCN	:	1-ethyl 3-methylimidazolium thiocyanate
N3	:	Cis-Bis(isothiocyanato)bis(4,40-dicarboxyl-2,20-bipyridine)-ruthenium(II), Ru(dcbpy)2 (NCS)2
N719	:	Cis-bis(isothiocyanato)bis(2,2'-bipyridy1-4,4'-dicarboxylato)-ruthenium(II)bis-tetrabutylammonium
AN	:	Acetonitrile
MOZ	:	3-methyl-2-oxazolidinone
(Pr)4NI	:	Tetrapropylammonium iodide
TBP	:	4-tertiary butylpyridine
MPIm-I	:	1-methyl-3-propylimidazolium iodide
DMHIm	:	1, 2-dimethyl-3-n-hexylimidazolium iodide

7. Acknowledgments

This work was supported by Department of Science & Technology project (DST/TSG/PT/2012/51-C) Government of India.

References

Abidin, S.Z.Z., A.M.M. Ali, O.H. Hassan and M.Z.A. Yahya. 2013. Electrochemical studies on cellulose acetate-LiBOB polymer gel electrolytes. Int. J. Electrochem. Sci. 8: 7320–7326.

Agrawal, R.C. and R.K. Gupta. 1999. Superionic solid: composite electrolyte phase-an overview. J. Mater. Sci. 34(6): 1131–1162.

Agrawal, R.C. and G.P. Pandey. 2008. Solid polymer electrolytes: materials designing and all-solid-state battery applications: an overview. J. Phys. D: Appl. Phys. 41: 223001.

Ahmad, K.A.S., R. Puteh and A.K. Arof. 2006. Characterizations of chitosan-ammonium triflate ($NH_4CF_3SO_3$) complexes by FTIR and impedance spectroscopy. Phys. Stat. Sol. (a) 203(3): 534–543.

Ahmad, Z. and M.I.N. Isa. 2012. Ionics conduction *via* correlated barrier hoping mechanism in CMC-SA solid biopolymer electrolytes. Int. J. Latest Research in Science and Technology 1(2): 70–75.

Ahmed, J., B.K. Tiwari, S.H. Imam and M.A. Rao. 2012. Starch-based polperic materials and nanocomposites chemistry, processing, and applications. Taylor & Francis Group, CRC Press.

Akhtar, M.S., J.M. Chun and O.B. Yang. 2007. Advanced composite gel electrolytes prepared with titania nanotube fillers in polyethylene glycol for the solid-state dye-sensitized solar cell. Electrochem. commun. 9: 2833.

Akhtar, M.S., K.K. Cheralathan, J.M. Chun and O.B. Yang. 2008. Composite electrolyte of heteropolyacid (HPA) and polyethylene oxide (PEO) for solid-state dye-sensitized solar cell. Electrochim. Acta. 53: 6623.

Al-Bat'hi Souad, A.M., I. Alaei and I. Sopyan. 2013. Natural photosensitizers for dye sensitized solar cells. Int. J. Renewable Energy Research 3(1): 138–143.

Ali, B.H., A. Ziada and G. Blunden. 2009. Biological effects of gum arabic: a review of some recent research. Food and Chemical Toxicology 47: 1–8.

Alias, S.S., S.M. Chee and A.A. Mohamad. 2014. Chitosan–ammonium acetate–ethylene carbonate membrane for proton batteries. Arab. J. Chem. 10: S3687–S3698.

Amelia, R.P.D., J.C. Tomic and W.F. Nirode. 2014. The determination of the solubility parameter (δ) and the mark-houwink constants (K & α) of food grade polyvinyl acetate (PVAc). J. Polymer and Biopolymer Physics Chemistry 2(4): 67–72.

Anandan, S., S. Pitchumani, B. Muthuraaman and P. Maruthamuthu. 2006. Heteropolyacid-impregnated PVDF as a solid polymer electrolyte for dye-sensitized solar cells. Sol. Energy Mater. Sol. Cells. 90: 1715.

Anandan, S., R. Sivakumar and P. Tharani. 2008. Solid-state dye-sensitized solar cells constructed with an electrochrome impregnated elastomeric electrolyte. Synth. Met. 158: 1067.

Andrade, J.R., E. Raphael and A. Pawlicka. 2009. Plasticized pectin-based gel electrolytes. Electrochim. Acta. 54(26): 6479–6483.

Armand, M.B., J.M. Chabagno and M.J. Duclot. 1979. *In*: Vashista, P. and Shenoy, G.K. (eds.). Fast Ion Transport in Solids. Elsevier, North Holland. 131.

Arof, A.K., N.E.A. Shuhaimi, N.A. Alias, M.Z. Kufian and S.R. Majid. 2010. Application of chitosan/iota-carrageenan polymer electrolytes in electrical double layer capacitor (EDLC). J Solid State Electrochem. 14(12): 2145–2152.

Arora, N., V. Sharma and R. Kumar. 2014. Xanthan gum based gel electrolyte containing NaOH i-manager's. J. Mater. Sci. 1(4): 20–23.

Augst, A.D., H.J. Kong and D.J. Mooney. 2006. Alginate hydrogels as biomaterials. Macromol. Biosci. 6: 623–633.

Ave, R.L. 2004. Biodegradable multiphase systems based on plasticized starch: a review. J. Macromol. Sci.: Part C: Polymer Rev. C 44(3): 231–274.

Aziz, N.A.N., N.K. Idris and M.I.N. Isa. 2010. Solid polymer electrolytes based on methylcellulose: FT-IR and ionic conductivity studies. Int. J. Polym. Anal. Ch. 15(5).

Bai, Y., Y. Cao, J. Zhang, M. Wang, R. Li, P. Wang, S.M. Zakeeruddin and M. Grätzel. 2008. High-performance dye sensitized solar cells based on solvent-free electrolytes produced from eutectic melts. Nat Mater. 7: 626–630.

Bakar, N.Y. and M.I.N. Isa. 2014. Potential of ionic conductivity and transport properties solid biopolymer electrolytes based carboxy methylcellulose/chitosan polymer blend doped with dodecyltrimethyl ammonium bromide. Res. J. Recent Sci. 3(10): 69–74.

Barker, R.E. and C.R. Thomas. 1964. Effects of moisture and high electric fields on conductivity in Alkali-Halide-Doped cellulose acetate. J. Appl. Phys. 35: 3203–3216.

Benedetti, J.E., M.A. De Paoli and A.F. Nogueira. 2008. Enhancement of photocurrent generation and open circuit voltage in dye-sensitized solar cells using Li+ trapping species in the gel electrolyte. Chem. Commun. 1121.

Bhattacharya, B., H.M. Upadhyaya and S. Chandra. 1996. Photoelectrochemical studies of an ion conducting polymer (PEO)/semiconductor (Si) junction. Solid State Commun. 98: 633.

Bonhote, P., A.P. Dias, N. Papageorgiou, K. Kalyanasundaram and M. Grätzel. 1996. Inorg. Hydrophobic, Highly Conductive Ambient-Temperature Molten Salts. Chem. 35: 1168.

Buraidah, M.H., L.P. Teo, S.R. Majid and A.K. Arof. 2010a. Characteristics of TiO_2/solid electrolyte junction solar cells with I^-/I^{3-} redox couple. Opt. Mater 32: 723–728.

Buraidah, M.H., L.P. Teo, S.R. Majid, R. Yahya, R.M. Taha and A.K. Arof. 2010b. Characterizations of chitosan-based polymer electrolyte photovoltaic cells. Int. J. Photoenergy. 2010: 7 Article ID 805836.

Buraidah, M.H. Bonhote A.K. Arof. 2011a. Characterization of chitosan/PVA blended electrolyte doped with NH_4I. J. Non-Cryst. Solids 357(16-17): 3261–3266.

Buraidah, M.H., L.P. Teo, S.N.F. Yusuf, M.M. Noor, M.Z. Kufian, M.A. Careem, S.R. Majid, R.M. Taha and A.K. Arof. 2011b. TiO_2/chitosan-$NH_4I(+I_2)$-BMII-based dye-sensitized solar cells with anthocyanin dyes extracted from black rice and red cabbage. Int. J. Photoenergy 2011: 11.

Campo, V.I., D.F. Kawano, D. Braz da Silva and I. Carvalho. 2009. Carrageenan: biological properties, chemical modifications and structural analysis—a review. Carbohyd. Polym. 77: 167–80.

Chai, M.N. and M.I.N. Isa. 2013. The oleic acid composition effect on the carboxymethyl cellulose based biopolymer electrolyte. J. Crystallization Process and Technology 3: 1–4.

Chamy, R. and F. Rosenkranz. 2013. Biodegradation - Life of Science. InTech. 378. ISBN 978-953-51-1154-2. doi: 10.5772/52777.

Chandra, R. and R. Rustgi. 1998. Biodegradable polymers. Prog. Polym. Sci. 23(7): 1273–1335.

Chandra, S. 1981. Superionic solids: principles and applications. Acta Cryst. A38: 878. doi: 10.1107/S0567739482001867.

Chatzivasiloglou, E., T. Stergiopoulos, N. Spyrellis and P. Falaras. 2005. Solid-state sensitized solar cells, using $[Ru(dcbpyH_2)_2Cl_2]\cdot 2H_2O$ as the dye and PEO/titania/I^-/I^{3-} as the redox electrolyte. J. Mater Process Tech. 161: 234.

Chelmecki, M., W.H. Meyer and G. Wegner. 2007. Effect of crosslinking on polymer electrolytes based on cellulose. J. Appl. Polym. Sci. 105(1): 25–29.

Chen, Z., Y. Tang, H. Yang, Y. Xia, F. Li, T. Yi and C. Huang. 2007. Nanocrystalline TiO_2 film with textural channels: exhibiting enhanced performance in quasi-solid/solid-state dye-sensitized solar cells. J. Power Sources 171: 990.

Croce, F., G.B. Appetecchi, L. Persi and B. Scrosati. 1998. Nanocomposite polymer electrolytes for lithium batteries. Nature 394: 456.

Dumitriu, S. 2004. Polysaccharides: Structural Diversity and Functional Versatility (ed.). Marcel Dekker, Inc. 1224.

Elizabeth, P. 1979. The polysaccharides of green, red and brown seaweeds: their basic structure, biosynthesis and function. Br. Phycol. J. 14: 103–117.

Ellis, A.B., S.W. Kaiser and M.S. Wrighton. 1976. Visible light to electrical energy conversion. Stable cadmium sulfide and cadmium selenide photoelectrodes in aqueous electrolytes. J. Am. Chem. Soc. 98: 1635.

Finkenstadt, V.L. 2005. Natural polysaccharides as electroactive polymers. Appl. Microbiol. Biotechnol. 67: 735–745.

Finkenstadt, V.L. and J.L. Willett. 2004. Electroactive materials composed of starch. J. Polym. Environ. 12(2): 43–46.

Finkenstadt, V.L. and J.L. Willett. 2005. Preparation and characterization of electroactive biopolymers. Macromol. Symp. 227(1): 367–372.

Flores, I.C., J.N. Freitas, C. Longo, M.A. De Paoli, H. Winnischofer and A.F. Nogueira. 2007. Dye-sensitized solar cells based on TiO_2 nanotubes and a solid-state electrolyte. J. Photochem. Photobiol. A: Chemistry 189: 153.

Fonseca, C.P., D.S. Rosa, F. Gaboardi and S. Neves. 2006. Development of a biodegradable polymer electrolyte for rechargeable batteries. J. Power Sources 155: 381–384.

Freitas, J.N., A.S. Goncalves, M.A. De Paoli, J.R. Durrant and A.F. Nogueira. 2008a. The role of gel electrolyte composition in the kinetics and performance of dye-sensitized solar cells. Electrochimica Acta 53: 7166.

Freitas, J.N., C. Longo, A.F. Nogueira and M.A. De Paoli. 2008b. Solar module using dye-sensitized solar cells with a polymer electrolyte. Sol. Energy Mater. Sol. Cells 92: 1110.

Freitas, J.N., A.F. Nogueira and M.A. De Paoli. 2009. New insights into dye-sensitized solar cells with polymer electrolytes. J. Mater. Chem. 19: 5279.

Ganesan, S., B. Muthuraaman, J. Madhavan, V. Mathew, P. Maruthamuthu and S.A. Suthanthiraraj. 2008a. The use of 2,6-bis (N-pyrazolyl) pyridine as an efficient dopant in conjugation with poly(ethylene oxide) for nanocrystalline dye-sensitized solar cells. Electrochim. Acta. 53: 7903.

Ganesan, S., B. Muthuraaman, V. Mathew, J. Madhavan, P. Maruthamuthu and S.A. Suthanthiraraj. 2008b. Performance of a new polymer electrolyte incorporated with diphenylamine in nanocrystalline dye-sensitized solar cell. Sol. Energy Mater. Sol. Cells. 92: 1718.

Gerischer, H. and J. Gobrecht. 1976. On the power characteristics of electrochemical solar cells. Ber. Bunsenges. Phys. Chem. 80: 327.

Gorlov, M. and L. Kloo. 2008. Ionic liquid electrolytes for dye-sensitized solar cells. Dalton Trans. 2655.

Grätzel, M. 2000. Perspectives for dye-sensitized nanocrystalline solar cells. Prog. Photovolt. Res. Appl. 8: 171.

Gratzel, M. 2001. Photoelectrochemical cells. Nature 414: 338–344.

Grätzel, M. 2003. Dye-sensitized solar cells. J. Photochem. Photobiol. C: Photochem. Rev. 4: 145.

Gratzel, M. 2009. Recent advances in sensitized mesoscopic solar cells. Acc. Chem. Res. 42: 1788–1798.

Gray, F.M. (ed.). 1991. Solid Polymer Electrolytes: Fundamentals and Technological Applications. VCH Publishers, New York.

Guo, X., P. Yi, Y. Yang, J. Cui, S. Xiao and W. Wang. 2013a. Effects of surfactants on agarose-based magnetic polymer electrolyte for dye-sensitized solar cells. Electrochim. Acta. 90: 524–529.

Guo, X., Y. Yi, F. Peng, W.J. Wang, S. Xiao and Y. Yang. 2013b. Effects of polyethylene glycol on agarose-based magnetic polymer electrolyte for dye-sensitized solar cell. Adv. Mater. Res. 860: 652–654.

Habibi, Y. and L.A. Lucia. 2012. Polysaccharide Building Blocks: A Sustainable Approach to the Development of Renewable Biomaterials. John Wiley and Sons Ltd. 430.

Hagfeldt, A. and M. Grätzel. 2000. Molecular photovoltaics. Acc. Chem. Res. 33: 269.

Halim, N.F.A., S.R. Majid, A.K. Arof, F. Kajzar and A. Pawlicka. 2012. Gellan gum-liI gel polymer electrolytes. Mol. Cryst. Liq. Cryst. 554: 232–238.

Hamdan, K.Z. and A.S.A. Khiar. 2014. Conductivity and dielectric studies of methylcellulose/chitosan-$NH_4CF_3SO_3$ polymer electrolyte. Key Eng. Mater. 594/595: 812.

Han, H., W. Liu, J. Zhang and X.Z. Zhao. 2005. A hybrid poly(ethylene oxide)/poly(vinylidene fluoride)/TiO_2 nanoparticle solid-state redox electrolyte for dye-sensitized nanocrystalline solar cells. Adv. Funct. Mater. 15: 1940.

Haque, S.A., E. Palomares, H.M. Upadhyaya, L. Otley, R.J. Potter, A.B. Holmes and J.R. Durrant. 2003. Flexible dye sensitised nanocrystalline semiconductor solar cells. Chem. Commun. 3008.

Hardin, B.E., H.J. Snaith and M.D. McGehee. 2012. Renaissance of dye-sensitized solar cells. Nat. Photonics 6: 162.

Harun, N.I., N.S. Sabri, N.H.A. Rosli, M.F.M. Taib, S.I.Y. Saaid, T.I.T. Kudin, A.M.M. Ali and M.Z.A. Yahya. 2010. Proton conductivity studies on biopolymer electrolytes. AIP Conf. Proc. 1250: 237.

Harun, N.I., R.M. Ali, A.M.M. Ali and M.Z.A. Yahya. 2012. Dielectric behaviour of cellulose acetate-based polymer electrolytes. Ionics 18(6): 599–606.

Hassan, F., H.J. Woo, N.A. Aziz, M.Z. Kufian and S.R. Majid. 2013. Synthesis of Al_2TiO_5 and its effect on the properties of chitosan–NH_4SCN polymer electrolytes. Ionics 19: 483–489.

Higgins, T.M., S.E. Moulton, K.J. Gilmore, G.G. Wallace and M. Panhuis. 2011. Gellan gum doped polypyrrole neural prosthetic electrode coatings. Soft Matter. 7(10): 4690–4695.

Hsu, H.L., W.T. Hsu and J. Leu. 2011. Effects of environmentally benign solvents in the agarose gel electrolytes on dye-sensitized solar cells. Electrochim. Acta. 56: 5904–5909.

Hsu, H.L., C.F. Tien, Yang Ya-Ting and Leu Jihperng. 2013. Dye-sensitized solar cells based on agarose gel electrolytes using allylimidazolium iodides and environmentally benign solvents. Electrochim. Acta. 91: 208–213.

Hsu, H.L., C.F. Tien and J. Leu. 2014. Effect of pore size/distribution in TiO_2 films on agarose gel electrolyte-based dye-sensitized solar cells. J. Solid State Electrochem. 18: 1665–1671.

Idris, N.H., H.B. Senin and A.K. Arof. 2007. Dielectric spectra of LiTFSI-doped chitosan/PEO blends. Ionics 13(4): 213–217.

Ileperuma, O.A., M.A.K.L. Dissanayake, S. Somasunderam and L.R.A.K. Bandara. 2004. Photoelectrochemical solar cells with polyacrylonitrile-based and polyethylene oxide-based polymer electrolytes. Sol. Energy Mater. Sol. Cells. 84: 117.

Isa, M.I.N. and A.S. Samsudin. 2013. Ionic conduction behavior of CMC based green polymer electrolytes. Adv. Mater. Res. 802: 194.

Iwakia, Y.O., E.M. Hernandez, J.R. Brionesc and A. Pawlicka. 2012. Sodium alginate-based ionic conducting membranes. Mol. Cryst. Liq. Cryst. 554(1): 221–231.

Jafirin, S., I. Ahmad and A. Ahamad. 2013. Potential use of cellulose from keaf in polymer electrolyte based on MG49 rubber composites. Bioresources 8(4): 5947–5964.

Jamaludin, A. and A.A. Mohamad. 2010. Application of liquid gel polymer electrolyte based on chitosan–NH_4NO_3 for proton batteries. J. Appl. Polym. Sci. 118(2): 1240–1243.

Johari, N.A., T.I.T. Kudin, A.M.M. Ali, T. Winie and M.Z.A. Yahya. 2009. Studies on cellulose acetate-based gel polymer electrolytes for proton batteries. Mater. Res. Innov. 13(3): 232–234.

Johari, N.A., T.I.T. Kudin, A.M.M. Ali and M.Z.A. Yahya. 2012. Electrochemical studies of composite cellulose acetate-based polymer gel electrolytes for proton batteries. P. Natl. A Sci. India A. 82(1): 49–52.

Jung, Y.S., A.R. Sathiya Priya, M.K. Lim, S.Y. Lee and K.J. Kim. 2010. Influence of amylopectin in dimethylsulfoxide on the improved performance of dye-sensitized solar cells. . J. Photoch. Photobio. A. Chemistry 209: 174–180.

Kadir, M.F.Z., S.R. Majid and A.K. Arof. 2010. Plasticized chitosan–PVA blend polymer electrolyte based proton battery. Electrochim. Acta. 55(4): 1475–1482.

Kadir, M.F.Z. and A.K. Arof. 2011a. Application of PVA-chitosan blend polymer electrolyte membrane in electrical double layer capacitor. Mater. Res. Innov. 15(2): S217–S220.

Kadir, M.F.Z., Z. Aspanut, R. Yahya and A.K. Arof. 2011b. Chitosan–PEO proton conducting polymer electrolyte membrane doped with NH_4NO_3. Mater. Res. Innov. 15: 164–167.

Kalaignan, G.P., M.S. Kang and Y.S. Kang. 2006. Effects of compositions on properties of PEO–KI–I2 salts polymer electrolytes for DSSC. Solid State Ionics 177 1091.

Kalia, S., B.S. Kaith and I. Kaur. 2011. Cellulose Fibers: Bio- and Nano-Polymer Composites. Green Chemistry and Technology, Springer Berlin Heidelberg. doi:10.1007/978-3-642-17370-7.

Kamarulzaman, N., Z. Osman, M.R. Muhamad, Z.A. Ibrahim, A.K. Arof and N.S. Mohamed. 2001. Performance characteristics of $LiMn_2O_4$/polymer/carbon electrochemical cells. J. Power Sources 97-98: 722–725.

Kang, J., W. Li, X. Wang, Y. Lin, X. Xiao and S. Fang. 2003. Polymer electrolytes from PEO and novel quaternary ammonium iodides for dye-sensitized solar cells. Electrochim Acta. 48: 2487.

Kang, M.S., J.H. Kim, Y.J. Kim, J. Won, N.G. Park and Y.S. Kang. 2005. Dye-sensitized solar cells based on composite solid polymer electrolytes. Chem. Commun. 889.

Kang, M.S., J.H. Kim, J. Won and Y.S. Kang. 2006. Dye-sensitized solar cells based on crosslinked poly(ethylene glycol) electrolytes. J. Photochem. Photobiol. A: Chemistry 183: 15.

Kang, M.S., J.H. Kim, J. Won and Y.S. Kang. 2007. Oligomer approaches for solid-state dye-sensitized solar cells employing polymer electrolytes. J. Phys. Chem. C 111: 5222.

Kang, M.S., K.S. Ahn, J.W. Lee and Y.S. Kang. 2008. Dye-sensitized solar cells employing non-volatile electrolytes based on oligomer solvent. J. Photochem. Photobiol. A: Chemistry. 195: 198.

Katsaros, G., T. Stergiopoulos, I.M. Arabatzis, K.G. Papadokostaki and P. Falaras. 2002. A solvent-free composite polymer/inorganic oxide electrolyte for high efficiency solid-state dye-sensitized solar cells. J. Photochem. Photobiol. A: Chemistry 149: 191.

Khanmirzaei, M.H. and S. Ramesh. 2013. Ionic transport and FTIR properties of lithium iodide doped biodegradable rice starch based polymer electrolytes. Int. J. Electrochem. Sci. 8: 9977–9991.

Khanmirzaei, M.H. and S. Ramesh. 2014a. Studies on biodegradable polymer electrolyte rice starch (RS) complexed with lithium iodide. Ionics 20: 691–695.

Khanmirzaei, M.H. and S. Ramesh. 2014b. Nanocomposite polymer electrolyte based on rice starch/ionic liquid/TiO_2 nanoparticles for solar cell application. Measurement 58: 68–72.

Khiar, A.S.A. and A.K. Arof. 2011. Electrical properties of starch/chitosan-NH_4NO_3 polymer electrolyte. International Journal of Mathematical, Computational, Physical, Electrical and Computer Engineering 5(11): 1662–1667.

Kim, J.H., M.S. Kang, Y.J. Kim, J. Won, N.G. Park and Y.S. Kang. 2004a. Dye-sensitized nanocrystalline solar cells based on composite polymer electrolytes containing fumed silica nanoparticles. Chem. Commun 1662–1663.

Kim, Y.J., J.H. Kim, M.S. Kang, M.J. Lee, J. Won, J.C. Lee and Y.S. Kang. 2004b. Supramolecular electrolytes for use in highly efficient dye-sensitized solar cells. Adv. Mater. 16: 1753.

Kim, J.H., M.S. Kang, Y.J. Kim, J. Won and Y.S. Kang. 2005. Poly(butyl acrylate)/NaI/I_2 electrolytes for dye-sensitized nanocrystalline TiO_2 solar cells. Solid State Ionics. 176: 579.

Koh, J.C.H. and Z.A. Ahmad. 2012. Bacto agar-based gel polymer electrolyte. Ionics 18(4): 359–364.

Kuang, D., C. Klein, Z. Zhang, S. Ito, J.E. Moser, S.M. Zakeeruddin and M. Grätzel. 2007. Stable, high-efficiency ionic-liquid-based mesoscopic dye-sensitized solar cells. Small. 3: 2094.

Kumar, M. and S.S. Sekhon. 2002. Role of plasticizer's dielectric constant on conductivity modification of PEO–NH4F polymer electrolytes. Eur. Polym. J. 38: 1297.

Kumar, M., T. Tiwari and N. Srivastava. 2012. Electrical transport behaviour of bio-polymer electrolyte system: potato starch + ammonium iodide. Carbohyd Polym. 88: 54–60.

Le, C.D., J. Bras and A. Dufresne. 2010. Starch nanoparticles: a review. Biomacromolecules 11: 1139–1153.

Lenz, R.W. 1993. Biodegradable polymer. Adv. Polym. Sci. 107: 1–40.

Leones, R., F. Sentanin, L.C. Rodrigues, I.M. Marrucho, J.M.S.S. Esperança, A. Pawlicka and M.M. Silva. 2012. Investigation of polymer electrolytes based on agar and ionic liquids. Express Polym. Lett. 6(12): 1007–1016.

Leones, R., M.B.S. Botelho, F. Sentanin, I. Cesarino, A. Pawlicka, A.S.S. Camargo and M.M. Silva. 2014. Pectin-based polymer electrolytes with Ir(III) complexes. Mol. Cryst. Liq. Cryst. 1: 604.

Li, B., L. Wang, B. Kang, P. Wang and Y. Qiu. 2006. Review of recent progress in solid-state dye-sensitized solar cells. Sol. Energy Mater. Sol. Cells. 90: 549.

Liew, C.W., S. Ramesh, K. Ramesh and A.K. Arof. 2012. Preparation and characterization of lithium ion conducting ionic liquid-based biodegradable corn starch polymer electrolytes. J. Solid State Electrochem. 16(5): 1869–1875.

Liew, C.W. and S. Ramesh. 2013. Studies on ionic liquid-based corn starch biopolymer electrolytes coupling with high ionic transport number. Cellulose 20: 3227–3237.

Longo, C. and M.A. De Paoli. 2003. Dye-sensitized solar cells: a successful combination of materials. J. Braz. Chem. Soc. 14: 889.

Lu, D.R., C.M. Xiao and S.J. Xu. 2009. Starch-based completely biodegradable polymer materials. Express Polym. Lett. 3(6): 366–375.

Ma, X., P.R. Chang, J. Yua and P. Lu. 2008. Characterizations of glycerol plasticized-starch (GPS)/carbon black (CB) membranes prepared by melt extrusion and microwave radiation. Carbohyd. Polym. 74(4): 895–900.

MacCallum, J.R. and C.A. Vincent. 1987 and 1989. Polymer Electrolyte Reviews. Elsevier, London, 1 and 2.

Majda, S. and L. Pierre. 1993. A spectroscopic investigation of the carrageenans and agar in the 1500–100 cm^{-1} spectral range. Spectrochim Acta. 49A (2): 209–221.

Majid, S.R. and A.K. Arof. 2005. Proton-conducting polymer electrolyte films based on chitosan acetate complexed with NH_4NO_3 salt. Physica B Condens Matter. 355(1-4): 78–82.

Majid, S.R. and A.K. Arof. 2009. Conductivity studies and performance of chitosan based polymer electrolytes in H_2/air fuel cell. Polym. Adv. Technol. 20(6): 524–528.

Mallick, H. and A. Sarkar. 2000. An experimental investigation of electrical conductivities in biopolymers. Bull. Mater. Sci. 23(4): 319–324.

Masao, K., H. Takayuki, K. Yuuki and U. Hirohito. 2004. Solid type dye-sensitized solar cell using polysaccharide containing redox electrolyte solution. J. Electroanal. Chem. 572: 21–27.

Mathew, S., A. Yella, P. Gao, B.R. Humphry, B.F.E. Curchod, N.A. Astani, I. Tavernelli, U. Rothlisberger, M.K. Nazeeruddin and M. Grätzel. 2014. Dye-sensitized solar cells with 13% efficiency achieved through the molecular engineering of porphyrin sensitizers. Nature Chemistry 6: 242–247.

Meyer, G.J. 2010. The 2010 millennium technology grand prize: dye-sensitized solar cells. ACS Nano 4: 4337–4343.

Mishra, R.K., A. Anis, S. Mondal, M. Dutt and A.K. Banthia. 2009. Preparation and characterization of amidated pectin based polymer electrolyte membranes. Chinese J. Polym. Sci. 27(5): 639–646.

Mishra, R.K., M. Datt, A.K. Banthia and A.B.A. Majeed. 2012. Development of novel pectin based membranes as proton conducting material. Int. J. Plastics Technology 16(1): 80–88.

Mohamad, S.A., R. Yahya, Z.A. Ibrahim and A.K. Arof. 2007. Photovoltaic activity in a ZnTe/PEO–chitosan blend electrolyte junction. Sol. Energ. Mat. Sol. C. 91(13): 1194–1198.

Mohamed, N.S., R H.Y. Subban and A.K. Arof. 1995. Polymer batteries fabricated from lithium complexed acetylated chitosan. J. Power Sources 56: 153–156.

Narsaiah, E.L., M.J. Reddy and U.V.S. Rao. 1995. Study of a new polymer electrolyte (PVP + KYF4) for solid-state electrochemical cells. J. Power Sources 55: 255.

Nemoto, J., M. Sakata, T. Hoshi, H. Ueno and M. Kaneko. 2007. All plastic dye-sensitized solar cell using a polysaccharide film containing excess redox electrolyte solution. J Electroanal Chem. 599: 23–30.

Ng, L.S. and A.A. Mohamad. 2006. Protonic battery based on a plasticized chitosan-NH_4NO_3 solid polymer electrolyte. J. Power Sources 163(1): 382–385.

Nijenhuis, K. 1997. Thermo reversible networks-viscoelastic properties and structure of gels. Adv. Polym. Sci. 130: 194–202.

Ning, W., Z. Xingxiang, L. Haihui and H. Benqiao. 2009a. 1-Allyl-3-methylimidazolium chloride plasticized-corn starch as solid biopolymer electrolytes. Carbohyd Polym. 76(3): 482–484.

Ning, W., Z. Xingxiang, L. Haihui and W. Jianping. 2009b. N, N-dimethylacetamide/lithium chloride plasticized starch as solid biopolymer electrolytes. Carbohyd Polym. 77(3): 607–611.

Nogueira, A.F., N.A. Vante and M.A. De Paoli. 1999. Solid-state photoelectrochemical device using poly(o-methoxy aniline) as sensitizer and an ionic conductive elastomer as electrolyte. Synth. Met. 105: 23.

Nogueira, A.F., J.R. Durrant and M.A. De Paoli. 2001a. Dye-sensitized nanocrystalline solar cells employing a polymer electrolyte. Adv. Mater. 13: 826.

Nogueira, A.F., M.A. De Paoli, I. Montanari, R. Monkhouse, J. Nelson and J.R. Durrant. 2001b. Electron transfer dynamics in dye sensitized nanocrystalline solar cells using a Polymer Electrolyte. J. Phys. Chem. B. 105: 7517.

Nogueira, A.F., M.A.S. Spinace, W.A. Gazotti, E.M. Girotto and M.A. De Paoli. 2001c. Poly(ethylene oxide-co-epichlorohydrin)/NaI: a promising polymer electrolyte for photoelectrochemical cells. Solid State Ionics 140: 327.

Nogueira, A.F., C. Longo and M.A. De Paoli. 2004. Polymers in dye sensitized solar cells: overview and perspectives. Coordin. Chem. Rev. 248: 1455.

Noor, I.S.M., S.R. Majid, A.K. Arof, D. Djurado, S.C. Neto and A. Pawlicka. 2012. Characteristics of gellan gum–$LiCF_3SO_3$ polymer electrolytes. Solid State Ionics 225: 649–653.

O'Regan, B. and M. Grätzel. 1991. A low-cost, high-efficiency solar-cell based on dye-sensitized colloidal TiO_2 films. Nature 353: 737–740.

Ohno, H. (ed). 2005. Electrochemical Aspects of Ionic Liquids. New Jersey, VCH–interscience.

Osman, Z., Z.A. Ibrahim and A.K. Arof. 2001. Conductivity enhancement due to ion dissociation in plasticized chitosan based polymer electrolytes. Carbohyd. Polym. 44: 167–173.

Özdemir, C. and A. Güner. 2007. Solubility profiles of poly(ethylene glycol)/solvent systems, I: qualitative comparison of solubility parameter approaches. Eur. Polym. J. 43(7): 3068–3093.

Ozer, S., J. Javorniczky and C.A. Angell. 2002. Polymer electrolyte photoelectrochemical cells with involatile plasticizers: I. The n Si $/I^-/I_2$ cell batteries and energy conversion. J. Electrochem. Soc. 149 A: 87.

Pang, S.C., C.L. Tay and S.F. Chin. 2014. Starch-based gel electrolyte thin films derived from native sago (Metroxylon sagu) starch. Ionics 20(10): 1455–1462.

Park, S.J., K. Yoo, J.Y. Kim, J.Y. Kim, D.K. Lee, B. Kim, H. Kim, J.H. Kim, J. Cho and M.J. Ko. 2013. Water-based thixotropic polymer gel electrolyte for dye-sensitized solar cells. ACS Nano 7(5): 4050–6.

Preechatiwong, W. and J.M. Schultz. 1996. Electrical conductivity of poly(ethylene oxide)—alkali metal salt systems and effects of mixed salts and mixed molecular weights. Polymer 37: 5109.

Ramesh, S., C.W. Liew and A.K. Arof. 2011a. Ion conducting corn starch biopolymer electrolytes doped with ionic liquid 1-butyl-3-methylimidazolium hexafluorophosphate. J. Non-Cryst. Solids. 357(21): 3654–3660.

Ramesh, S., R. Shanti, E. Morris and R. Durairaj. 2011b. Utilisation of corn starch in production of 'green' polymer electrolytes. Mater Res. Innov. 15(2): s13–s8.

Ramesh, S., R. Shanti and E. Morris. 2012a. Discussion on the influence of DES content in CA-based polymer electrolytes. J. Mater. Sci. 47(4): 1787–1793.

Ramesh, S., R. Shanti and E. Morris. 2012b. Studies on the plasticization efficiency of deep eutectic solvent in suppressing the crystallinity of corn starch based polymer electrolytes. Carbohyd. Polym. 87(1): 701–706.

Ramesh, S., R. Shanti and E. Morris. 2012c. Studies on the thermal behavior of CS:LiTFSI:[Amim] Cl polymer electrolytes exerted by different [Amim] Cl content. Solid State Sci. 14(1): 182–186.

Ramesh, S., R. Shanti and E. Morris. 2013. Employment of [Amim] Cl in the effort to upgrade the properties of cellulose acetate based polymer electrolytes. Cellulose 20(3): 1377–1389.

Ramlli, M.A. and M.I.N. Isa. 2014. Conductivity study of carboxyl methyl cellulose solid biopolymer electrolytes (SBE) doped with ammonium fluoride. Res. J. Recent Sci. 3(6): 59–66.

Rani, M.S.A., S. Rudhziah, A. Ahmad and N.S. Mohamed. 2014. Biopolymer electrolyte based on derivatives of cellulose from kenaf bast fiber. Polymers 6(9): 2371–2385.

Raphael, E., C.O. Avellaneda, B. Manzolli and A. Pawlicka. 2010. Agar-based films for application as polymer electrolytes. Electrochim. Acta. 55: 1455–1459.

Rathod, S.G., R.F. Bhajantri, V. Ravindrachary, P.K. Pujari and T. Sheela. 2014. Ionic conductivity and dielectric studies of $LiClO_4$ doped poly(vinylalcohol) (PVA)/chitosan (CS) composites. J. Adv. Dielectr. 4(4): 1450033.

Rees, D.A. 1977. Polysaccharide Shapes, Outline Studies in Biology. Springer, Netherlands.

Ren, Y., Z. Zhang, S. Fang, M. Yang and S. Cai. 2002. Application of PEO based gel network polymer electrolytes in dye-sensitized photoelectrochemical cells. Sol. Energ. Mat. Sol. C. 71: 253.

Rinaudo, M. 2006. Chitin and chitosan: properties and applications. Prog. Polym. Sci. 31: 603–632.

Robertson, N. 2006. Optimizing dyes for dye-sensitized solar cells. Angew. Chem. Int. Ed. Engl. 45: 2338–2345.

Saaid, S.I.Y., T.I.T. Kudin, A.M.M. Ali, A.H. Ahmad and M.Z.A. Yahya. 2009. Solid state proton battery using plasticised cellulose–salt complex electrolyte. Mater. Res. Innov. 13(3): 252–254.

Salvi, D.T.B.D., H.S. Barud, A. Pawlicka, R.I. Mattos, E. Raphael, Y. Messaddeq and S.J.L. Ribeiro. 2014. Bacterial cellulose/triethanolamine based ion-conducting membranes. Cellulose 21: 1975–1985.

Samsudin, A.S., E.C.H. Kuan and M.I.N. Isa. 2011. Investigation of the potential of proton-conducting biopolymer electrolytes based methyl cellulose-glycolic acid. Int. J. Polym. Anal. Ch. 16(7): 477–485.

Samsudin, A.S. and M.I.N. Isa. 2012 Structural and ionic transport study on CMC doped NH_4Br: A new types of biopolymer electrolytes. J. Appl. Sci. 12: 174–179.

Sankri, A., A. Arhaliass, I. Dez, A.C. Gaumont, Y. Grohens, D. Lourdin, I. Pillin, S.A. Rolland and E. Leroy. 2010. Thermoplastic starch plasticized by an ionic liquid. Carbohyd. Polym. 82(2): 256–263.

Sequeira, C. and D. Santos. 2010. Polymer Electrolytes Fundamentals and Applications. Woodhead Publishing Limited.

Shuhaimi, N.E.A., N.A. Alias, S.R. Majid and A.K. Arof. 2008. Electrical double layer capacitor with proton conducting K-carrageenan–chitosan electrolytes. Funct. Mater. Lett. 01(03): 195–201.

Shuhaimi, N.E.A., L.P. Teo, S.R. Majid and A.K. Arof. 2010a. Transport studies of NH_4NO_3 doped methyl cellulose electrolyte. Synth. Met. 160 (9-10): 1040–1044.

Shuhaimi, N.E.A., N.A. Alias, M.Z. Kufian, S.R. Majid and A.K. Arof. 2010b. Characteristics of methyl cellulose-NH_4NO_3-PEG electrolyte and application in fuel cells. J. Sol. State Elecchem. 14(12): 2153–2159.

Shukur, M.F., Y.M. Yusof, S.M.M. Zawawi, H.A. Illias and M.F.Z. Kadir. 2013a. Conductivity and transport studies of plasticized chitosan-based proton conducting biopolymer electrolytes. Phys. Scr. 014050.

Shukur, M.F., F.M. Ibrahim, N.A. Majid, R. Ithnin and M.F.Z. Kadir. 2013b. Electrical analysis of amorphous corn starch-based polymer electrolyte membranes doped with LiI. Phys. Scr. 88: 025601.

Shukur, M.F., N.A. Majid, R. Ithnin and M.F.Z. Kadir. 2013c. Effect of plasticization on the conductivity and dielectric properties of starch–chitosan blend biopolymer electrolytes infused with NH_4Br. Phys. Scr. T157: 014051.

Shukur, M.F., R. Ithnin and M.F.Z. Kadir. 2014a. Protonic transport analysis of starch-chitosan blend based electrolytes and application in electrochemical device. Mol. Cryst. Liq. Cryst. 603: 52–65.

Shukur, M.F., R. Ithnin and M.F.Z. Kadir. 2014b. Electrical characterization of corn starch-LiOAc electrolytes and application in electrochemical double layer capacitor. Electrochim. Acta. 136: 204–216.

Shukur, M.F., R. Ithnin and M.F.Z. Kadir. 2014c. Electrical properties of proton conducting solid biopolymer electrolytes based on starch–chitosan blend. Ionics 20(7): 977–999.

Shukur, M.F. and M.F.Z. Kadir. 2015. Electrical and transport properties of NH_4Br-doped cornstarch-based solid biopolymer electrolyte. Ionics 21(1): 111–124.
Singh, P.K., K.I. Kim, N.G. Park and H.W. Rhee. 2007. Dye sensitized solar cell using polymer Electrolytes based on Poly(ethylene oxide) with an ionic liquid. Macromol. Symp. 162: 249–250.
Singh, P.K., K.W. Kim and H.W. Rhee. 2009. Ionic liquid (1-methyl 3-propyl imidazolium iodide) with polymer electrolyte for DSSC application. Polym. Eng. Sci. 49: 862.
Singh, P.K., B. Bhattacharya, R.K. Nagarale, K.W. Kim and H.W. Rhee. 2010a. Synthesis, characterization and application of biopolymer-ionic liquid composite membranes. Synthetic Metals 160: 139–142.
Singh, P.K., B. Bhattacharya, R.K. Nagarale, S.P. Pandey, K.W. Kim and H.W. Rhee. 2010b. Ionic liquid doped poly(N-methyl 4-vinylpyridine iodide) solid polymer electrolyte for dye sensitized solar cell. Synthetic Metals 160: 950.
Singh, N.B. and D.S. Saran. 2011. Introduction to Polymer Science and Technology. New Age International Limited. New Delhi.
Singh, P.K., R.K. Nagarale, S.P. Pandey, H.W. Rhee and B. Bhattacharya. 2012. Present status of solid state photoelectrochemical solar cells and dye sensitized solar cells using PEO-based polymer electrolytes. Adv. Nat. Sci: Nanosci. Nanotechnol. 2: 023002.
Singh, R., N.A. Jadhav, S. Majumder, B. Bhattacharya and P.K. Singh. 2013. Novel biopolymer gel electrolyte for dye-sensitized solar cell application. Carbohyd. Polym. 91: 682–685.
Singh, R., B. Bhattacharya, H.W. Rhee and P.K. Singh. 2014a. New biodegradable polymer electrolyte for dye sensitized solar cell. Int. J. Electrochem. Sci. 9: 2620–2630.
Singh, R., J. Baghel, S. Shukla, B. Bhattacharya, H.W. Rhee and P.K. Singh. 2014b. Detailed electrical measurements on sago starch biopolymer solid electrolyte. Phase Transit. 87(12): 1237–1245.
Singh, R., A.R. Polu, B. Bhattacharya, Hee-Woo Rhee, C. Varlikli and P.K. Singh. 2016a. Perspectives for solid biopolymer electrolytes in dye sensitized solar cell and battery application. Renew. Sust. Energ. Rev. 65: 1098–1117.
Singh, R., P.K. Singh, S.K. Tomar and B. Bhattacharya. 2016b. Synthesis, characterization and dye sensitized solar cell fabrication using solid biopolymer electrolyte membranes. High Perform. Polym. 28: 47–54.
Singh, R., B. Bhattacharya, S.K. Tomar, V. Singh and P.K. Singh. 2017. Electrical, optical and electrophotochemical studies on agarose based biopolymer electrolyte towards dye sensitized solar cell application. Measurement 102: 214–219.
Siti, S.A. and A.M. Ahmad. 2013. Effect of NH4I and I2 concentration on agar gel polymer electrolyte properties for a dye-sensitized solar cell. Ionics 19: 1185–1194.
Skotheim, T.A. 1981. Tandem photovoltaic cell using a thin-film polymer electrolyte. Appl. Phys. Lett. 38: 712.
Skotheim, T.A. and O. Inganas. 1985. Polymer solid electrolyte photoelectrochemical cells with n-Si-polypyrrole photoelectrodes. J. Electrochem. Soc. 132: 2116.
Steinbuchel, A. 2003. Biopolymers, General Aspects and Special Applications. Wiley-VCH, Weinheim. 10. 526.
Stephan, A.M. 2006. Review on gel polymer electrolytes for lithium batteries. Eur. Polym. J. 42(1): 21–42.
Stephen, A.M., G.O. Phillips and P.A. Williams. 2006. Food Polysaccharides and their Applications. Second Edition. Taylor & Francis Group LLC.
Stergiopoulos, T., I.M. Arabatzis, G. Katsaros and P. Falaras. 2002. Binary polyethylene oxide/titania solid-state redox electrolyte for highly efficient nanocrystalline TiO_2 photoelectrochemical cells. Nano Lett. 2: 1259.
Subban, R.H.Y., A.K. Arof and S. Radhakrishna. 1996. Polymer batteries with chitosan electrolyte mixed with sodium perchlorate. Mater Sci. Eng. B. 38: 156–160.

Suzuki, K., M. Yamaguchi, M. Kumagai, N. Tanabe and S. Yanagida. 2006. Dye-sensitized solar cells with ionic gel electrolytes prepared from imidazolium salts and agarose. Comptes Rendus Chimie. 9(5-6): 611–616.

Tarascon, J.M. and M. Armand. 2001. Issues and challenges facing rechargeable lithium batteries. Nature 414: 359–367.

Teoh, K.H., S. Ramesh and A.K. Arof. 2012. Investigation on the effect of nanosilica towards corn starch–lithium perchlorate-based polymer electrolytes. J. Solid State Electr. 16(10): 3165–3170.

Teoh, K.H., C.S. Lim and S. Ramesh. 2014. Lithium ion conduction in corn starch based solid polymer electrolytes. Measurement 48: 87–95.

Tiwari, T., K. Pandey, N. Srivastava and P.C. Srivastava. 2011a. Effect of glutaraldehyde on electrical properties of arrowroot starch + NaI electrolyte system. Journal of Applied Polymer Science 121(1): 1–7.

Tiwari, T., N. Srivastava and P.C. Srivastava. 2011b. Electrical transport study of potato starch-based electrolyte system. Ionics 17: 353–360.

Tiwari, T., N. Srivastava and P.C. Srivastava. 2013. Ion dynamics study of potato starch + sodium salts electrolyte system. Int. J. Electrochemistry. 2013: 8.

Tiwari, T., M. Kumar, N. Srivastava and P.C. Srivastava. 2014. Electrical transport study of potato starch-based electrolyte system-II. Mater Sci. Eng. B. 182: 6–13.

Toyoda, T., T. Sano, J. Nakajima, S. Doi, S. Fukumoto, A. Ito, T. Tohyama, M. Yoshida, T. Kanagawa, T. Motohiro, T. Shiga, K. Higuchi, H. Tanaka, Y. Takeda, T. Fukano, N. Katoh, A. Takeichi, K. Takechi and M. Shiozawa. 2004. Outdoor performance of large scale DSC modules. J. Photochem. Photobiol. A: Chemistry 164: 203.

Ummartyotin, S. and H. Manuspiya. 2015. A critical review on cellulose: from fundamental to an approach on sensor technology. Renew. Sust. Energ. Rev. 41: 402–412.

Ummartyotin, S. and H. Manuspriya. 2015. An overview of feasibilities and challenge of conductive cellulose for rechargeable lithium based battery. Renew. Sust. Energ. Rev. 50: 204–213.

Wang, G.X., L. Yang, J.Z. Wang, H.K. Liu and S.X. Dou. 2005. Enhancement of ionic conductivity of PEO based polymer electrolyte by the addition of nanosize ceramic powders. J. Nanosci. Nanotechnol. 5: 1135.

Wang, Y., K.S. Chen, J. Mishler, S.C. Cho and X.C. Adroher. 2011. A review of polymer electrolyte membrane fuel cells: Technology, applications, and needs on fundamental research. Appl. Energy 88(4): 981–1007.

Weijia, W., G. Xueyi and Y. Ying. 2011. Lithium iodide effect on the electrochemical behavior of agarose based polymer electrolyte for dye-sensitized solar cell. Electrochim. Acta. 56: 7347–7351.

Winie, T. and A.K. Arof. 2006. Hexanoyl chitosan-based gel electrolyte for use in lithium-ion cell. Polymer. Adv. Tech. 17(7-8): 552–555.

Winie, T. and A.K. Arof. 2006. Transport properties of hexanoyl chitosan-based gel electrolyte. Ionics 12: 149–152.

Winie, T., S. Ramesh and A.K. Arof. 2009. Studies on the structure and transport properties of hexanoyl chitosan-based polymer electrolytes. Physica B. 404: 4308–4311.

Xiao, S.Y., Y.Q. Yang, M.X. Li, F.X. Wang, Z. Chang, Y.P. Wu and X. Liu. 2014. A composite membrane based on a biocompatible cellulose as a host of gel polymer electrolyte for lithium ion batteries. J. Power Sources 270: 53–58.

Yahya, M.Z.A. and A.K. Arof. 2002. Characteristics of chitosan-lithium acetate-palmitic acid complexes. J. New. Mat. Electrochem. Systems 5: 123–128.

Yahya, M.Z.A. and A.K. Arof. 2003. Effect of oleic acid plasticizer on chitosan–lithium acetate solid polymer electrolytes. Eur. Polym. J. 39(5): 897–902.

Yahya, M.Z.A. and A.F. Arof. 2004. Conductivity and X-ray photoelectron studies on lithium acetate doped chitosan films. Carbohyd. Polym. 55(1): 95–100.

Yahya, M.Z.A., A.M.M. Ali, M.F. Mohammat, M.A.K.M. Hanafiah, M. Mustaffa, S.C. Ibrahim, Z.M. Darus and M.K. Harun. 2006. Ionic conduction model in salted chitosan membranes plasticized with fatty acid. Appl. Energy. 6(6): 1287–1291.

Yang, Y., J. Cui, P. Yi, X. Zheng, X. Guo and W. Wang. 2014. Effects of nanoparticle additives on the properties of agarose polymer electrolytes. J. Power Sources 248: 988–993.

Yang, Y., X.Y. Guo and X.Z. Zhao. 2011a. Influence of polymer concentration on polysaccharide electrolyte for quasi-solid-state dye-sensitized solar cell. Mater Sci. Forum. 685: 76–81.

Yang, Y., C. Zhou, S. Xu, H. Hu, B. Chen, J. Zhang, S. Wu, W. Liu and X.Z. Zhao. 2008. Improved stability of quasi-solid-state dye-sensitized solar cell based on poly(ethylene oxide)–poly(vinylidene fluoride) polymer-blend electrolytes. J. Power Sources 185: 1492.

Yang, Y., H. Hu, C.H. Zhou, S. Xu, B. Sebo and X.Z. Zhao. 2011b. Novel agarose polymer electrolyte for quasi-solid state dye-sensitized solar cell. J. Power Sources 196: 2410–2415.

Yella, A., H.W. Lee, H.N. Tsao, C. Yi, A.K. Chandiran, M.K. Nazeeruddin, E.W.G. Diau, C.Y. Yeh, S.M. Zakeeruddin and M. Grätzel. 2011. Porphyrin-sensitized solar cells with cobalt (II/III)-based redox electrolyte exceed 12% efficiency. Science 334: 629–634.

Yohannes, T. and O. Iganas. 1998. Photoelectrochemical studies of the junction between poly[3-(4-octylphenyl)thiophene] and a redox polymer electrolyte. Sol. Energy Mater. Sol. Cells 51: 193.

Yusof, Y.M., H.A. Illias and M.F.Z. Kadir. 2014. Incorporation of NH_4Br in PVA-chitosan blend-based polymer electrolyte and its effect on the conductivity and other electrical properties. Ionics 20: 1235–1245.

Yusof, Y.M., M.F. Shukur, H.A. Illias and M.F.Z. Kadir. 2014. Conductivity and electrical properties of corn starch–chitosan blend biopolymer electrolyte incorporated with ammonium iodide. Phys. Scr. 89: 035701.

Yusof, Y.M., N.A. Majid, R.M. Kasmani, H.A. Illias and M.F.Z. Kadir. 2014. The effect of Plasticization on Conductivity and Other Properties of Starch/Chitosan Blend Biopolymer Electrolyte Incorporated with Ammonium Iodide. Mol. Cryst. Liq. Cryst. 603: 73–88.

Zhang, C., S. Gamble., D. Ainsworth, A.M.Z. Slawin, Y.G. Andreev and P.G. Bruce. 2009. Alkali metal crystalline polymer electrolytes. Nat. Mater. 8: 580–584.

Zhang, J., H. Han, S. Wu, S. Xu, C. Zhou, Y. Yang and X.Z. Zhao. 2007a. Ultrasonic irradiation to modify the PEO/P(VDF–HFP)/TiO2 nanoparticle composite polymer electrolyte for dye sensitized solar cells. Nanotechnology 18: 295606.

Zhang, J., H. Han, S. Wu, S. Xu, Y. Yang, C. Zhou and X.Z. Zhao. 2007b. Conductive carbon nanoparticles hybrid PEO/P(VDF-HFP)/SiO_2 nanocomposite polymer electrolyte type dye sensitized solar cells. Solid State Ionics 178: 1595.

Zhao, S., C.Y. Wang, M.M. Chen, J. Wang and Z.O. Shi. 2009. Potato starch-based activated carbon spheres as electrode material for electrochemical capacitor. J. Phys. Chem. Solids 70: 1256–1260.

3

Application of Nanomaterials for Lithium Ion Batteries

Masashi Kotobuki

1. Introduction

Lithium batteries have been widely used as a power source for mobile electronic devices like mobile phones and laptop computers since their first appearance in the market in 1991. The morphology, size and pore structure of electrode active materials largely influence the performance of lithium batteries. Additionally, nanocomposite of electrode materials and electrolytes and surface coating of electrode materials in nano-scale are also keys to determine the performance of batteries. Therefore, the design of the batteries and battery components in nano-scale are very important for developing next generation lithium batteries as well as developing novel materials.

In this chapter, overview of lithium battery such as its principle, structure, and feature is described first. Then, the application of nanomaterials for each component of lithium batteries, that is, electrode materials (cathode and anode) and electrolyte are described in detail.

2. Overview of Lithium Battery

Due to the global efforts towards the application of clean and renewable energy, a major breakthrough for energy storage devices is required

Department of Mechanical Engineering, National University of Singapore, 9, Engineering Drive 1, Singapore 117546.
E-mail: mpemako@nus.edu.sg

(Chu and Majumdar 2012). Among the various energy storage devices, electrochemical ones, such as rechargeable batteries and electrochemical capacitors are especially attractive to be used as a power source in portable electronic devices, plug-in hybrid vehicles, electric vehicles and so on (Tarascon and Armand 2001, Armand and Tarascon 2008). The fundamental difference between the batteries and capacitors is the charge storage mechanism. The batteries store charges inside the electrodes through the Faradic reactions, whereas the capacitors store charges in the vicinity of electrode surface (electrochemical double layer) as shown in Fig. 1. Some Faradic reactions proceed very fast at the surface of the electrodes. This is called a pseudo-capacitor. Therefore, capacitors can allow flow of larger current, but batteries cannot because of the limitation of the reaction rate of Faradic reactions. On the contrary, batteries can store large amount of charges while capacitors cannot since the charges are stored only near the surface of the electrodes.

The performance of energy storage systems is evaluated by energy and power densities. The energy density is expressed by electrical energy that can be stored in the systems, either per unit of volume (Wh l^{-1}) or weight (Wh kg^{-1}) of the systems. The energy density is calculated by multiplying the operation voltage (V) by capacity (Ah kg^{-1}), both of which are directly connected to the chemistry of the system. On the contrary, the power density is expressed by electrical power that the system can provide, either per unit of volume (W l^{-1}) or weight (W kg^{-1}) of the systems. The power density is calculated by multiplying operation voltage (V) by current (A) per unit, which is more connected to kinetic rather than the chemistry. In general, batteries possess high energy density and the capacitors provide high power density because of the reasons mentioned above.

Figure 2 shows the energy densities of various battery systems. As can be seen, Li ion batteries (LIBs) surpass other batteries in terms of energy density mainly due to their high operation voltage (Table 1) (Cheng et al. 2011, Gonzalez et al. 2012, Mukherjee et al. 2012, Ji et al. 2009, Ji et al. 2011a, Arico et al. 2005). Therefore, LIBs have been widely used as a power source especially for mobile electronic devices since their first appearance in 1991

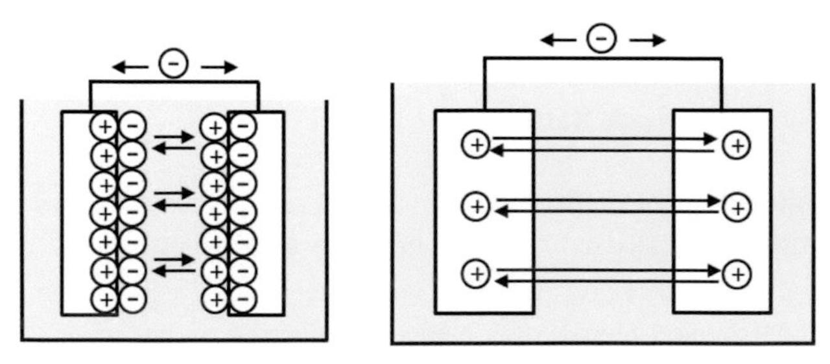

Figure 1. Structure of battery and capacitor.

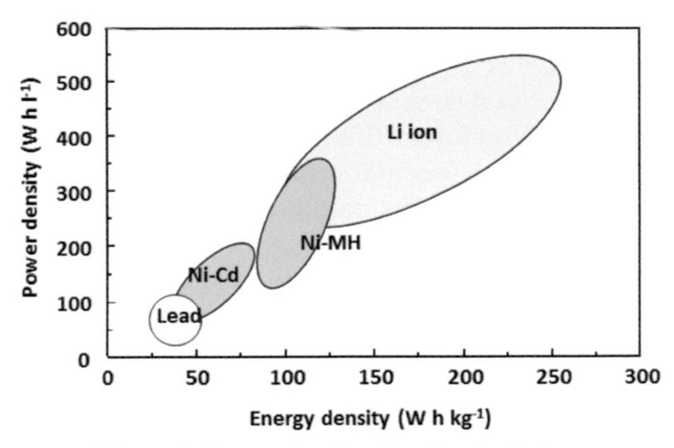

Figure 2. Energy density of various batteries.

Table 1. Operation voltage and energy density of various batteries.

Types	Operation voltage (V)	Energy density (Wh kg^{-1})
Lead acid	2.0	30–50
Ni-MH	1.2	60–120
Li-ion	3.7	110–250

by Sony (Kanamura and Kotobuki 2009, Nagaura and Tozawa 1990). A typical structure of LIBs is schematically shown in Fig. 3. LIBs do not contain lithium metal as an anode. In general, the battery containing the lithium metal is called "lithium battery", whereas the battery without the lithium metal is called "lithium ion battery". The LIBs are like a lithium ion device composed of a non-aqueous electrolyte (because aqueous electrolyte is not stable at low potentials), and lithium intercalated anode and cathode. In the first appearance of LIBs, a graphite anode and a layered $LiCoO_2$ were used (see Fig. 3).

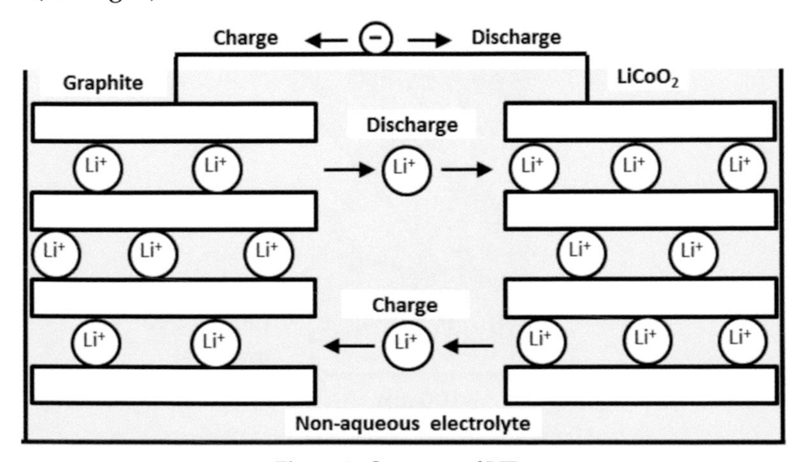

Figure 3. Structure of LIBs.

The electrodes in LIBs are generally composed of three components: an active material like $LiCoO_2$ or graphite, an electronic conducting material (normally carbon material such as Ketjen black or acetylene black), and a binder (normally polyvinylidene fluoride, PVdF). These three components are mixed with a solvent (normally N-methyl-2-pyrrolidone, NMP) to prepare a slurry, which is then painted onto a current collector. Al and Cu foils are typically used as current collectors for the cathode and anode, respectively. In general, the thickness of the electrode is approximately below 100 μm (see Fig. 4). In the charge process, the Li ion is deintercalated from $LiCoO_2$ and moves toward the graphite anode through the non-aqueous electrolyte. Then, the Li ion is intercalated into the graphite anode. At same time, Co^{3+} in $LiCoO_2$ releases an electron and is oxidized to Co^{4+}. The released electron runs in the external circuit, where we can detect a current flow. A reverse reaction proceeds in the discharge process.

The battery reactions of LIBs using $LiCoO_2$ cathode and graphite anode can be described as below.

$$\text{Cathode}: \ LiCoO_2 \leftrightarrow xLi^+ + Li_{1-x}CoO_2 + xe^-$$

$$\text{Anode}: \ 6C + xLi^+ + xe^- \leftrightarrow Li_xC_6$$

$$\text{Overall}: \ LiCoO_2 + 6C \leftrightarrow Li_{1-x}CoO_2 + Li_xC_6$$

As can be seen, the electrolyte works only as a Li ion conduction media and is not concerned with the battery reaction. For this feature, the LIBs are sometimes called "Rock-chair structure". The lesser the amount of electrolyte, the better is the battery performance in terms of energy and power densities. This feature is totally different from Lead-acid batteries in which the Pb ion reacts with water molecule and sulfate ions in the electrolyte (see below).

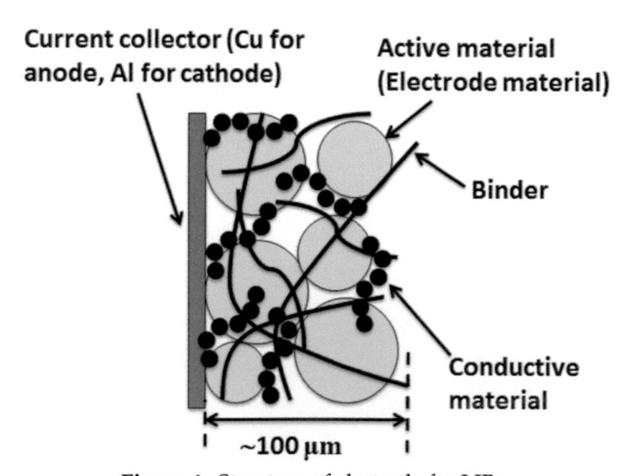

Figure 4. Structure of electrode for LIBs.

The reactions of the lead-acid batteries are

Cathode : $PbSO_4 + 2H_2O \leftrightarrow PbO_2 + 4H^+ + SO_4^{2-} + 2e^-$

Anode : $PbSO_4 + 2e^- \leftrightarrow Pb + SO_4^{2-}$

Overall : $2PbSO_4 + 2H_2O \leftrightarrow Pb + PbO_2 + 4H^+$

In this case, certain amount of electrolyte must be needed to complete the battery reaction. This is one of the reasons for poor energy density of the lead-acid batteries. In the commercial LIBs, the electrolyte is soaked in porous organic membrane with thickness of 20 ~ 30 μm used as a separator to prevent a short circuit (see Fig. 5).

Before describing about the non-aqueous electrolytes, SEI (solid electrolyte interphase) must be explained. SEI is *adhoc* protective layer on the graphite anode surface formed by decomposition of the electrolyte in the first cycle of charging in the electrolyte containing ethylene carbonate (EC). Depending upon the nature of the graphite anode, the type and distribution of surface functional groups can either enhance the chemical bonding of the SEI layer, thus stabilizing the layer (Novak et al. 2010), or enhance the decomposition of the electrolyte, thereby increasing the irreversible capacity (Buiel and Dahn 1999, Banks et al. 2005). Therefore, surface functional groups on the graphite anode should be carefully chosen. Figure 6 shows the typical discharge curve of graphite anode in electrolyte containing EC in the first Li intercalation into graphite (Flandrois and Simon 1999). In the first intercalation, a plateau was observed at 0.8–0.9V vs. Li/Li^+. This is attributed to the SEI formation. The intercalation of Li into the graphite is proceeding below 0.3 V that is shown at the other plateau in Fig. 6.

Thanks to the SEI, the graphite can stably intercalate and deintercalate Li ion. Almost theoretical capacity of the graphite electrode could be

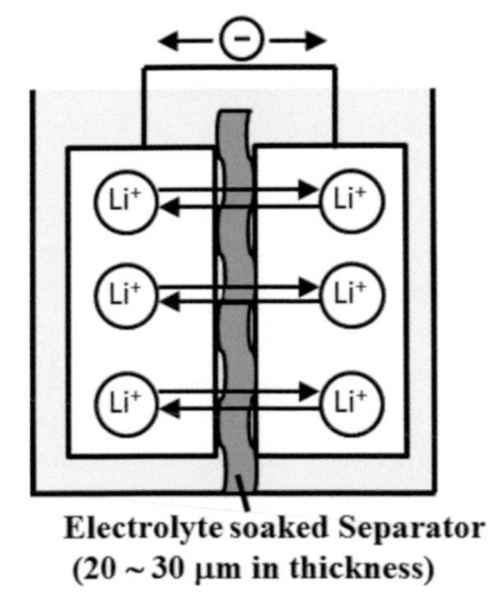

Figure 5. Structure of LIBs with separator.

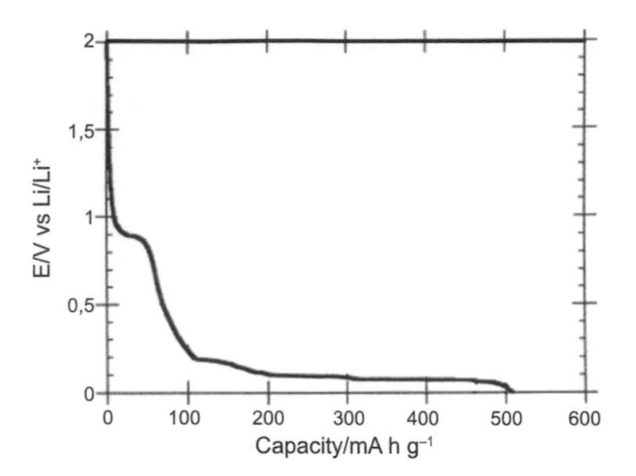

Figure 6. Typical discharge curve of graphite anode in the electrolyte containing EC.

obtained at 3 C (Kottegoda et al. 2002). Here, C means a current which can be charged or discharged on the active (cathode or anode) material for 1 h. For example, when the electrode is charged at 5 C, it takes 12 min (= 60 min/5) to charge the electrode completely. However, except thin film batteries which are composed of thin film electrodes and electrolyte with a thickness of several μm, the electrode cannot be fully charged at 5 C and the charge process usually finished within 12 min. Once a stable layer has been deposited on the electrode surface, the inorganic and organic insoluble Li salts that comprise the layer prevent further solvent reduction reactions, while still allowing for the intercalation of Li ions into the carbon matrix and *vice versa*. Therefore, the formation of a stable SEI layer is a key factor that determines the performance of LIBs.

Although much research has been performed on determining the reaction pathways that the electrolyte undergoes during the reductive decomposition and formation of the SEI layer, the precise mechanisms on those reactions have not been firmly established yet. It has been proposed that the decomposition of EC that gives rise to the common components of the SEI layer and results in generation of gasses can be attributed to a 2-electron reduction of single molecule or two one-electron reductions of two molecules (Aurbach et al. 1999a, Ostrovskii et al. 2001, Moshkovich et al. 2001, Aurbach et al. 1999b). In addition, computational studies have reported that the energy barrier associated with the initial reduction of a cyclic carbonate (i.e., EC) would be reduced to essentially eliminated level, if the solvent molecules were coordinated with a Li ion (Li and Balbuena 2000, Wang et al. 2001, Wang et al. 2002a, Wang and Balbuena 2002b). Those results indicated that it is the Li-coordinated molecules that are preferentially reduced at the electrode surface. The substantial efforts of

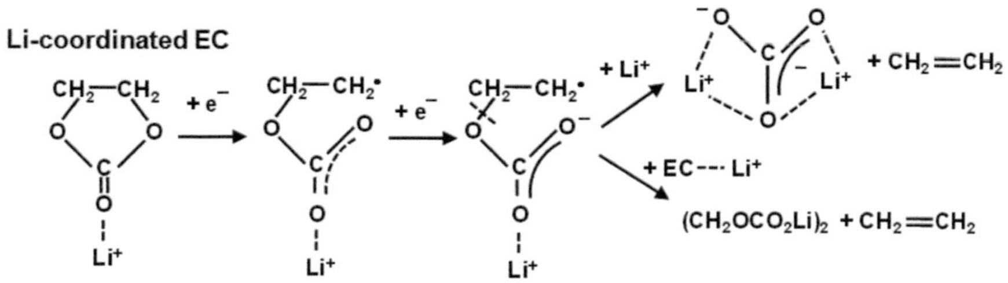

Figure 7. Reduction pathway for EC.

Aubach et al. 1999a have led to a detailed representation of the probable reduction pathways for EC which is shown in Fig. 7 (Naji et al. 1996, Wang et al. 2001, Wang and Balbuena 2004, Marom et al. 2010).

The 'low-population' and 'high-population' conditions generally refer to the concentration of EC in the electrolyte relative to other solvent components (The detailed explanation of other solvent components is described later). However, it can also refer to a lower or higher concentration of other Li-coordinated solvent molecules within the immediate environment relative to the reacting solvent molecules. It is thought that Li ion is solvated by one or more solvent molecule in the electrolyte. Coordination of a solvent molecule with a Li ion facilitates its reduction because this results in a negative-free energy change for the first reduction reaction. There is an energy barrier associated with the ring-opening step that has to be overcome to form the SEI layer. When the Li-coordinated cyclic carbonate adsorbs on the carbon surface, that additional coordination gives strain to the surface ring structure. This facilitates the reduction of the cyclic molecule and the ring-opening step by lowering the activation energy barrier for reaction (Wang et al. 2003, Marom et al. 2011). The Li-coordination would stabilize the radical anion that would then form, which undergoes a second 1-electron reduction. The affinity of electron for the anion is predicted to be much lower than the uncharged molecule, so it is reasonable to expect that the second reaction may not immediately proceed after the first reduction. Because these reactions should take place at the carbon surface, the radical anion can be stabilized by the active surface of the carbon. Once the second reduction occurs, the molecule can either react with a Li ion under 'low-population' conditions or with another Li-coordinated molecule under 'high-population' conditions. Both pathways result in the generation of ethene and either lithium carbonate or the alkyl carbonate. These lithium carbonate and/or alkyl carbonate are thought to be the main components of SEI.

In order to develop a stable SEI layer, additives have also been employed, most of which are functional derivatives of the cyclic carbonates (Sun and Dai 2010, Ein-Eli et al. 1997, Xing et al. 2009, Yao et al. 2009). Structures of common electrolyte additives are depicted in Fig. 8.

Among them, vinylene carbonate (VC) is the most successful additive employed for developing the stable SEI layer (Yun et al. 2011, Tsubouchi et al. 2012, Hongyou et al. 2013, Kim et al. 2014). The addition of VC to electrolytes enhances long-term stability of graphite anode. VC reacts with the Li-carbon electrode surface and is reduced to species such as poly-Li-alkyl carbonate, $ROCO_2Li$ species and polycarbonates. Once carbon anode is polarized to low potential, VC is reduced to more stable intermediates than one from EC (Wang et al. 2002). The reduced VC undergoes decomposition to form a radical anion, which then might follow several termination reactions to produce alkyl dicarbonates, Li-carbides, and so on. These reduction products, which contain unsaturated double bonds, may further polymerize on the electrode surface and therefore have the potential to build up an effective SEI film. Due to this promising nature, VC is expected to be an effective additive in non-EC based electrolytes.

The non-aqueous liquid electrolytes in the commercial LIBs are mainly composed of Li salt and organic electrolyte. As only EC can form the SEI on the surface of graphite by the decomposition as mentioned above, EC is present in almost all commercial batteries. Representative organic solvents are revealed in Table 2 with dielectric constant, melting point and so on, while their structures are shown in Fig. 9.

High dielectric constant of EC promotes a dissociation of Li salt and increases Li ion concentration, leading to high Li ion conductivity. The conductivity of Li ion in the non-aqueous liquid electrolyte can be described as below

$$\sigma = zenu \qquad \text{(Equation 1)}$$

where σ, z, e, n, and u are conductivity, valence of Li ion (= +1), elementary charge (= 1.602×10^{-19} C), density and mobility of Li ion, respectively.

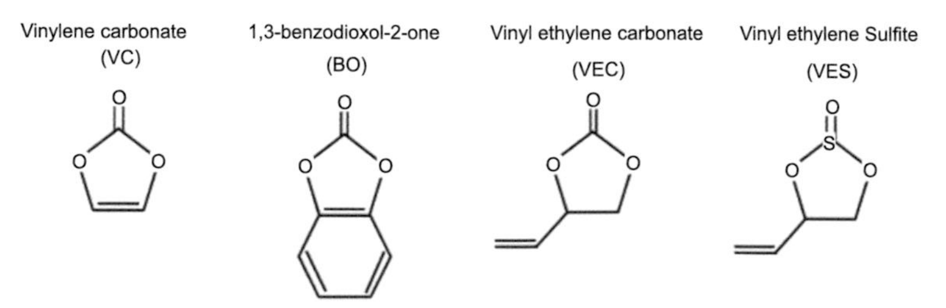

Figure 8. Structures of common electrolyte additive.

Table 2. Properties of various organic solvents.

Solvent	Molecular weight	Melting point (°C)	Boiling point (°C)	Dielectric constant	Viscosity (cp)
EC	88.0	36.0	248.0	89.6	1.92
PC	102.0	–49.0	242.0	64.4	2.53
DMC	90.0	0.5	90.2	3.1	0.59
DEC	118.0	–43.0	126.0	2.8	0.75
EMC	104.0	–53.0	110.0	3.0	0.65

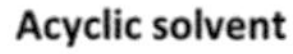

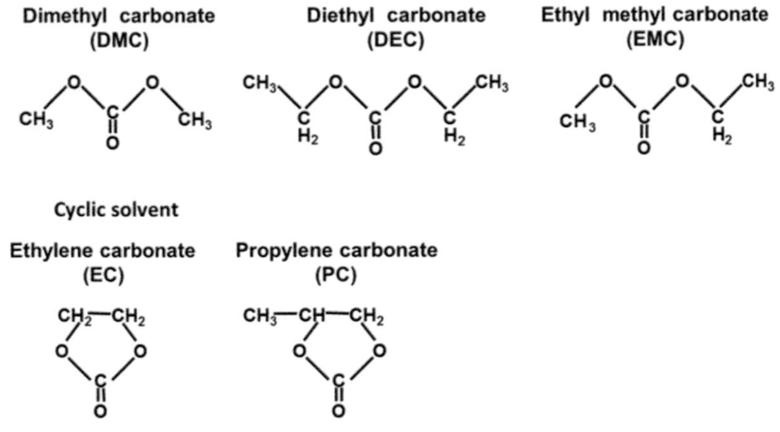

Figure 9. Structure of organic solvents used in LIBs.

The mobility, u, can be written by the Nernst-Einstein equation

$$u - Dze/k_BT \quad \text{(Equation 2)}$$

Here, D is diffusion constant and k_B and T are Boltzmann constant (= 1.381×10^{-23} J K^{-1}) and absolute temperature, respectively.

When the Li ion moves in the liquid electrolyte, D can be expressed by the Stokes-Einstein equation

$$D = k_BT/6\pi\eta r \quad \text{(Equation 3)}$$

where η and r are viscosity of the liquid and radius of the Li ion, respectively.

Overall, the Li ion conductivity in the liquid electrolyte can be defined as:

$$\sigma = n(ze)^2/6\,\pi\eta r$$

It can be clearly seen that the density of Li ion (= concentration of Li ion) and viscosity of the electrolyte determine the conductivity. Therefore, EC is a promising electrolyte due to its high dielectric constant which increases the Li ion concentration and is able to dissolve many lithium salts as well as the SEI formation. However, as shown in Table 2, high viscosity of EC is unfavorable for the Li ion conduction. This high viscosity of EC also causes low wettability of the electrolyte with porous electrode. The electrode reaction proceeds only at electrode/electrolyte interface. The low wettability of electrolyte forms limited electrode/electrolyte interface and provides low reaction area. In addition, high melting point of EC has made a difficulty to use EC for the liquid electrolyte alone. On the other hand, propylene carbonate (PC) possesses high dielectric constant and low melting point. Therefore, PC has been considered a promising organic solvent. However, PC is not involved in the commercial batteries. Figure 10 shows the charge and discharge curves of the graphite anode in EC and PC, both of which dissolved $LiClO_4$ as the Li salt (Inaba et al. 1997).

Compared with EC, a large plateau at 0.8 V was observed in PC electrolyte. This is attributed to an exfoliation of the graphite. The Li intercalation into graphite anode does not proceed in PC due to the absence of the SEI protective layer. Thus, PC cannot be used for Li batteries as electrolyte alone.

In order to circumvent the low viscosity of EC, acyclic alkyl carbonates (see Fig. 9) are usually added because they offer low viscosities and melting points. However, their dielectric constant is low and their boiling and flash points are also low, thus limiting the safety margin of the practical cell. Given the lack of better alternatives, mixtures of organic carbonates are the solvents of choice for the commercial cells such as mixture of EC/DEC, EC/DMC, and EC/DMC/DEC.

It is essential to realize that the structure and properties of electrode-electrolyte interfaces, which influence the battery performance, depend

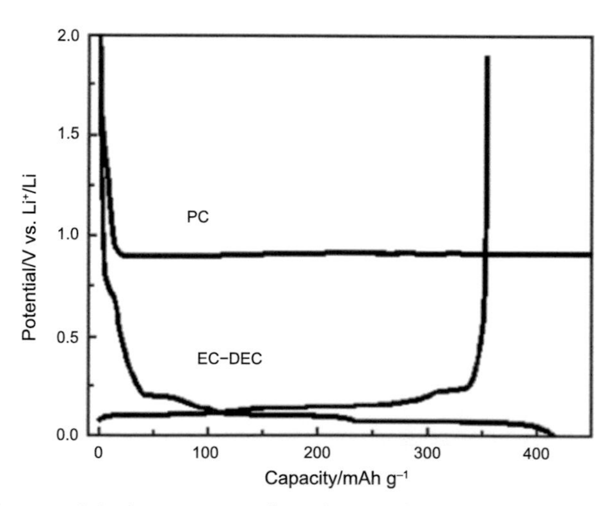

Figure 10. Charge and discharge curves of graphite anode in PC and EC/DEC electrolytes.

not only on the electrolyte solvent but also vastly on the salt. This makes lithium salt the other critical element of the electrolyte and one of the key elements of the entire LIBs. The ideal properties of lithium salts for battery application are described below:

1) The salts should completely dissolve into the solvent (mixture of carbonates as mentioned above) at desired concentration and ions (particularly Li ions) should be able to transfer through the solution.
2) The counteranion should be stable towards oxidative or reductive decomposition on both electrodes.
3) The salts can promote a formation of ionically conducting and stable interfacial layers (SEI) on both electrodes.
4) The counteranion should be non-toxic and remain thermally stable at the battery working conditions.
5) Cheap and easy to synthesize in large quantities.
6) The counteranion should be stable with electrolyte solvent and other cell components.
7) Preferentially stable against water and ambient air.
8) Non-toxic and non-flammable.

$LiPF_6$ has been widely used in commercial LIBs. However, low thermal stability and chemical stability of $LiPF_6$ have always caused safety concern especially in high temperature operation and water contamination in which $LiPF_6$ reacts with water and forms toxic HF. Table 3 shows the property comparison of various lithium salts (Marcus 1985) and their corresponding structures are shown in Fig. 11.

As can be seen, lithium bis(trifluoro methanesulfonyl) imide (LiTFSI) possesses a good thermal stability and chemical stability and is paid much attention as a novel Li salt.

Herein, structure and components of lithium batteries were briefly explained. In order to further improve the performance of batteries, usage of nano-sized electrode materials is one of the good strategies. Interestingly, even though same electrode materials are used, the performance of batteries

Table 3. Properties of various Li salts.

Property	**From best → to worst**					
Chemical inertness	LiTf	LiTFSI	$LiAsF_6$	$LiBF_4$	$LiPF_6$	
Solubility	LiTFSI	$LiPF_6$	$LiAsF_6$	$LiBF_4$	LiTf	
Ion pair dissociation	LiTFSI	$LiAsF_6$	$LiPF_6$	$LiClO_4$	$LiBF_4$	LiTf
Thermal stability	LiTFSI	LiTf	$LiAsF_6$	$LiBF_4$	$LiPF_6$	
Ion mobility	$LiBF_4$	$LiClO_4$	$LiPF_6$	$LiAsF_6$	LiTf	LiTFSI
SEI formation	$LiPF_6$	$LiAsF_6$	LiTFSI	$LiBF_4$		

LiTf = lithium triflate, LiTFSI = lithium bis(trifluoro methanesulfonyl) imide

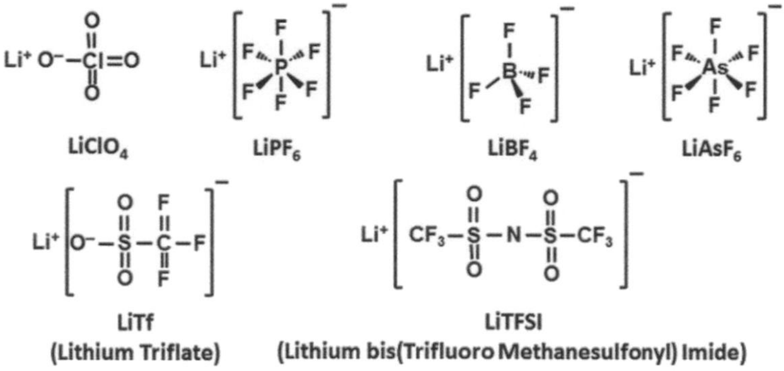

$LiClO_4$ $LiPF_6$ $LiBF_4$ $LiAsF_6$

LiTf (Lithium Triflate) LiTFSI (Lithium bis(Trifluoro Methanesulfonyl) Imide)

Figure 11. Structure of various Li salts.

varies by the size of electrode materials. In the next section, recent research on nano-sized electrode for battery application is described.

3. Application of Nanomaterials for Electrodes of Lithium Batteries

3.1 Electrode materials

Electrode materials should fulfill these three fundamental requirements:

1) High specific charge and charge density, i.e., a high number of charge carriers per mass and volume unit of the material are available.
2) High cell voltage, that is, electrode redox reaction proceeds at high (cathode) or low (anode) potential.
3) High reversibility: electrochemical reactions at both cathodes and anodes to maintain the specific charge for hundreds to thousands of charge-discharge cycles.

These requirements can be satisfied by using the electrode composed of nanomaterials. Some advantages and disadvantages appear when the nanomaterials are applied for the electrode materials of LIBs. Herein, the advantages and disadvantages of nanomaterials in the application of the electrode are summarized and itemized.

Advantages

1. Some reactions can proceed only on nanopowders. For example, reversible lithium intercalation into mesoporous β-MnO_2 without destruction of the rutile structure (Jiao and Bruce 2007).
2. The nanostructured electrodes, i.e., grain size is in the nano-size range, providing a short ion and electron transport length and a

large surface area, are beneficial to improve LIB's performances (see Fig. 12) (Guo et al. 2008b, Wang and Cao 2008). The characteristic time constant for diffusion is described in $t = L^2/D$, where L and D are the diffusion length and the diffusion constant, respectively. The time t for Li ion intercalation into the electrode materials decreases with the square of the diffusion length, therefore, replacing micrometer with nanometer particles gives a large impact to the time (Arico et al. 2005).

3. Electron transport in the particles is also enhanced by nanometer-sized particles due to short length of diffusion, as mentioned above. These reduction of transport distances of both electron and Li ion increase the rate of lithium insertion/deintercalation significantly, resulting in improved LIB's performances, especially under high rate.
4. A high surface area permits a high contact area with the electrolyte. This enlargement of reaction fields causes a high lithium-ion flux across the interface.
5. For very small particles, especially nano-sized particles, the chemical potentials for lithium ions and electrons may be modified probably by the quantum effect, which probably affects the electrode potential (thermodynamics of the reaction) and operation voltage of the LIBs (Balaya et al. 2006).
6. The strain due to volume change of electrode materials during the electrode reaction can be absorbed well by the existing large free space among the nanoparticles.

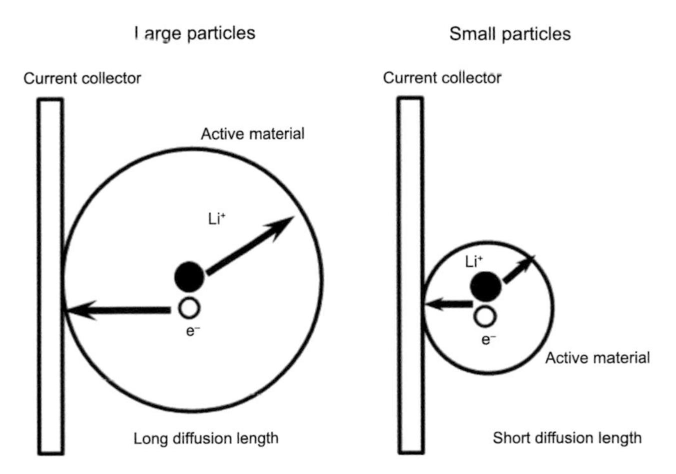

Figure 12. Diffusion length of electron and Li ion in the electrode.

Disadvantages

1. Nanoparticles may be more difficult to be synthesized than large particles. The synthesis of nanoparticles with uniform particle size distribution would make it more difficult. Furthermore, the nanoparticles also increase synthesis cost.
2. Nanoparticles possess a large surface area. If the materials are not stable in air where they react with air (oxygen, moisture, and so on), this large surface area promotes the reaction, resulting in decay of the materials. In the worst case, safety issue like explosion may be caused.
3. The density of nanopowder is generally less than the micro-sized powder even though they are composed of the same composition. The volume of the electrode composed of the nanopowders becomes larger than that comprised of large particles, resulting in reduced volumetric energy density.
4. High electrode/electrolyte surface area may lead to more significant side reactions with the electrolyte and reduce the battery performance.

As described above, the application of nanomaterials for LIBs is thought to bring a lot of advantages, but disadvantages as well. A lot of researchers have paid efforts to emphasise the advantages and to suppress the disadvantages. Hereafter, some representative researches and progresses on application of the nanomaterials for the electrodes of the LIBs are reviewed.

3.2 *Cathode*

Most of the lithium intercalation compounds can be candidates as electrodes in rechargeable lithium batteries depending on their redox potential. This gives us high motivation to find new electrode materials. Figure 13 depicts the intercalation voltage and capacity of the possible cathode materials. Among them, $LiCoO_2$, $LiFePO_4$, $LiMn_2O_4$ and $LiCo_{1/3}Mn_{1/3}Ni_{1/3}O_2$ have been already loaded in the commercial LIBs.

Intercalation electrodes in batteries are electroactive materials and serve as a host solid where guest species can be reversibly intercalated from electrolyte. Intercalation compounds are a special family of materials. Intercalation refers to the reversible intercalation of mobile guest species (atoms, molecules, or ions) into a crystalline host lattice that contains an interconnected system of empty lattice sites of appropriate size, while the structural integrity of the host lattice is formally conserved (O'Hare 1991). Although various host lattices reported so far are metal dichalcogenides, metal oxyhalides, metal phosphorous trisulphides, metal oxides, metal phosphates, hydrogen phosphates and phosphonates, graphite and the

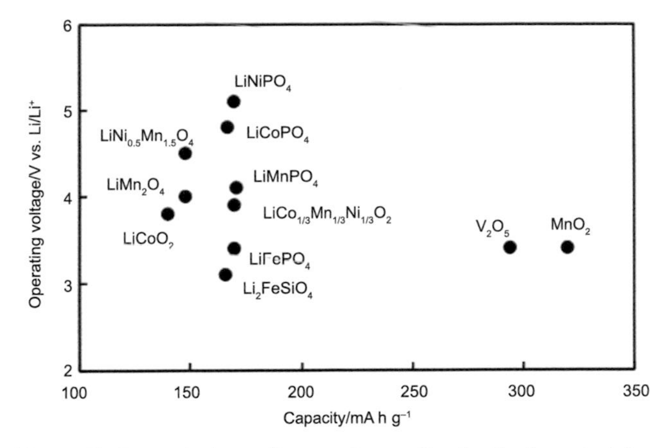

Figure 13. Intercalation voltage and capacity of cathode materials.

transition metal oxides have been exclusively used for the LIBs. When guest species are intercalated into host lattices, various structural changes will occur.

1) Change in interlayer spacing.
2) Change in stacking mode of the layers.
3) Formation of intermediate phases at low guest concentrations (Schoellhorn 1987).

There are two categories of cathode materials. One comprises layered compounds with an anion close-packed lattice; transition metal cations occupy alternate layers between the anion layers, and lithium ions are intercalated into remaining empty layers between the anion layers. $LiCoO_2$, $LiNi_{1-x}Co_xO_2$, and $LiNi_xMn_xCo_{1-2x}O_2$ belong to this group. The spinels with transition metal cations ordered in all the layers can be considered to be in this group as well. This class of materials has the inherent advantage of higher energy density (high energy per unit of volume) owing to their more compact lattices. Another group of cathode materials has more open structures, such as vanadium oxides, the tunnel compounds of manganese oxides, and transition metal phosphates (e.g., the olivine $LiFePO_4$, $LiMnPO_4$). These materials generally provide the advantages of better safety and lower cost compared to the first group.

Lithium transition metal oxides

Many lithium transition metal oxides have been researched thus far. Among them, $LiCoO_2$ is the most popular owing to the convenience and simplicity of preparation. This material can be synthesized by conventional solid state reaction, sol-gel method and so on (Reimers and Dahn 1992, Kumta et al. 1998). $LiCoO_2$ system has been studied extensively (Ohzuku et al. 1993, Amatucci et al. 1996). Li_xCoO_2 exhibits excellent cyclability at room temperature for $1 > x > 0.5$. Therefore, the specific capacity of the material is limited to the range of 137 to 140 mA h g^{-1}, although the theoretical capacity of $LiCoO_2$ is 273 mA h g^{-1} (Ohzuku and Ueda 1994). On the other hand, Li_xCoO_2 is expensive and highly toxic, which is unfortunate when considering its good electrochemical properties and easy synthesis. $LiMn_2O_4$ is also a popular cathode material and has been studied for long time. In principle, $Li_xMn_2O_4$ permits the intercalation/deintercalation of lithium ions in the range of $0 < x < 2$. In $1 < x < 2$. The material consists of two different phases: cubic in bulk and tetragonal at the surface. Simultaneously, the intercalation of lithium ions effectively decreases the average valence of manganese ions ($Mn^{4+} \rightarrow Mn^{3+}$) and leads to a pronounced cooperative Jahn–Teller effect, in which the cubic spinel crystal becomes distorted tetragonal with a $c/a \approx 1.16$, and the volume of the unit cell increases by 6.5% (Thackeray 1995). The transition metal oxides usually show poor rate capability due to low electronic conductivity and Li diffusion constant. To improve the rate capability, that is, the intercalation-deintercalation kinetics of the material, it is necessary to downsize the material to obtain short diffusion distance and large surface area. Hereafter, a few results using nano-transition metal oxides are described.

A particle size of $LiMn_2O_4$ prepared by a sol-gel method can be controlled by calcination temperature (Curtis et al. 2004). The nanoparticles having the size of 10 nm were prepared *via* a low temperature calcination at 350°C, whereas sub-micrometer sized particles were obtained through a high temperature calcination at 550°C. The prepared nanoparticle cathode showed improved capacity and cyclability. For RF sputtered $LiCoO_2$ thin film heating at 300°C, the average grain size was increased but still within nanosize order, while lattice distortion was reduced by the heating (Whitacre et al. 2001). The heated thin film $LiCoO_2$ demonstrated better electrochemical performance. It can be said the lattice distortion is also an important factor to determine the electrochemical performance.

$LiNi_{0.5}Mn_{1.5}O_4$ with spinel structure, same as $LiMn_2O_4$ is a promising cathode material for the next generation LIBs. The redox potential of $LiNi_{0.5}Mn_{1.5}O_4$ is 4.7 V vs. Li/Li^+ (Kim et al. 2004, Hoshina et al. 2012). This high redox potential compared with $LiCoO_2$ and $LiMn_2O_4$ is expected to improve energy and power densities of LIBs. Nanostructured $LiNi_{0.5}Mn_{1.5}O_4$ containing both ordered $P4_332$(p) spinel and disordered Fd3m spinel revealed high capacity retention of 80% at 6 C (Kunduraci and Amatucci 2006). The disordered spinel contained a small amount of Mn^{3+} and the distorted spinel has higher electronic conductivity than that of the ordered spinel. This high electronic conductivity contributes to the high capacity retention. These are good examples that nanoparticles can improve the electrochemical performance of the electrodes due to short diffusion distance and large surface area.

Nanostructured metal oxides

Vanadium oxide (V_2O_5) is a typical intercalation compound as a result of its layered structure. For Li-ion intercalation applications, vanadium oxide offers the essential advantages such as low cost, abundant source, easy synthesis, and high energy densities. Tuning the morphology or texture of V_2O_5 to obtain high-surface-area composite electrodes which can compensate the slow electrochemical kinetics and slow diffusion has been investigated (Le et al. 1996). V_2O_5 aerogels, which are mesoporous materials where the nanometer-sized domains are networked through a continuous, highly porous volume of free space, were reported to have electroactive capacities up to 100% greater than polycrystalline non-porous V_2O_5 powders (Dong et al. 2000). In another example, V_2O_5 nanoroll was prepared by a combination of sol-gel method and hydrothermal treatment (Krumeich 1999). The nanoroll V_2O_5 possesses four different sites, that is, tube opening, outer surface, inner surface and interstitial sites, compared to other tubular system and it showed better electrochemical performance. Also, defects in V_2O_5 nanoroll influence its electrochemical performance. It was reported that the specific capacity of defect-rich nanorolls was higher than that of well-ordered nanorolls (Muhr et al. 2000).

The olivine materials such as $LiFePO_4$ and $LiMnPO_4$ are generally recognized as a strong candidate for the cathode material owing to its low cost, safety and good cyclability. In fact, $LiFePO_4$ has already been successfully implemented in numerous commercial batteries for the automotive market. The shortcomings of the olivine materials are low Li ion and electronic conductivities. To solve these problems, reduction of particle size, nanostructures (carbon coating, and porous structure) have been applied for the olivine materials. Kotobuki used the ball-milling to reduce the particle size of $LiMnPO_4$ prepared by hydrothermal synthesis and found higher specific capacity of the milled sample than that of non-milled sample (Kotobuki 2013b). Meanwhile, a 3D porous hierachical $LiFePO_4$/graphene composite was prepared by the sol-gel method. The composite showed stable cyclability, 146 mA h g^{-1} after 100 cycle (Yang et al. 2012b). Graphene can absorb volume change of $LiFePO_4$ during charge and discharge and provide a good cyclability. Therefore, $LiFePO_4$ nanograin/graphene (Liu et al. 2014), graphene encapsulated $LiFePO_4$ (Luo et al. 2014, Shi et al. 2012), graphene nanoribbon-wrapped $LiFePO_4$ (Li et al. 2014) and so on have been investigated. The graphene composite was also applied for $LiMnPO_4$ (Zong and Liu 2014, Wang et al. 2014, Jiang et al. 2013). In the case of $LiMnPO_4$, heteroatom doping such as Mg and Fe was also applied to improve its electronic conductivity (Xu et al. 2014, Kotobuki et al. 2011a, Wi et al. 2014). $LiNiPO_4$ and $LiCoPO_4$ with olivine structure have also been expected as new cathode materials due to their high operation voltage (operation voltage of $LiNiPO_4$ and $LiCoPO_4$ are 4.8 V (Kotobuki 2013) and 5.1 V (Amine et al. 2000) vs. Li/Li^+, respectively). However, their high operation voltage decomposes the organic liquid electrolyte and makes difficult to evaluate their properties precisely (Kotobuki 2013). Further development including the electrolytes is needed for these materials.

3.3 *Anode*

The increase in the energy density of lithium batteries can be achieved by developing high capacity anode. As mentioned above, graphite allows intercalation of only one Li-ion with six carbon atoms, with a resulting stoichiometry of LiC_6 and thus an equivalent reversible capacity of 372 mA h g^{-1} (Kotobuki et al. 2011b). This low capacity confines further enhancement of the energy density of the next generation lithium batteries. In addition, the diffusion rate of lithium into carbon materials is between 10^{-12} and 10^{-6} cm^2 s^{-1} (for graphite it is between 10^{-9} and 10^{-7} cm^2 s^{-1}), which results in batteries with low power density (Persson et al. 2010, Kaskhedikar and Maier 2009). Therefore, development of new anode with high capacity, energy and power densities is an urgent issue. Although lithium metal has the highest capacity among the anode materials (3862 mA h g^{-1}) (Kotobuki et al. 2011b), safety issue like dendrite Li formation which can cause short circuit between anode and cathode, prevents it from practical use (Tarascon and Armand 2001, Liu 2013, Orsini 1999). Fortunately, we can find both carbon and non-carbon materials with high capacity such as carbon nanotubes (1100 mA h g^{-1}) (Landi et al. 2009), carbon nanofibers (450 mA h g^{-1}) (Kim et al. 2006), graphene (960 mA h g^{-1}) (Hou et al. 2011), porous carbon (800–1100 mA h g^{-1}) (Zhou et al. 2003), SiO (1600 mA h g^{-1}) (Yang et al. 2002), silicon (4200 mA h g^{-1}) (Ge et al. 2012), germanium (1600 mA h g^{-1}) (Hwang et al. 2012), tin (994 mA h g^{-1}) (Zhuo et al. 2013) and transition metal oxides (500–1000 mA h g^{-1}) (Jiang et al. 2012, Wang et al. 2012, Prosini et al. 2000). Furthermore, metal sulphides, phosphides and nitrides are also expected as the anode for LIBs (Lai et al. 2012, Boyanov et al. 2008, Ji et al. 2011b). However, high volume expansion, poor electron transport, capacity fading and low coulombic efficiency of these materials are the main limitations that have to be overcome before they can be used as effective anodes. In this sense, promising results and a bright perspective are offered by nanostructuring the above listed materials with nano-size and tailored morphology (Sun et al. 2012, Chen et al. 2012, Ueda et al. 2013, Hwang et al. 2011).

In the next section, the state of the art of anode materials for LIBs using nanotechnology to enhance their compatibility with practical use is described. For the purpose of simplicity, anode materials are classified into three main groups, depending on their reaction mechanism with Li:

1. Intercalation/de-intercalation materials, such as carbon based materials, porous carbon, carbon nanotubes, graphene, TiO_2, $Li_4Ti_5O_{12}$, etc.;
2. Alloy/de-alloy materials such as Si, Ge, Sn, Al, Bi, etc.;
3. Conversion materials like transition metal oxides (Mn_xO_y, NiO, Fe_xO_y, CuO, Cu_2O, MoO_2, etc.), metal sulphides, metal phosphides and metal nitrides (M_xX_y; here X = S, P, N). Some conversion materials such as In, Si, Ge, Sn, Pb, Sb, and Bi show remarkably large capacities due to the alloy formation (Fiordiponti et al. 1978, Huggins 1997, Idota et al. 1997, Courtney 1997, Li 1999, Martos et al. 2003, Pereira et al. 2003a, Kim et al. 2008, Guo et al. 2008a). These materials are included in this group.

Intercalation/de-intercalation materials

Carbon based materials. Carbon materials have been the most widely investigated anode materials due to its ease of availability, stability in thermal, chemical and electrochemical environment, low cost, and good lithium intercalation and de-intercalation reversibility (Marom et al. 2011, Girishkumar et al. 2010, Scrosati and Garche 2010, Li and Zhou 2012). The variety of carbon based materials used as active anode in LIBs are classified into two categories, according to the degree of crystallinity and carbon atoms stacking (Park et al. 2013): (i) SOFT carbon (graphitizable carbons) where crystallites are stacked almost in the same direction and (ii) HARD carbon (non-graphitizable carbons) where crystallites have disordered orientation (see Fig. 14).

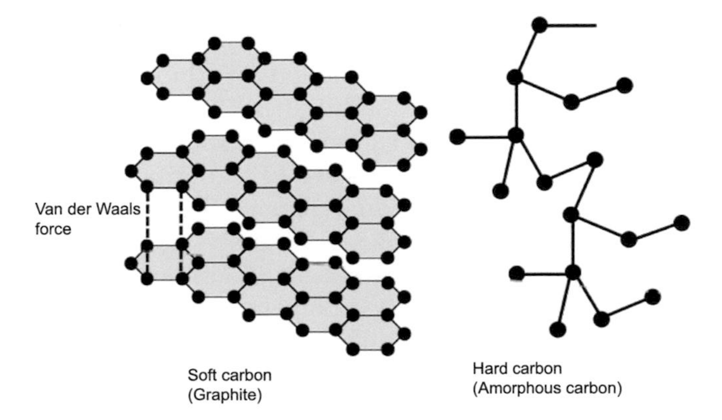

Figure 14. Structure of soft carbon and hard carbon.

The soft carbon consists of layered structure. In each layer, the carbon atoms are arranged in a honeycomb lattice. The carbon atoms are connected to neighborhood carbon atoms through strong covalent bond in the layers, while each layer is connected by weak van der Waals force. The Li ion is accommodated between the layers. The soft carbon is quite popular in the battery community. In fact, it shows stable reversible capacity (i.e., 350–370 mA h g^{-1}), long cycling life and good coulombic efficiency (Marom et al. 2011, Li and Wang 2013a, Boyanov et al. 2008, Yashio et al. 2004, Haik et al. 2011, Ganin et al. 2011, Yoshio et al. 2002). These properties can be enhanced by the preparation of coating layer on the graphite particles. For example, Lux et al. coated hydrophobic fumed silica particles on the graphite by dry-coating method (Lux et al. 2012). The coated graphite exhibited 3.5 time longer life than uncoated one when the fumed silica with 16 nm in diameter was coated. This enhancement is related to the protection of vulnerable site of the graphite to reaction with electrolyte. Also, SEI formation is influenced by the nature of the coating layer. The SEI film for carbon coated natural graphite spheres was found to be quite compact with a thickness from 60 nm to 150 nm, much thinner than the SEI film found on uncoated graphite spheres, showing thickness from 450 nm to 980 nm. The conclusion was that carbon coating can reduce the decomposition of the electrolyte and it can lead to the formation of a thin SEI on the electrode surface (Zhang et al. 2006). This is quite an important result because the SEI formation causes irreversible capacity mostly in the first cycle, resulting in energy loss. The energy loss in the first cycle can be reduced by the thin SEI formation. Additionally, the coating layer can reduce the van der Waals force between the particles. This effect leads to better flowability of the graphite powder which is important for further processing like preparation of electrode slurry.

Despite their massive production and the relative low cost of the industrial processes, these classes of carbon materials have, as a major issue, a low specific capacity (i.e., theoretical capacity of 372 mA h g^{-1}), especially for electric vehicle (EV) and hybrid vehicle (HV) applications. On the other hand, hard carbons have high reversible capacity (more than 500 mAh g^{-1}) in the potential range 0–1.5 V vs. Li/Li^+, therefore, they represent a valid alternative to soft carbons. Hard carbon has disordered orientation. This orientation breaks stoichiometry between Li and carbon atom (1:6 in soft carbon), being able to provide the high reversible capacity. Hard carbons were developed in 1991 by Kureha Corporation (Japan) and used as the negative electrode materials in the first built lithium-ion secondary battery, but they were subsequently dismissed from the electronic industry because lithium diffusion inside the hard carbons is very slow, namely very poor rate capacity. To overcome this shortcoming, nanoporous hard carbons are focused on. The performance of the hard carbon electrode is strongly affected by the porous structure (Yang et al. 2012a). The hard carbons in three different pore structures are shown in Fig. 15, in which non-porous carbon, mesoporous carbon and hierarchical porous carbon are composed of mesopores and micropores.

Charge and discharge test of these carbons were examined in EC-DES electrolyte dissolving $LiPF_6$ salt. Table 4 shows the discharge capacity and Li ion diffusion constant of hard carbons with different pore structure. As can be clearly seen, the pore structure strongly influences the discharge capacity and the hierarchical structure shows the highest discharge capacity, 503.5 mA h g^{-1} at 0.2 C. This surpasses the theoretical capacity of soft carbon (372 mA h g^{-1}). Also, this high capacity is retained even under high C rate condition. The discharge capacity of the hierarchical hard carbon is 332.8 mA h g^{-1} at 5 C, indicating a good candidate for EV and HV applications. This enhanced performance is attributed to the shorter diffusion pathways as evidenced by high Li ion diffusion constant of the hierarchical hard carbon. Additionally, the porous structure is expected to provide better structural flexibility that can absorb volume change during Li intercalation/ deintercalation, resulting in better cycle stability.

Carbon nanotubes (CNTs) are allotropes of carbon with a cylindrical nanostructure. Normally, CNTs are constructed with length-to-diameter ratio of up to 132,000,000:1 (Wang et al. 2009) significantly larger than that for any other material. These cylindrical carbon molecules have unusual properties, which are valuable for nanotechnology, electronics, optics and other fields of materials science. CNTs can be classified into single (SWCNTs) and multiwall carbon nanotubes (MWCNTs) depending on the thickness and on the number of coaxial layers (see Fig. 16).

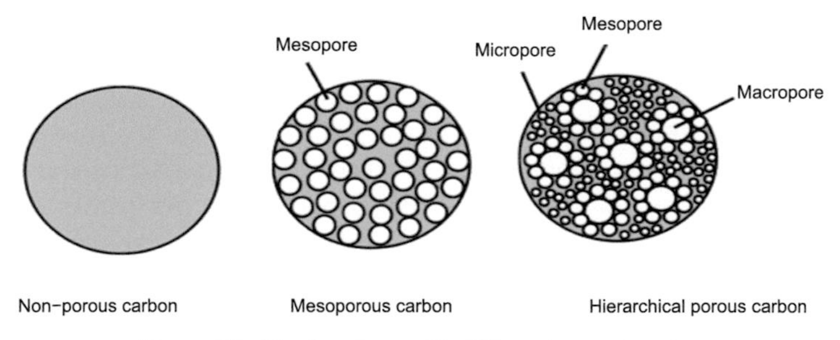

Figure 15. Hard carbon with different pore structure.

Table 4. Discharge capacity and diffusion constant of hard carbon with different pore structure.

Sample	**Discharge capacity at 0.2 C (mA h g^{-1})**	**Discharge capacity at 5 C (mA h g^{-1})**	**Diffusion constant (cm^2 s^{-1})**
Non-porous	223.9	41.8	1.03×10^{-9}
Mesoporous	433.4	267.5	1.66×10^{-8}
Hierarchiral	503.5	332.8	7.55×10^{-8}

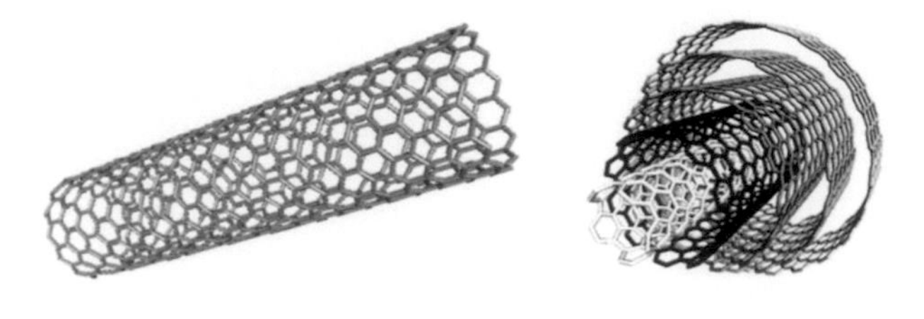

Figure 16. Structure of carbon nanotube.

Since their discovery in 1991, both SWCNTs and MWCNTs have been extensively investigated as anode materials and composites due to their superior electronic conductivity, good mechanical and thermal stabilities, adsorption and transport properties (Yu et al. 2013). The reversible capacity of SWCNTs is estimated at 1116 mA h g^{-1} in LiC_2 stoichiometry (Meunier et al. 2002, Schauerman et al. 2012, Nishidate and Hasegawa 2005, Zhao et al. 2000). This is the highest value among the carbon materials and it makes them as promising anode material for the next generation lithium batteries. This high capacity is considered to be due to the intercalation of lithium into stable sites on the surface of the tubes which have pseudo-graphitic layers as well as inside the central tube. The high capacity (1050 mA h g^{-1}) was reported in purified SWCNTs prepared by laser vaporization procedure (DiLeo et al. 2010). However, obtaining close value to the theoretical value is still a major challenge. In addition, low coulomb efficiency in CNT electrodes was observed very often due to a lot of surface defects on the large surface area and high voltage hysteresis. In order to overcome these issues, some attempts such as surface modification (Meunier et al. 2002), and adjustment of tube dimensions (diameter, wall thickness and porosity) (Lv et al. 2008, Zhou et al. 2010) have been done. Among the attempts, nano-structured CNTs are recognized as a good strategy (Oktaviano et al. 2012). Small holes can be prepared on the MWCNTs. Cobalt oxide nanoparticles were wired onto the CNT surface and an oxidation process was performed. Finally, 4 nm size holes on the MWCNTs were formed by removal of the oxide particles (see Fig. 17).

The performance of MWCNT with the holes as anode for lithium batteries has exhibited higher capacity, better cycle life and higher coulomb efficiency than the pristine one. It is confirmed that the CNT is a promising anode material and we can find a lot of interesting results. However, from the battery industry's point of view, CNT technology is not mature enough. In fact, production cost of CNT has hindered their usage in LIBs applications. A breakthrough to suppress the production cost is needed.

Graphene is also a promising anode material for the next generation of lithium battery. Graphene consists in a honeycomb network of sp^2 carbons bonded into two dimensional sheet with nano-meter thickness (single atom thickness). It is like a single layer of graphite (Fig. 18).

Since their introduction in 1987, this material has gained much attention because of its admirable properties and versatility in a number of fields such as chemical, physical, biological and engineering sciences (Hou et al. 2011, Huang et al. 2012). Among its astonishing properties, its good electrical conductivity, mechanical strength, high value of charge mobility and high surface area make it a suitable material for LIB electrode (Cui et al. 2008, Liang et al. 2009, Brownson et al. 2011). In fact, high quality graphite with few graphene layers (~ 4 layers) and a surface area over 490 m^2 g^{-1} showed large reversible capacities close to 1200 mA h g^{-1} at the initial cycles, with values around 848 mA h g^{-1} even at the 40th cycle (Lian et al. 2010). The nano-scale porous graphene is also a favorable structure for anode like CNTs. The graphene electrode with nano-pores exihibited high reversible

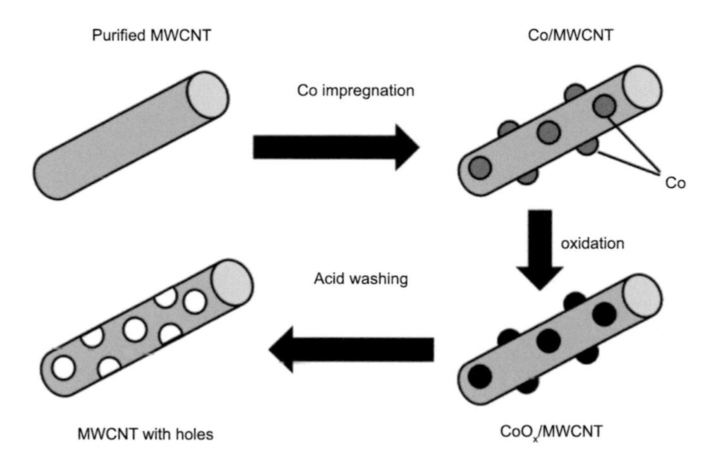

Figure 17. Preparation of MWCNT with small holes.

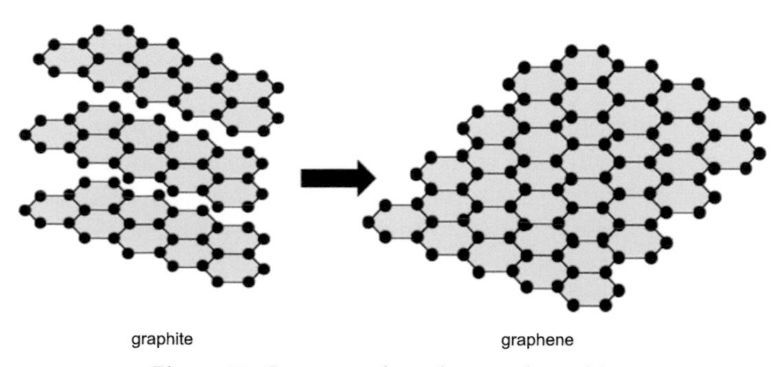

Figure 18. Structure of graphene and graphite.

capacity with long cyclability, ~ 3000 cycle. These results were attributed to the synergic effect of porous structure and good conducting network which facilitate the mass-transport and speed up the electrochemical reactions. In another strategy, graphene/metal or metal oxide hybrid is also an attractive anode. SnO_2 is well known as a good anode material which possesses high capacity. However, its volume change causes capacity fading while cycling. The graphene/SnO_2 hybrid can absorb the volume change of SnO_2 by structural flexibility of the graphene and additionally, the graphene can provide electrical conductive network. In fact, the hybrid system formed by 2 ~ 3 nm SnO_2 particle/graphene showed high capacity and stability (1220 mA h g^{-1} over 100 cycle) (Vinayan and Ramaprabhu 2013). Similarly, Si/graphene and Fe_3O_4/graphene hybrids were also reported and showed higher reversible capacity than present graphite anode (Wang et al. 2013a, Hu et al. 2013). The graphene/metal or metal oxide hybrids are also promising system for the anode. However, their production cost currently restricts their use in LIBs application.

Titanium based oxides. The titanium based oxides have been gaining much attention for their inexpensiveness, low toxicity, low volume change (2–3%) during both lithium insertion and de-insertion, along with an excellent cycling life (Wagemaker and Mulder 2013, Chen et al. 2013, Hong and Wei 2013, Moretti et al. 2013). The electrochemical performance and the lithium insertion/removal capacity of titanium based oxides mainly depend on their structure, morphology and size. To date, among the titanium based oxides, spinel $Li_4Ti_5O_{12}$ (LTO) has been extensively studied for anode purposes.

Spinel LTO is considered the most appropriate titanium based oxide as anode material for LIBs. In fact, TOSHIBA developed the lithium battery with LTO anode (SCiB) and has successfully commercialized it in 2008. LTO exhibits excellent Li-ion reversibility at the operating potential of 1.55 V vs. Li/Li^+. Lithium insertion/extraction in LTO occurs by the lithiation of spinel $Li_4Ti_5O_{12}$ yielding rock salt type $Li_7Ti_5O_{12}$. During the insertion process, the spinel symmetry and its structure remain almost unaltered (Li and Wang 2013b, Martha et al. 2011). This stable structure is the origin of stable charge-discharge profile of LTO. As shown in Fig. 19, typical charge-discharge curve of LTO anode reveals a long plateau at 1.5 V vs. Li/Li^+.

The high operation voltage guarantees safety conditions. In fact the formation of the solid electrolyte interface (SEI) is mitigated and the development of dendrites, a typical issue in carbon based anodes, can be avoided especially under high current condition (Mahmoud et al. 2013). The high operation voltage of the LTO anode reduces the operation voltage of batteries and this is unfavorable in terms of energy density of the batteries. The operation voltage will be improved by developing high voltage cathode

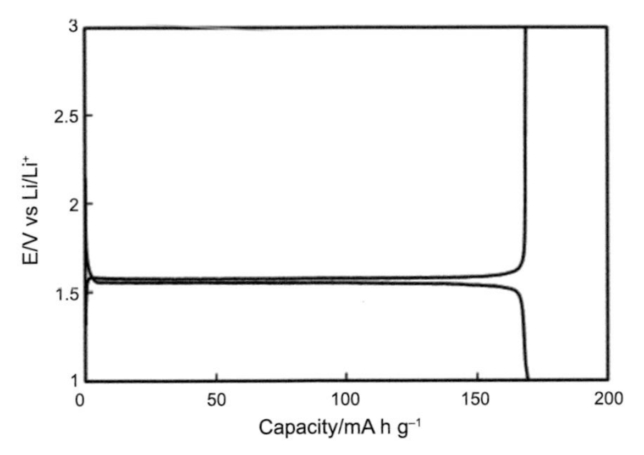

Figure 19. Charge and discharge curves of LTO.

materials such as $LiNi_{0.5}Mn_{1.5}O_4$ and $LiCoPO_4$ and novel electrolytes which can tolerate high voltage (present electrolyte is deteriorated over 4.5 V (Arrebola et al. 2006, Aurbach et al. 1999a)). The low theoretical capacity of 175 mA h g^{-1} and the low electronic conductivity ~ 10^{-13} S cm^{-1} of LTO limit the full capacity at high charge-discharge rates and reduce the Li-ion diffusion (Zhu et al. 2013a). In order to overcome this issue, two strategies have been performed. One is to enhance the Li ion diffusion by downsizing the LTO particles to nanoscale. The other is to improve the electrical conductivity of the LTO by doping heteroatoms. Nanocrystalline LTO with particle sizes between 20 to 50 nm has showed remarkable performance as anode (Prakash et al. 2010). Discharge capacity of 170 mA h g^{-1}, which was close to the theoretical capacity was obtained at 0.5 C. In addition, stable capacities of 140 and 70 mA h g^{-1} were observed under high charge-discharge rate of 10 and 100 C, respectively. Nanostructured LTO electrode had also showed a good result. LTO nanowires were grown directly on the Ti substrate. The LTO nanowire was hydrogenated to enhance electrical conductivity by introducing Ti^{3+} ion (see Fig. 20) (Shen et al. 2012). These nanowires containing Ti foil substrate could be used as electrode and current collector without binders and any conductive additives. They revealed good rate performance, i.e., 173 mA h g^{-1} at 0.2 C and 121 mA h g^{-1} at 30 C, and good cycle life.

As mentioned above, the electrochemical performance of LTO anode can be totally changed by its structure, particle size and doping. LTO is considered the most promising anode material for the next generation of lithium batteries for high power density application, especially with its high rate capability and stability. However, the structure of LTO anode has not been optimized. Further research on this topic is required.

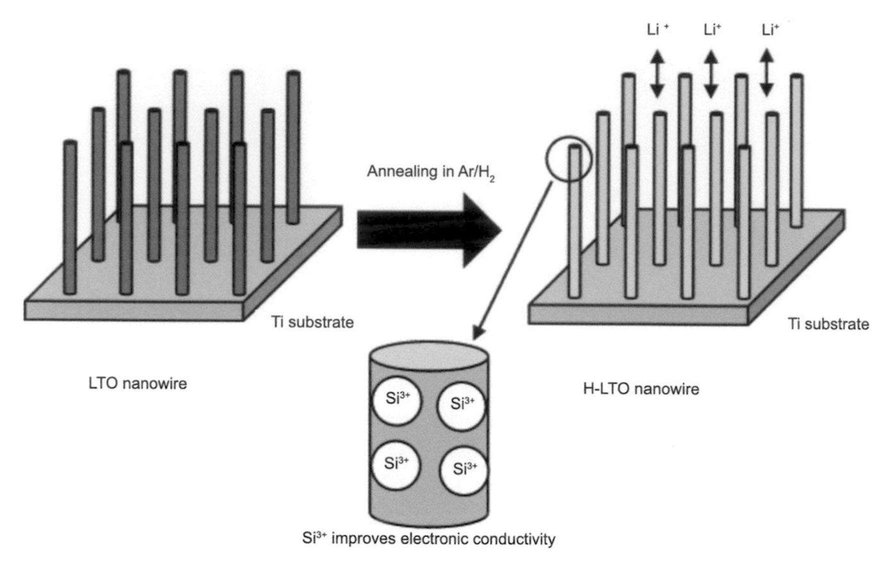

Figure 20. Preparation of LTO nanowires grown on the Ti substrate.

Alloy anode

Owing to ability to store large amounts of lithium and to solve safety issues related to lithium deposition (Shukla and Kumar 2008, Winter et al. 1998), lithium metal alloys, Li_xM_y, are of great interest as high capacity anode materials in LIBs. Table 5 compares the electrochemical properties of the alloy anodes, graphite and lithium metal. The theoretical capacity of the alloy anodes is 2 ~ 10 times higher than that of present graphite anode. It must be noted that the density of the host metals is higher than that of the graphite which leads to reduce volume of anode, resulting in increase of volumetric energy density of the battery. The second advantage of the alloy anodes is moderate operation voltage vs. lithium. This moderate operation voltage can avoid Li metal deposition on the anode especially under high current density, i.e., fast charge and discharge conditions, and solve the safety concerns.

Table 5. Electrochemical properties of various alloy anodes.

Materials	**Li**	**Graphite**	**Si**	**Sn**	**Sb**	**Al**
Density (g cm^{-3})	0.53	2.25	2.33	7.29	6.7	2.7
Lithiated phase	Li	LiC_6	$Li_{4.4}Si$	$Li_{4.4}Sn$	Li_3Sb	LiAl
Theoretical capacity (mA h g^{-1})	3862	372	4200	994	660	993
Volume change	100	12	320	260	200	96
Potential vs. Li (V)	0	0.05	0.4	0.6	0.9	0.3

The main challenge for the implementation of alloy anodes is their large volume change (up to 300%) during lithium insertion and extraction, which often leads to pulverization of the active alloy particles and poor cycle stability (Benedek and Thackeray 2002, Kasavajjula et al. 2007). In addition, the first-cycle irreversible capacity loss of alloy anodes is too high for practical application. Extensive research has been carried out to address these two issues and significant progress has been made during the last two decades. Among them, the downsizing from micro to nanoscale particle and the fabrication of composites with both lithium active and inactive material, are the most promising ways. In the latter case, the lithium inactive material serves as a conductive buffer matrix between the alloy materials and lithium source (Kasavajjula et al. 2007, Woo et al. 2010). Nanostructured alloy materials with different morphologies like nanowires and nanotubes have been considered as an implementable path to achieve high capacity with a good rate capability and long cycling life (Zhang and Braun 2012, Gu et al. 2012). Many metals are reactive towards lithium, e.g., Si, Sn, Sb, Al, Mg, Bi, In, Zn, Pb, Ag, Pt, Au, Cd, As, Ga and Ge (Winter and Besenhard 1999). However, only the first two elements have been widely investigated because they are cheap, abundant and environmentally friendly.

Silicon has been extensively investigated as an anode material because of its high theoretical energy capacity (Table 5). In addition, the discharge (lithiation) potential of silicon is almost close to graphite, i.e., 0.4 V vs. Li/Li^+. Si is the 2nd most abundant element on earth, hence it is inexpensive and eco-friendly. It is then understandable why silicon and its derivatives are considered the most promising and interesting materials for the realization of the future generation of LIBs, which justifies the strong academic and industrial interest for their development as anode active materials. The electrochemical lithiation of Si electrodes has been deeply investigated by many groups. It has been clarified that the high specific capacity value is due to the formation of intermetallic Li-Si binary compounds such as $Li_{12}Si_7$, Li_7Si_3, $Li_{13}Si_4$, $Li_{22}Si_5$ (Szczech and Jin 2011). To overcome the poor cyclability and large first cycle irreversible capactiy, nanostructured Si, especially on the morphology aspect has been intensively studied. Reducing the particle size from 10 μm to 10–40 nm slightly decreased the irreversible capacity in the first cycle because a lot of free spaces between nano-Si particles can absorb the volume change during Li insertion/extraction (Luo et al. 2009, Chan et al. 2008, Li et al. 2009). Si nanowires (NWs) has been also intensively studied due to its unique structure. For example, the Si NWs directly prepared on the metallic current collector showed charge-discharge capacity of 3500 mA h g^{-1} over 20 cycles at 0.2C rate and 2100 mA h g^{-1} at 1C (Chan et al. 2008). This result was due to the good electronic contact between the current collector and the Si NWs, to an efficient charge transport mechanism and to the small diameter of Si

NWs, which allows better accommodation of the volume change without any initiation of fractures. It is well known that carbon addition to Si powders also significantly improves their electrochemical performance by the enhancement of electronic conductivity (Saint et al. 2007, Si et al. 2009, Wang et al. 1998, Gu et al. 2010). This can be applied for Si NWs as well. Improved performance of the carbon coated Si NWs was reported (Chan et al. 2010). Also, porous Si NW was prepared and it reported superior electrochemical performance (Ge et al. 2012). Si nanotubes also demonstrated remarkable electrochemical performance (Song et al. 2010, Park et al. 2009). These superior electrochemical properties were attributed to available free volume to accommodate the volume change. Recently, it was reported that Si nanotube with silicon oxide on its surface showed stable charge-discharge property. The SiO_x on the Si nanotube allows Li diffusion, but prevents direct contact Si with electrolyte. This influences SEI formation on the electrode surface, resulting in improvement of electrode stability. The Si nanotubes with SiO_x surface layer showed outstanding stable anode performance with cycle life of over 6000 cycles (Wu et al. 2012). Despite the mentioned nanostructure electrodes showing good electrochemical performances, their preparation is not cost-effective. Many research groups are looking for new cost-effective preparation methods such as hydrothermal and solvothermal techniques.

Tin was also widely studied as an alternative anodic material to carbon for lithium ion batteries. Although the theoretical capacity of Sn is 994 mA h g^{-1} (Ke et al. 2007), which is lower than that of Si, its environmental benignity, low cost and ease of processing makes it an attractive anode material. In the case of Sn anode, extensive attention has been paid to Sn based intermetallic compounds to improve the performance of cyclability, i.e., Sn_xM_y (M: inactive element) such as Sn-Fe (Mao et al. 1999a, Mao et al. 1999b, Larcher et al. 2000b), SnCu (Kepler et al. 1999, Larcher et al. 2000a, Wang et al. 2000, Tamura et al. 2002), Sn-Mn (Beaulieu et al. 2000), Sn-Co (Mao et al. 1999c) and so on. Ni is a typical element which does not react with Li and Sn-Ni anode has been most widely studied so far. One of the promising features of Sn-Ni alloy is that it can be easily prepared by plating. Electrochemical performance of Sn-Ni thin film electroplated on a Cu sheet which is 3 ~ 4 μm in thickness was investigated by Mukaibo et al. (Mukaibo et al. 2005a). They reported that the best composition of the Sn-Ni alloy was $Sn_{62}Ni_{38}$ and its discharge capacity was 650 mA h g^{-1} at 70th cycle (Mukaibo et al. 2003). XRD study of structural change during charge-discharge cycling of Sn-Ni alloy with different composition revealed Ni_3Sn_4 phase which could reversibly react with Li, was observed in $Sn_{62}Ni_{38}$, whereas only metastable phase close to structure of SnNi was observed in $Sn_{54}Ni_{46}$ and pure Sn and Sn rich metastable phases were observed in $Sn_{84}Ni_{16}$ (Mukaibo et al. 2005b). Nanostructured Sn-Ni alloy anode was also prepared by applying the electroplating technique. A three dimensional ordered microporous structure (3DOM) Sn-Ni alloy was prepared by electroplating using a colloidal crystal template consisting of monodisperse polystyrene (PS) latex. Figure 21 shows the SEM image of 3DOM Sn-Ni electrode prepared by a combination of the PS template with electroplating (Nishikawa et al. 2009).

The initial discharge capacity of the 3DOM Sn-Ni electrode was 455 mA h g^{-1}. However, the discharge capacity decreased rapidly after 30 charge-discharge cycles owing to the collapse of the porous structure caused by the volume change during lithiation and delithiation. Furthermore, they prepared highly patterned cylindrical Ni–Sn alloy anode with 3DOM structure by using a photoresist substrate which has holes with 20 μm in diameter and 30 μm of depth (see Fig. 22).

The discharge capacity of the highly patterned cylindrical Ni-Sn alloy was 632 mA h g^{-1} and it exhibited stable discharge capacity until 200th cycle (Woo et al. 2010). This improved performance of the patterned 3DOM Sn-Ni alloy anode was attributed to the reduction of volume change of the anode held by the photo resist substrate. In addition, they examined other patterns such as square and hexagon types to improve the areal capacity because the cylinder type has lots of void spaces which cannot relate the electrochemical reaction (refer Fig. 23) (Kotobuki et al. 2011c).

In the result, the hexagon pattern exhibited the best performance of 1.9 mA h cm^{-2}, followed by square (1.6 mA h cm^{-2}) and cylinder (1.3 mA h cm^{-2}). The void space was in the order of hexagon < square < cylinder. Therefore, the smaller void space the electrode had, the larger the capacity was obtained. The authors concluded that the void space was needed to absorb the stress during the volume change, but it must be as small as possible. Wan et al. prepared nanoporous Sn-Ni electrode without the PS template (Wan et al. 2015). They first deposited Ni-Cu alloy electrochemically and then dissolved Cu selectively by adjusting the anodic potential. Finally, Sn was electrodeposited on the porous Ni and the nanoporous Sn-Ni anode was obtained. The electrode revealed stable charge-discharge behavior.

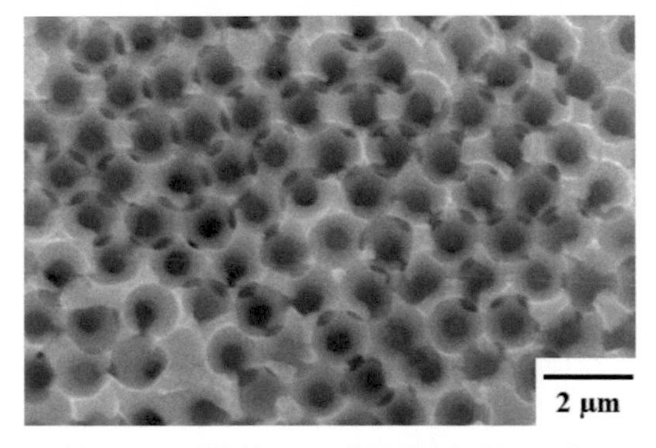

Figure 21. SEM image of 3DOM Sn-Ni anode.

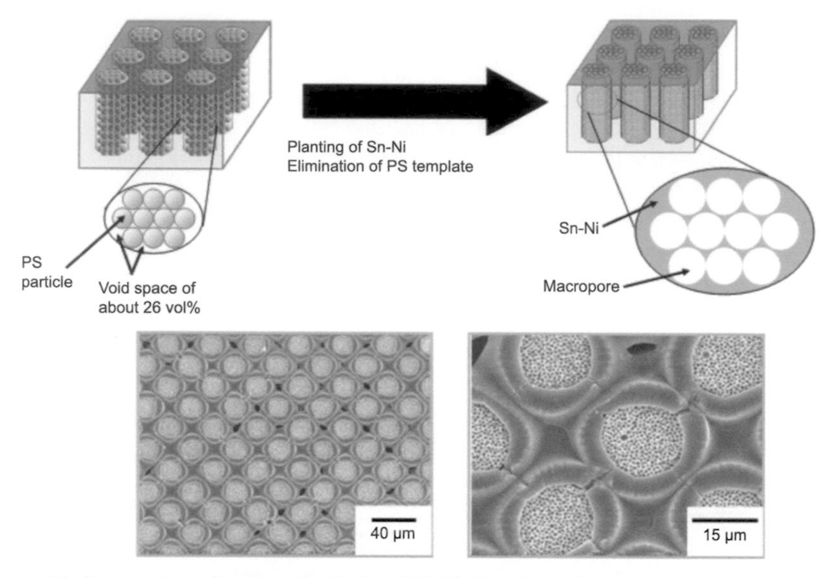

Figure 22. Preparation of patterned cylindrical 3DOM Sn-Ni anode using photoresist substrate.

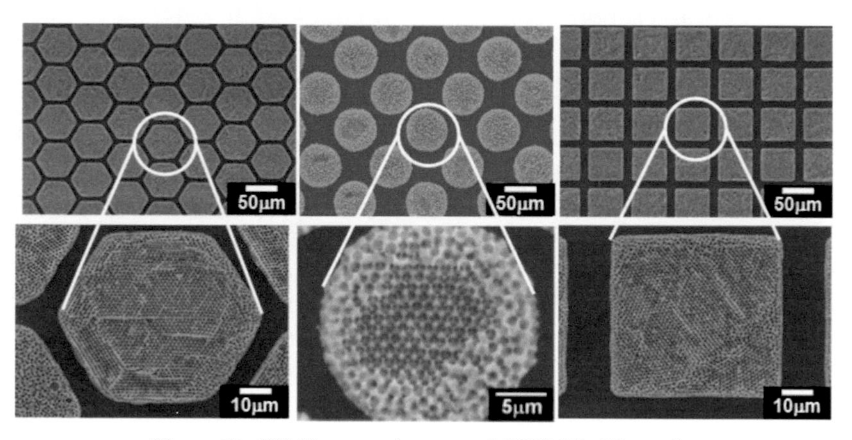

Figure 23. SEM images of patterned 3DOM Sn-Ni anode.

The discharge capacity was 537 mA h g^{-1} at 200th cycle. Enhancement of performance of Sn-Ni anode by compositing with CNTs due to its superior electrical conductivity and accommodation to volume change during charge and discharge was also reported (Guo et al. 2005). Nanocrystalline Sn-Ni/MWCNT prepared by ultrasonic-pulse electrodeposition technique exhibited stable charge and discharge behavior compared with pristine one (Uysal et al. 2015). Other useful preparation method for Sn-Ni alloy is the co-precipitation method. The co-precipitation method makes the materials react uniformly at molecular level and has the advantages of lower polycrystalline synthesized temperature and shorter sintering time (Li et al. 2000). Sn-Ni alloy nanoparticles prepared by the co-precipitation method are also extensively researched (Guan et al. 2014). The advantage of this method is the ease of mass production of Sn-Ni nanopowder and this makes the co-precipitation method an attractive method for the implementation of Sn-Ni alloy anode.

Conversion anode

A new reactivity concept with reversible electrochemical reaction of lithium with transition metal oxides was brought by Poizot et al. in the advent of 21st century (Poizot et al. 2000). This is called "conversion reaction" which is generalized as follows:

$$M_aX_b + (b \cdot n)\,Li \leftrightarrow a\,M + b\,Li_nX$$

where M, X and n are transition metal, anion and formal oxidation state of X, respectively.

Many transition metal compounds that do not have any vacant sites in the structure were disregarded as a possible candidate for active materials because of the impossibility of intercalating lithium. After it was proven that several oxides can deliver stable gravimetric capacities as much as three times that of carbon, these phases started to be considered as promising alternatives in rechargeable batteries. Since then, the conversion reaction turns out to be widespread; since the original discovery, many other examples of conversion reactions including sulfides, nitrides, fluorides, and phosphides have been reported (Pereira et al. 2003a, Li et al. 2003, Badway et al. 2003, Bervas et al. 2005, Silva et al. 2003, Pralong et al. 2002). For example, the simple binary transition metal oxides with rock salt structure (CoO, CuO, NiO, FeO) can react reversibly with lithium, according to the general reaction (refer Fig. 24):

$$MO + 2Li^+ + 2e^- \leftrightarrow Li_2O + M^0$$

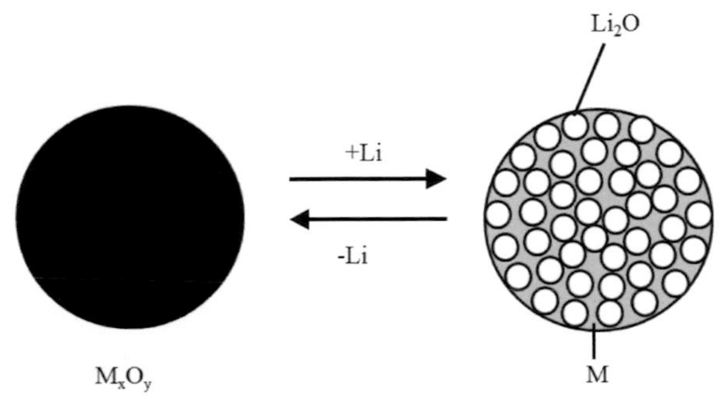

Figure 24. Conversion reaction.

Their full reduction leads to composite materials consisting of nanometric metallic particles (2–8 nm) dispersed in an amorphous Li_2O matrix. The nanometric character of the metal particles has been shown to be maintained even after several reduction-oxidation cycles (Grugeon et al. 2003). Li_2O has been always reported to be electrochemically inactive. The nanometric metal particles are believed to enhance their electrochemical activity toward formation/decomposition of Li_2O. As can be expected, this phenomenon is directly correlated with the covalence of M-O (M-X) band. The potential at which the conversion reaction proceeds is shown in Table 6. The potential depends on the transition metal and the anionic species. In other words, the reaction potential can be tuned easily by choosing suitable transition metal and anionic species (Goodenough and Kim 2010). As shown in Table 6, almost all conversion reaction occurs at low potential except CuF_2. Therefore, the conversion electrode is considered as an alternative of commercial graphite anode.

Table 6. Potential at which the conversion reaction proceeds in binary transition metal compounds.

	X = O		X = S		X = N		X = P		X = F	
	Phase	**E_{conv}**	**Phase**	**E_{conv}**	**Phase**	**E_{conv}**	**Phase**	**E_{conv}**	**Phase**	**E_{conv}**
M = Cr	Cr_2O_3	0.2	CrS	0.85	CrN	0.2			CrF_3	1.8
M = Mn	MnO_2	0.4	MnS	0.7			MnP_4	0.2		
	Mn_2O_3	0.3								
	MnO	0.2								
M = Fe	Fe_2O_3	0.8	FeS_2	1.5			FeP_2	0.3	FeF_3	2.0
	Fe_3O_4	0.8	FeS	1.3	Fe_3N	0.7	FeP	0.1		
	FeO	0.75								
M = Co	Co_3O_4	1.1	CoS_2	1.65–1.3	CoN	0.8	CoP_3	0.3	CoF_2	2.2
	CoO	0.8	$Co_{0.92}S$	1.4	Co_3N	1.0				
			Co_9S_8	1.1						
M = Cu	CuO	1.5	CuS	2.0–1.7			CuP_2	0.7	CuF_2	3.0
	Cu_2O	1.4	Cu_2S	1.7			Cu_3P	0.8		
M = Mo	MoO_3	0.45	MoS_2	0.6						
M = Ni	NiO	0.6	NiS_2	1.6	Ni_3N	0.6	NiP_3	0.7	NiF_2	1.9
			NiS	1.5			NiP_2	0.5–0.3		
			Ni_3S_2	1.4						
M = Ru	RuO_2	0.9								
M = W			WS_2	0.8–0.6						
M = Ti									TiF_3	0.95
M = V									VF_3	0.4

The conversion materials are promising anode materials for the lithium ion batteries due to high capacity, for example, the theoretical capacities of Cr_2O_3 and CoS_2 are 1058 and 870 mA h g^{-1}, respectively. However, they are still far from commercial application although intensive research has been done. Common issues of the conversion materials are induction of large volume changes like the alloy anodes mentioned above by structural re-organization that takes place to accommodate the chemical changes, resulting in particle decohesion and unsatisfactory cycling performance. In order to deal with these issues, nanostructured conversion materials have been firmly studied. Hereafter, application of nanomaterials for the conversion material and other approaches for the issues are described.

Transition metal oxides

Transition metal oxides are, so far, the family of compounds that react through conversion reactions which deserve the most attention. Iron based oxides have been extensively studied as the anode material because of their low cost, non-toxicity and high natural abundance. Iron oxides, both haematite (α-Fe_2O_3) and magnetite (Fe_3O_4), are capable of participating in reversible conversion reactions with lithium, providing a theoretical capacity of 1007 and 926 mA h g^{-1}, respectively (Xu and Zhu 2012). However, low electrical conductivity, low diffusion rate of Li ions, high volume change and iron aggregation of iron oxides during charging and discharging provide poor cycle performance. As a result, many papers can be found to date in which the electrochemical activity is evaluated for samples prepared by numerous synthetic methods of forming iron oxide nanomaterials in various sizes, shapes (Koo et al. 2012, Kang et al. 2012, Ma et al. 2013, Mitra et al. 2006) and porosities as well as carbon coating (Zhu et al. 2013b, Hwang et al. 2013, Kim et al. 2013). It has been reported that α-Fe_2O_3 hollow sphere (Wang et al. 2011), α-Fe_2O_3 nanotube (Liu et al. 2009), α-Fe_2O_3 nanoparticles with 300 to 500 nm (Wu et al. 2006) and spindle-like porous α-Fe_2O_3 (Xu et al. 2012b) had showed high reversible capacity, more than 700 mA h g^{-1}. Regarding Fe_3O_4, although its behavior received less attention than α-Fe_2O_3, carbon-decorated crystalline Fe_3O_4 nanowires (Muraliganth et al. 2009), nanocomposites Fe_3O_4 with porous carbon-silicate layers (Sohn et al. 2013), Fe_3O_4 deposited onto Cu nanorod current collector (Taberna et al. 2006) had showed remarkable performance, both in terms of cycle life and rate capability, making this material to be taken into consideration for practical use. γ-Fe_2O_3 has not received much attention like α-Fe_2O_3 thus far. However, it was proven that γ-Fe_2O_3 behaved in a similar way like α-Fe_2O_3 (Abraham et al. 1990). This material is also a valid alternative to the graphite anode.

Cobalt oxides (CoO and Co_3O_4) are among the most explored transition metal oxides for their application as negative electrodes in LIBs. The theoretical capacities of Co_3O_4 and CoO are 890 and 715 mA h g^{-1}, respectively (Barreca et al. 2010, Zhang et al. 2011). These values are slightly lower than those of iron oxide, but still much larger than that of present graphite anode. Similarly to other kinds of conversion materials, a number of different forms of cobalt oxides have been studied. Porous nanostructures, nanosheets, nanocubes, nanowires and nanotubes have been prepared by various synthetic routes such as wet chemical, solid-state, hydrothermal and microwave, therefore allowing the tailoring of their electrochemical performance (Fei et al. 2013, Yuan et al. 2013, Wang et al. 2013b, Li et al. 2011, Binotto et al. 2007). Also, carbon composites of cobalt oxides such as CoO/graphene (Sun et al. 2012, Peng et al. 2012) and Co_3O_4/graphene (Wu et al. 2010) exhibited a good performance. Comparing with CoO, electrochemical reaction of Co_3O_4 proceeded in 1 or 2 moles of lithium and formed $Li_xCo_3O_4$ or Li_2O and CoO depending on the crystal size and surface area of the Co_3O_4 and applied current (Larcher et al. 2002, Binotto et al. 2007). The CoO was further reduced to Co and Li_2O (Shaju et al. 2007, Pralong et al. 2004, Li et al. 2008, Fu et al. 2004a). This CoO became active material instead of Co_3O_4 in prolonged cycle (Connor and Irvine 2002). Therefore, CoO may be more close to the practical use.

Various other metal oxides, which show a conversion reaction mechanism with lithium, have also been studied. Most of the studies focused on the 1st row transition metal in the periodic table like CrO_x, MnO_x, NiO and CuO_x and they showed miscellaneous physicochemical properties and a large reversible capacity around 500 mA h g^{-1} (Dupont et al. 2007, Zhong et al. 2010, Hosono et al. 2006).

In the 2nd row transition metal, MoO_3 has been intensively investigated due to its high theoretical capacity (1117 mA h g^{-1}). It was reported that the first discharge capacity was more than 1000 mA h g^{-1} and the high capacity was sustained in several cycles (Lee et al. 2008, Jung et al. 2009, Riley et al. 2010, Hassan et al. 2010). The 2nd row transition metal conversion anode is not popular compared to the 1st row metal, but it is a potential alternative to present graphite anode. Studies on the 2nd row transition metal conversion anode will be popular and it is expected that a lot of results will come out soon.

Transition metal sulfides. Transition metal sulfides were initially used as lithium primary cell materials (Gabano et al. 1972). Layered phases such as TiS_2 had been long considered as candidates for positive electrode materials in rechargeable lithium batteries (Whittingham 1978) and high temperature cells based on iron sulfide electrodes were considered as a possible energy storage option for the electric vehicle in the 1980s (Henriksen and Jansen 2002).

The technical feasibility of the electrochemical reaction of alkali metals with sulfur is well-known. Particularly, the reaction of lithium with sulfur has received much attention in the development of Li-sulfur batteries which have a theoretical capacity of 1675 mA h g^{-1} and a theoretical specific energy of 2500 Wh kg^{-1}, assuming complete reaction to form Li_2S. However, the reaction is complicated by the formation of a variety of lithium polysulfide intermediates, which are partially soluble in the electrolyte, reducing its conductivity, and loss of active material, resulting in severe inefficiencies (Yamin and Peled 1983, Mikhaylik and Aldridge 2004).

To suppress the dissolution of the intermediates into the electrolyte, usage of carbon matrix has showed a good result. Ultra-thin carbon coated FeS nanosheets were prepared by surfactant assisted solution method (Xu et al. 2012a). Here the carbon coating served to avoid lithium sulphides (Li_xS) from dissolving and then improve the cycle life of LIBs. A similar improvement was observed in graphene/CoS composite (Gu et al. 2013). Graphene plays an important role to not only suppress the dissolution, but also improve the electronic conductivity. Another approach to avoid the loss of active material is the use of solid electrolyte. Improvement of cycle life of FeS_2 and FeS was reported for all-solid-state cell with $LiCoO_2$ cathode and grassy-type electrolytes (Takada et al. 1999, Kim et al. 2005a). Moreover, NiS in batteries using PEO (poly-ethylene oxide)-based polymer electrolyte operating at 80°C showed discharge capacity of 540 mA h g^{-1} after 100 cycle (theoretical capacity of NiS is 591 mA h g^{-1}) (Han et al. 2003). Similar results were reported in CuS using PEO-based polymer electrolyte and glass ceramics electrolytes (Hayashi et al. 2004). These results suggested that the interaction between the organic solvents used in liquid cells and the lithium polysulfide intermediates may be a major issue. To use the transition metal sulfide as the anode for LIBs, selection of electrolyte is very important (Marmorstein et al. 2000, Shim et al. 2002, Kim et al. 2005).

Transition metal phosphides. The transition metal phosphides have also been studied as alternative anode to the graphite (Lai et al. 2012, Boyanov et al. 2008, Pralong et al. 2002, Villevieille et al. 2008, Boyanov et al. 2009, Li et al. 2013a, Kim and Cho 2009, Stan et al. 2013). They can react with Li in both conversion reaction and intercalation/de-intercalation reaction depending on the nature of the transition metal and phosphorous bond stability upon electrochemical environment. While conversion reaction has not been observed for phosphides of early transition metals such as Ti or V (Kim et al. 2005, Woo et al. 2006, Gillot et al. 2007), the conversion reaction, that is, formation of Li_3P and nano-sized metallic particles has been found in other first row of transition metals like Mn, Fe, Co, Ni or Cu. Transition metal phosphides can deliver high capacities between 500 and 1800 mA h g^{-1} and lower conversion potentials compared with the transition metal oxides and sulfides (refer Table 6). However, their low electrical conductivity and large volume change upon charge-discharge cycle are the problems which are hindering their development. To overcome these problems, the design of electrode is very important. For example, a monoclinic NiP_2 was grown on Ni foam which acted as a current collector, provided high discharge capacity (1300 mA h g^{-1}) and remarkably enhanced rate capabilities (Gillot et al. 2005). Moreover, Cu_3P grown on copper foil by vapor phase reaction also improved the cycle life and rate capability significantly (Villevieille et al. 2008). While further improvements are needed to meet the minimum standards of cycle life and rate capability, these innovative works with NiP_2 and Cu_3P show that a lot of things must be done in order to optimize the electrochemical performances of the phosphide compounds.

Transition nitride and fluoride. Conversion reactions have been reported to take place in transition metal nitrides when prepared as crystallized (e.g., VN, CoN, Co_3N, Fe_3N, Mn_4N) or amorphous (e.g., Ni_3N) thin films with a few hundred nanometers in thickness (Das et al. 2009, Wang et al. 2004, Fu et al. 2004b). The discharge capacity of the transition metal nitrides depends on a ratio of nitrogen to metal (N/M). In $N/M < 1$, the discharge capacity of 350–400 mA h g^{-1} was usually obtained whereas much larger capacity of > 1000 mA h g^{-1} was observed in $N/M = 1$. CrN showed the discharge capacity of 1800 and 1200 mA h g^{-1} in the first and second cycle, respectively, and the value remained relatively stable in prolonged cycles (Sun and Fu 2007, Sun and Fu 2008a, Sun and Fu 2008b). However, upon prolonged cycling, additional redox couples sometimes appeared in the transition metal nitride anode (Pereira et al. 2003a). This complexity of reaction which takes place in the transition metal nitride is still unclear and must be characterized.

Transition metal fluorides constitute a special group within the family of compounds that react with lithium through conversion reactions. The very high ionicity of the M-F bond results in the reduction potentials to LiF and metal nanoparticles that can be around or even above 2 V (Table 6), which is in stark contrast with the potentials below 1.5 V as observed in oxides, sulfides, nitrides, and phosphides. This high potential is preferred in terms of SEI formation and safety concern. The fluorides of Ti, V, Cr, Mn, Ni, Fe Co, Cu and so on have been reported. Nanocomposites of carbon and the transition metal fluorides showed better reversibility of the conversion reaction (Badway et al. 2003, Badway et al. 2007), compared to other conversion materials.

In conversion materials, a volume change during charge and discharge processes and low conductivity of compounds of Li and counter anion (Li_2O, Li_2S and so on) make them difficult for practical use. Reduction of particle size of the conversion materials to nano-scale shortens the Li ion diffusion path and addition of conductive material helps to improve the electronic conductivity by ensuring inter-particle contact. This is a good strategy and some of such attempts have reported promising results. For practical application of the conversion anode, more precise control of structure of the electrode and reduction of processing cost are the critical issues that need to be solved at the moment.

3.4 Electrolytes containing nanomaterials

As mentioned above, the non-aqueous liquid electrolyte dissolving Li salt has been used for LIBs and its flammability sometimes causes serious safety issue. A good strategy that has been recognized to solve this issue is by replacement of the nonflammable solid electrolyte for the non-aqueous liquid electrolyte. There are two kinds of solid electrolytes: ceramics and polymer electrolytes. The polymer electrolytes, that is, ionically conducting polymer membranes, are expected to enhance lithium-battery technology by replacing the liquid electrolyte and thereby enabling the fabrication of flexible, compact, laminated solid-state structures without leakage and they are available in various geometries (Kotobuki et al. 2010). Furthermore, it is supposed that the introduction of polymer electrolytes would allow practical utilization of lithium metal anodes, with considerable gain in terms of cell energy density. Therefore, strenuous efforts have been made for over 20 years. Polymer electrolytes are commonly composed of complexes of lithium salt (LiX) with a high-molecular-weight polymer such as polyethylene oxide (PEO). The basic structure of PEO-LiX polymer electrolytes involves oxygen in PEO chains coordinated with Li^+ cations, thereby separating them from the X-counteranions (refer Fig. 25) (Lightfoot et al. 1993, Vincent and Scrosati 1997).

Polymer electrolytes therefore require local relaxation and segmental motion of the solvent (PEO) chains to allow ion (Li^+) transport, and this condition can only be obtained when the polymer is in the amorphous state. But PEO crystallizes below 60°C. So the conductivity of PEO-LiX electrolytes reaches practically useful values (about ~ 10^{-4} S cm^{-1}) only in the amorphous

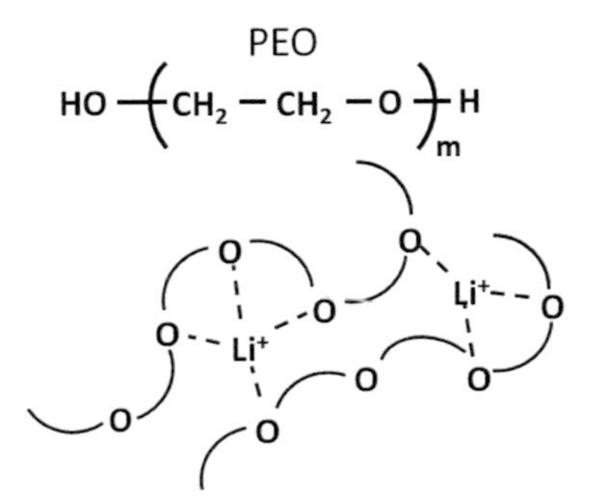

Figure 25. Structure and interaction between PEO and Li salt.

state, i.e., temperatures of 60 ~ 80°C. Therefore, most of the research has focused on lowering the crystallization temperature to keep the amorphous state and high conductivity at room temperature. When the polymer is in crystal state, the Li ion conduction was reported only in a ratio of PEO/Li = 6 (Gadjourova et al. 2001). In this ratio, pairs of PEO chains fold to form cylindrical tunnels within which the Li^+ ions are located and coordinated by the six ether oxygens. The anions are located outside these tunnels in the interchain space and do not coordinate with the cations (see Fig. 26).

The structure suggests that Li^+ ion transport along the tunnels may be possible in the crystalline 6:1 complexes. In fact, when the Li ion conduction was reported, however, the value was $10^{-6} \sim 10^{-7}$ S cm^{-1} at room temperature, which was less for practical use. Furthermore, transference number, which is a fraction of the current transported by Li cation, of the crystalline polymer electrolyte was reported at ≈ 1. Nevertheless, the amorphous polymer electrolyte has gained more attention because of high Li ion conductivity and a lot of research has been conducted to enhance the Li ion conductivity of the amorphous polymer electrolytes.

The most common approach for lowering the crystalization temperature of the amorphous polymer electrolyte is to add liquid plasticizers, but this approach promotes deterioration of the electrolyte's mechanical properties and increases its reactivity towards the lithium metal anode. Therefore, the development of solid plasticizers is required. In several cases, polymer electrolytes have been mixed with plasticizers, such as succinonitrile (Fan et al. 2008), polysquarate (Itoh et al. 2008), carbonates, or poly (methylmethacrylate) (Ghosh and Kofinas 2008). Insulative nanoparticulate oxide powders (ceramic filler) such as TiO_2, Al_2O_3 and SiO_2 have been reported to enhance the conductivity when the powders are dispersed into the polymer matrix (Croce et al. 1989, Capuano et al. 1991,

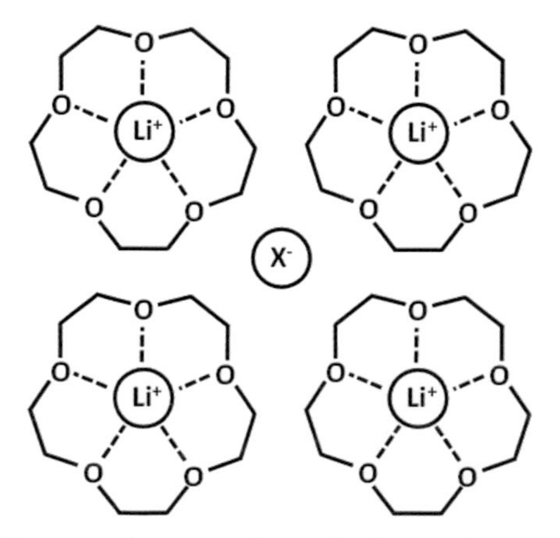

Figure 26. Structure of crystal polymer electrolyte.

Croce and Scrosati 1993, Kumer and Scanlon 1994). Furthermore, addition of the inorganic oxide particles to PEO-based electrolyte suppresses lithium dendrite formation, which is one of the main obstacles in the usage of lithium metal anode (Liu et al. 2010). Two mechanisms were originally suggested to explain the improvement of ionic conductivity by the addition of the inorganic oxide particles: (a) the creation of interfaces between the polymer and the inorganic nanopowders which act as preferential ion-conduction pathways; and (b) the inhibition of the crystallinity of the polymeric matrix, a well-known obstacle to the long-range ion migration in the polymer electrolytes. As a representative example of suggestion (a), Bhattacharyya and Maier reported that the improvement of Li ion conductivity by addition of the ceramic filler was also observed even in non-aqueous liquid electrolyte dissolving lithium salt (Bhattacharyya and Maier 2004). They used the non-aqueous electrolyte to neglect the amorphization of polymer. This is because increase of Li ion concentration due to breaking up of ion pairs of Li salt through adsorption of anion on acidic surface of the ceramic fillers, results in the improvement of Li ion conductivity. On the other hand, Croce et al. reported that the ceramic fillers hindered the reorganization of PEO chain due to large surface area and Lewis acid character of the fillers, resulting in amorphization of the polymer electrolyte (Croce et al. 1998). They also reported that the structural properties of the PEO-LiX electrolyte containing dispersed and nano-sized ceramic fillers were similar to those of liquid electrolytes (Croce et al. 1992). Therefore, it can be considered that both suggestions contribute to the enhancement of Li ion conductivity.

Furthermore, the ceramic filler promotes the enhancement of the lithium-ion transference number, which is associated with the Lewis acid-base interactions occurring between the surface of the ceramic and both the X anion of the salt and the segments of the PEO chain (Arico et al. 2005, Bruce et al. 2005, Appetecchi et al. 2000). Transference number is defined as the fraction of current transported by Li cation. The transference number is described as below.

$$t^+ = i^-/I$$

where t^+, i^- and I are transference number of cation (Li^+), current generated by the movement of cation and total current, respectively.

The high transference number is desired for electrolyte used in batteries, because a low lithium ion transference number results in mixed cation and anion conductivity, leading to the formation of salt concentration gradients in the electrolyte during the battery operation. Such concentration gradients give rise to voltage loss due to concentration polarization. Furthermore, in the electrolyte with high transference number, side-reactions caused by migration of anion can be suppressed. The transference number of ceramic electrolyte and crystalline polymer electrolyte is considered to be ≈ 1. In these electrolytes, only Li ion can move in the conduction channel of their crystal structure, while the anion is fixed in the crystal, resulting in unity of transference number. Generally, in ceramic-free polymer electrolyte, transference number is lower than that transported by anion, usually, 0.2 ~ 0.3 (Gray 1991). The transference number of ceramic-containing polymer electrolyte was ≈ 0.6, in TiO_2 containing polymer electrolyte, the peculiar high transport properties (transference number = 0.8) was observed (Tominaga and Yamazaki 2014).

As can be seen, surface functional groups of the ceramic fillers influence the properties of the ceramic-containing polymer electrolyte. In other words, the use of surface functionalized inorganic particles can improve and tune particular properties of the polymer electrolyte. The hypergrafted silica can improve the conductivity to over 10^{-4} S cm^{-1} by addition of (vinyl alcohol)/ $LiClO_4$ electrolytes (Hu et al. 2012). In other examples, it was reported that poly(methyl methacrylate) grafted titanium dioxide (PMMA-g-TiO_2), lithiated silica, lithiated core-shell SiO_2, lithiated fluorine-modified titania and so on could enhance the conductivity, stability and transference number of polymer electrolyte and suppress lithium dendrite formation (Cui et al. 2013, Xiao et al. 2012, Li et al. 2013b, Lee et al. 2013, Chinnam and Wunder 2013, Bertasi et al. 2014a, Bertasi et al. 2014b).

The addition of ceramic fillers in nanometer size can modify the properties of the polymer electrolyte. The development of polymer electrolyte with high-mechanical strength, high Li ion conductivity, and high transference number and application of it for the lithium battery technology would cause a drastic change in our life. This is when the polymer electrolyte is expected to apply for flexible micro battery which can be embedded in smart cards, sensors, and so on, due to the flexibility of the polymer electrolyte. To realize this, the ceramics containing polymer electrolyte has been strenuously studied. The commercialization of the polymer electrolyte would be achieved soon.

4. Summary

Some examples of the application of nanomaterials to lithium battery components, that is, electrode materials and electrolyte are introduced and explained in this chapter. The nanomaterials such as nano particles, nano-composite materials, and materials structured in nano-scale largely influence the battery performance and battery life as well as safety. Even though same materials are used in the batteries, the performance completely varies depending on the size and structure of the materials. In order to develop the Li battery with high energy and power densities, it not only requires research on new materials with high performance, but design of battery structure and electrode structure controlled in nano-scale are also important. In other words, a development of process technology which can produce nano-sized material and nano-structured material is also needed. "Nanomaterial" is a keyword for next generation Li battery.

References

Abraham, K.M., D.M. Pasquariello and E.B. Willstaedt. 1990. Preparation and characterization of some lithium insertion anodes for secondary lithium batteries. J. Electrochem. Soc. 137: 743–749.

Amatucci, G.G., J.M. Tarascon and L.C. Klein. 1996. CoO_2, The end member of the Li_xCoO_2 solid solution. J. Electrochem. Soc. 143: 1114–1123.

Amine, K., H. Yasuda and M. Yamauchi. 2000. Olivine $LiCoPO_4$ as 4.8V electrode material for lithium batteires. Electrochem. and Solid-State Lett. 3(4): 178–179.

Appetecchi, G.B., F. Croce, L. Persi, L. Ronci and B. Scrosati. 2000. Transport and interfacial properties of composite polymer electrolytes. Electrochim. Acta 45: 1481–1490.

Arico, A.S., P. Bruce, B. Scrosati, J.-M. Tarascon and W. Van Schalkwijc. 2005. Nanostructured materials for advanced energy conversion and storage devices. Nat. Mater. 4: 366–377.

Armand, M. and J.M. Tarascon. 2008. Building better batteries. Nature 451: 652–657.

Arrebola, J.C., A. Caballero, L. Hernan, M. Melero, J. Morales and E.R. Castellon. 2006. Electrochemical properties of $LiNi_{0.5}Mn_{1.5}O_4$ films prepared by spin-coating deposition. J. Power Sources 162: 606–613.

Aurbach, D., B. Markovsky, M.D. Levi, E. Levi, A. Schechter, M. Moshkovich et al. 1999a. New insights into the interactions between electrode materials and electrolyte solutions for advanced nonaqueous batteries. J. Power Sources 81-82: 95–111.

Aurbach, D., B. Markovskya, I. Weissmana, E. Levia and Y. Ein-Eli. 1999b. On the correlation between surface chemistry and performance of graphite negative electrodes for Li ion batteries. Electrochim Acta 45(1-2): 67–86.

Badway, F., A.N. Mansour, N. Pereira, J.F. Al-Sharab, F. Cosandey, I. Plitz et al. 2007. Electrochemical reactions of lithium with transition metal nitride electrodes. Chem. Mater. 19: 4129–4141.

Badway, F., N. Pereira, F. Cosandey and G.G. Amatucci. 2003. Carbon-metal fluoride nanocomposites: structure and electrochemistry of FeF_3:C. J. Electrochem. Soc. 150: A1209–A1218.

Balaya, P., A.J. Bhattacharyya, J. Jamnik, Y.F. Zhukovskii, E.A. Kotomin and J. Maier. 2006. Nano-ionics in the context of lithium batteries. J. Power Sources 159: 171–178.

Banks, C.E., T.J. Davies, G.G. Wildgoose and R.G. Compton. 2005. Electrocatalysis at graphite and carbon nanotube modified electrodes: edge-plane sites and tube ends are the reactive sites. Roy. Soc. Chem. 0(7): 829–41.

Barreca, D., M. Cruz-Yusta, A. Gasparotto, C. Maccato, J. Morales, A. Pozza et al. 2010. Cobalt oxide nanomaterials by vapor-phase synthesis for fast and reversible lithium storage. J. Phys. Chem. C 114: 10054–10060.

Beaulieu, L., D. Larcher, R.A. Dunlap and J.R. Dahn. 2000. Nanocomposites in the Sn-Mn-C system produced by mechanical alloying. J. Alloys Compd. 297: 122–128.

Benedek, R. and M.M. Thackeray. 2006. Lithium reactions with intermetallic-compound electrodes. J. Power Sources 110: 406–411.

Bertasi, F., K. Vezzu, E. Negro, S. Greenbaum and V. DiNoto. 2014a. Single-ion-conducting nanocomposite polymer electrolytes based on PEG400 and anionic nanoparticles: Part 1. Synthesis, structure and properties. Int. J. Hydrog. Energy 39: 2872–2883.

Bertasi, F., K. Vezzu, G.A. Giffin, T. Nosach, P. Sideris, S. Greenbaum et al. 2014b. Single-ion-conducting nanocomposite polymer electrolytes based on PEG400 and anionic nanoparticles: Part 2. Electrical characterization. Int. J. Hydrog. Energy 39: 2884–2895.

Bervas, B., F. Badway, L.C. Klein and G.G. Amatucci. 2005. Bismuth fluoride nanocomposite as a positive electrode material for rechargeable lithium batteries. Electrochem. Solid-State Lett. 8: A179–183.

Bhattacharyya, A.J. and J. Maier. 2004. Second phase effects on the conductivity of non-aqueous salt solutions: soggy sand electrolytes. Adv. Mat. 16: 811–814.

Binotto, G., D. Larcher, A.S. Prakash, R. Herrera Urbina, M.S. Hegde and J.M. Tarascon. 2007. Synthesis, characterization, and Li-electrochemical performance of highly porous Co_3O_4 powders. Chem. Mater. 19: 3032–3040.

Bodenes, L., A. Darwiche, L. Monconduit and H. Martinez. 2015. The solid electrolyte interphase a key parameter of the high performance of Sb in sodium-ion batteries: comparative X-ray photoelectron spectroscopy study of Sb/Na-ion and Sb/Li-ion batteries. J. Power Sources 273: 14–24.

Boyanov, S., K. Annou, C. Villevieille, M. Pelosi, D. Zitoun and L. Monconduit. 2008. Nanostructured transition metal phosphide as negative electrode for lithium-ion batteries. Ionics 14: 183–190.

Boyanov, S., D. Zitoun, M. Ménétrier, J.C. Jumas, M. Womes and L. Monconduit. 2009. Comparison of the electrochemical lithiation/delithiation mechanisms of FeP_x ($x = 1, 2, 4$) based electrodes in Li-ion batteries. J. Phys. Chem. C. 113: 21441–21452.

Brownson, D.A.C., D.K. Kampouris and C.E. Banks. 2011. An overview of graphene in energy production and storage applications. J. Power Sources 196: 4873–4885.

Buiel, E. and J.R. Dahn. 1999. Li-insertion in hard carbon anode materials for Li-ion batteries. Electrochim Acta 45: 121–30.

Capuano, F., F. Croce and B. Scrosati. 1991. Composite polymer electrolytes. J. Electrochem. Soc. 52: 1918–1922.

Chan, C.K., H. Peng, G. Liu, K. Mcilwrath, X.F. Zhang, R.A. Huggins and Y. Cui. 2008. High-performance lithium battery anodes using silicon nanowires. Nat. Nanotechnol. 3: 31–35.

Chan, C.K., R.N. Patel, M.J. O'Connell, B.A. Korgel and Y. Cui. 2010. Solution-grown silicon nanowires for lithium-ion battery anodes. ACS Nano 4: 1443–1450.

Chen, J., X. Xia, J. Tu, Q. Xiong, Y.Yu, X.Wang et al. 2012. Co_3O_4-C core-shell nanowire array as an advanced anode material for lithium ion batteries. J. Mater. Chem. 22: 15056–15061.

Chen, Z., I. Belharouak, Y.K. Sun and K. Amine. 2013. Titanium-based anode materials for safe lithium-ion batteries. Adv. Funct. Mater. 23: 959–969.

Cheng, F.Y., J. Liang, Z.L. Tao and J. Chen. 2011. Functional materials for rechargeable batteries. Advanced Materials 23: 1695–1715.

Chinnam, P.R. and S.L. Wunder. 2013. Self-assembled janus-like multi-ionic lithium salts form nano-structured solid polymer electrolytes with high ionic conductivity and li + ion transference number. J. Mater. Chem. A. 1: 1731–1739.

Chu, S. and A. Majumdar. 2012. Opportunities and challenges for a sustainable energy future. Nature 488: 294–303.

Connor, P.A. and J.T.S. Irvine. 2002. Combined X-ray study of lithium (tin) cobalt oxide matrix negative electrodes for Li-ion batteries. Electrochim. Acta 47: 2885–2892.

Courtney, A. and J.R. Dahn. 1997. Behavior of Bi_2O_3 as a cathode for lithium cells. J. Electrochem. Soc. 144: 2045–2052.

Croce, F, F. Bonino, S. Panero and B. Scrosati. 1989. Properties of mixed polymer and crystalline ionic conductors. Phil. Mag. B. 59: 161–168.

Croce, F., B. Scrosati and G. Mariotto. 1992. Electrochemical and spectroscopic study of the transport properties of composite polymer electrolytes. Chem. Mater. 4: 1134–1136.

Croce, F. and B. Scrosati. 1993. Composite polymer ionics: advanced electrolyte materials for thin-film batteries. Polym. Adv. Techn. 4: 198–204.

Croce, F., G.B. Appetecchi, L. Persi and B. Scrosati. 1998. Nanocomposite polymer electrolytes for lithium batteries. Nature 394: 456–458.

Cui, G., L. Gu, L. Zhi, N. Kaskhedikar, P.A. Aken, K. Mullen et al. 2008. A germanium–carbon nanocomposite material for lithium batteries. Adv. Mater. 20: 3079–3083.

Cui, W.W., D.Y. Tang and Z.L. Gong. 2013. Electrospun poly(vinylidene fluoride)/poly(methyl Methacrylate) grafted TiO_2 composite nanofibrous membrane as polymer electrolyte for lithium-ion batteries. J. Power Sources 223: 206–213.

Curtis, C.J., J. Wang and D.L. Schulz. 2004. Preparation and characterization of $LiMn_2O_4$ spinel nanoparticles as cathode materials in secondary Li batteries. J. Electrochem. Soc. 151: A590–598.

Das, B., M.V. Reddy, P. Malar, T. Osipowicz, G.V.S. Rao and B.V.R. Chowdari. 2009. Nanoflake CoN as a high capacity anode for Li-ion batteries. Solid State Ionics 180: 1061–1068.

DiLeo, R.A., A. Castiglia, M.J. Ganter, R.E. Rogers, C.D. Cress, R.P. Raffaelle et al. 2010. Enhanced capacity and rate capability of carbon nanotube based anodes with titanium contacts for lithium ion batteries. ACS Nano 4: 6121–6131.

Dong, W., D.R. Rolison and B. Dunn. 2000. Electrochemical properties of high surface area vanadium oxides aerogels. Electrochem. Solid State Lett. 3: 457–459.

Dupont, L., S. Grugeon, S. Laruelle and J.M. Tarascon. 2007. Structure, texture and reactivity versus lithium of chromium-based oxides films as revealed by TEM investigations. J. Power Sources 164: 839–848.

Ein-Eli, Y., S.F. McDevitt, D. Aurbach, B. Markovsky and A. Schechter. 1997. Methyl propyl carbonate: a promising single solvent for li ion battery electrolytes. J. Electrochem. Soc. 144(7): L180–L184.

Fan, L.Z., X.L. Wang, F. Long and X. Wang. 2008. Enhanced ionic conductivities in composite polymer electrolytes by using succinonitrile as a plasticizer. Solid State Ionic 179: 1772–1775.

Fei, L., Q. Lin, B. Yuan, M. Naeemi, Y. Xu, Y. Li et al. 2013. Controlling morphology and enhancing electrochemical performance of cobalt oxide by addition of graphite. Mater. Lett. 98: 59–62.

Fiordiponti, P., G. Pistoia and C. Temperoni. 1978. Behavior of Bi_2O_3 as a cathode for lithium cells. J. Electrochem. Soc. 125: 14–17.

Flandrois, S. and B. Simon. 1999. Carbon materials for lithium-ion rechargeable batteries. Carbon 37: 165–180.

Fu, Z.W., Y. Wang, Y. Zhang and Q.Z. Qin. 2004a. Electrochemical reaction of nanocrystalline Co_3O_4 thin film with lithium. Solid State Ionics. 170: 105–109.

Fu, Z.W., Y. Wang, X.L. Yue, S.L. Zhao and Q.Z. Qin. 2004b. Electrochemical reactions of lithium with transition metal nitride electrodes. J. Phys. Chem. B. 108: 2236–2244.

Gabano, J.P., V. Dechenaux, G. Gerbier and J. Jammet. 1972. D-size lithium cupric sulfide cells. J. Electrochem. Soc. 119: 459–461.

Gadjourova, Z., Y. Andreev, D.P. Tunstall and P.G. Bruce. 2001. Ionic conductivity in crystalline polymer electrolytes. Nature 412: 520–522.

Ge, M., J. Rong, X. Fang and C. Zhou. 2012. Porous doped silicon nanowires for lithium ion battery anode with long cycle life. Nano Lett. 12: 2318–2323.

Ghosh, A. and P. Kofinas. 2008. Nanostructured block copolymer dry electrolyte. J. Electrochem. Soc. 155: A428–A431.

Gillot, F., S. Boyanov, L. Dupont, M.L. Doublet, A. Morcrette, L. Monconduit et al. 2005. Electrochemical reactivity and design of NiP_2 negative electrodes for secondary Li-ion batteries. Chem. Mater. 17: 6327–6337.

Gillot, F., M. Menetrier, E. Bekaert, L. Dupont, M. Morcrette, L. Monconduit et al. 2007. Vanadium diphosphides as negative electrodes for secondary Li-ion batteries. J. Power Sources 172: 877–885.

Girishkumar, G., B. McCloskey, A.C. Luntz, S. Swanson and W. Wilcke. 2010. Lithium-air battery: promise and challenges. J. Phys.Chem. Lett. 1: 2193–2203.

Gonzalez, Z., S. Vizireanu, G. Dinescu, C. Blanco and R. Santamaria. 2012. Carbon nanowalls thin films as nanostructured electrode materials in vanadium redox flow batteries. Nano Energy 1: 833–839.

Goodenough, J.B. and Y. Kim. 2010. Challenges for rechargeable Li batteries. Chem. Mater. 22: 587–603.

Gray, F.M. 1991. Solid polymer electrolytes-fundamentals and technical applications. VCH, Wenheim.

Grugeon, S., S. Laruelle, L. Dupont and J.M. Tarascon. 2003. An update on the reactivity of nanoparticles Co-based compounds towards Li Solid State Sci. 5: 895–904.

Gu, J., S.M. Collins, A.I. Carim, X. Hao, B.M. Bartlett and S. Maldonado. 2012. Template-free preparation of crystalline Ge nanowire film electrodes *via* an electrochemical liquid–liquid–solid process in water at ambient pressure and temperature for energy storage. Nano Lett. 12: 4617–4623.

Gu, P., R. Cai, Y. Zhou and Z. Shao. 2010. Si/C composite lithium-ion battery anodes synthesized from coarse silicon and citric acid through combined ball milling and thermal pyrolysis. Electrochim. Acta 55: 3876–3883.

Gu, Y., Y. Xu and Y. Wang. 2013. Graphene-wrapped CoS nanoparticles for high-capacity lithium-ion storage. ACS Appl. Mater. Interfaces. 5: 801–806.

Guan, D., J. Li, X. Gao and C. Yuan. 2014. A comparative study of enhanced electrochemical stability of tin–nickel alloy anode for high-performance lithium ion battery. J. Alloy Comp. 617: 464–471.

Guo, B.K., J. Shu, Z.X. Wang, H. Yang, L.H. Shi, Y.N. Liu et al. 2008a. Electrochemical reduction of nano-SiO_2 in hard carbon as anode material for lithium ion batteries. Electrochem. Commun. 10: 1876–1878.

Guo, Y.G., J.S. Hu and L.J. Wan. 2008b. Nanostructured materials for electrochemical energy conversion and storage devices. Adv. Mater. 20: 2878–2887.

Guo, Z.P., Z.W. Zhao, H.K. Liu and S.X. Dou. 2005. Electrochemical lithiation and de-lithiation of MWNT-Sn/SnNi nanocomposites. Carbon 43: 1392–1399.

Haik, O., S. Ganin, G. Gershinsky, E. Zinigrad, B. Markovsky, D. Aurbach et al. 2011. On the thermal behavior of lithium intercalated graphites. J. Electrochem. Soc. 158: A913–A923.

Han, S.C., K.W. Kim, H.J. Ahn, J.H. Ahn and J.Y. Lee. 2003. Charge–discharge mechanism of mechanically alloyed NiS used as a cathode in rechargeable lithium batteries. J. Alloys Compd. 361: 247–251.

Hassan, M.F., Z.P. Guo, Z. Chen and H.K. Liu. 2010. Carbon-coated MoO_3 nanobelts as anode materials for lithium-ion batteries. J. Power Sources 195: 2372–2376.

Hayashi, A., T. Ohtomo, F. Mizuno, K. Tadanaga and M. Tatsumisago. 2004. Rechargeable lithium batteries, using sulfur-based cathode materials and Li_2S–P_2S_5 glass-ceramic electrolytes. Electrochim. Acta 50: 893–897.

Henriksen, G.L. and A.N. Jansen. 2002. Handbook of batteries, McGraw-Hill, New York.

Hong, Z. and M. Wei. 2013. Layered titanate nanostructures and their derivatives as negative electrode materials for lithium-ion batteries. J. Mater. Chem. A. 1: 4403–4414.

Hongyou, K., T. Hattori, Y. Nagai, T. Tanaka, H. Nii and K. Shoda. 2013. Dynamic *in situ* Fourier transform infrared measurements of chemical bonds of electrolyte solvents during the initial charging process in a Li ion battery. J. Power Sources 243: 72–77.

Hoshina, K., K. Yoshima, M. Kotobuki and K. Kanamura. 2012. Fabrication of $LiNi_{0.5}Mn_{1.5}O_4$ thin film cathode by PVP sol-gel process and its application of all-solid-state lithium ion batteries using $Li_{1+x}Al_xTi_{2-x}(PO_4)_3$ solid electrolyte. Solid State Ionics 209-210: 30–35.

Hosono, E., S. Fujihara, I. Honma and H.S. Zhou. 2006. The high power and high energy densities Li ion storage device by nanocrystalline and mesoporous Ni/NiO covered structur. Electrochem. Commun. 8: 284–288.

Hou, J., Y. Shao, M.W. Ellis, R.B. Moore and B. Yi. 2011. Graphene-based electrochemical energy conversion and storage: fuel cells, supercapacitors and lithium ion batteries. Phys. Chem. Chem. Phys. 13: 15384–15402.

Hu, A., X. Chen, Y. Tang, Q. Tang, L. Yang and S. Zhang. 2013. Self-assembly of Fe_3O_4 nanorods on graphene for lithium ion batteries with high rate capacity and cycle stability. Electrochem. Commun. 28: 139–142.

Hu, X.L., G.M. Hou, M.Q. Zhang, M.Z. Rong, W.H. Ruan and E.P. Giannelis. 2012. A new nanocomposite polymer electrolyte based on poly(vinyl alcohol) incorporating hypergrafted nano-silica. J. Mater. Chem. 22: 18961–18967.

Huang, X., X. Qi, F. Boey and H. Zhang. 2012. Graphene-based composites. Chem. Soc. Rev. 41: 666–686.

Huggins, A. 1997. Alloy negative electrodes for lithium batteries formed in situ from oxides. Ionics. 3: 245–255.

Hwang, H., H. Kim and J. Cho. 2011. MoS_2 nanoplates consisting of disordered graphene-like layers for high rate lithium battery anode materials. Nano Lett. 11: 4826–4830.

Hwang, I.S., J.C. Kim, S.D. Seo, S. Lee, J.H. Lee and D.W. Kim. 2012. A binder-free Ge-nanoparticle anode assembled on multiwalled carbon nanotube networks for Li-ion batteries. Chem. Commun. 48: 7061–7063.

Hwang, J.K., H.S. Lim, Y.K. Sun and K.D. Suh. 2013. Monodispersed hollow carbon/Fe_3O_4 composite microspheres for high performance anode materials in lithium-ion batteries. J. Power Sources 244: 538–543.

Idota, Y., T. Kubota, A. Matsufuji, Y. Maekawa and T. Miyasaka. 1997. Tin-based amorphous oxide: a high-capacity lithium-ion-storage material. Science 276: 1395–1397.

Inaba, M., Z. Siroma, Y. Kawatate, A. Funabiki and Z. Ogumi. 1997. Electrochemical scanning tunneling microscopy analysis of the surface reactions on graphite basal plane in ethylene carbonate-based solvents and propylene carbonate. J. Power Sources 68: 221–226.

Itoh, T., K. Hirai, T. Uno and M. Kubo. 2008. Solid polymer electrolytes based on squaric acid structure. Ionics 14: 1–6.

Ji, L., Z. Lin, M. Alcoutlabi and X. Zhang. 2011a. Recent developments in nanostructured anode materials for rechargeable lithium-ion batteries. Energy Environ. Sci. 4: 2682–2699.

Ji, X., S. Evers, R. Black and L.F. Nazar. 2011b. Stabilizing lithium–sulphur cathodes using polysulphide reservoirs. Nature Communications 2: 325.

Ji, X.L., K.T. Lee and L.F. Nazar. 2009. A highly ordered nanostructured carbon–sulphur cathode for lithium-sulphur batteries. Nature Materials 8: 500–506.

Jiang, J., Y. Li, J. Liu, X. Huang, C. Yuan and X.W. Lou. 2012. Recent advances in metal oxide-based electrode architecture design for electrochemical energy storage. Adv. Mater. 24: 5166–5180.

Jiang, Y., R.Z. Liu, W.W. Xu, Z. Jiao, M.H. Wu, Y.L. Chu et al. 2013. A novel graphene modified $LiMnPO_4$ as a performance-improved cathode material for lithium-ion batteries. J. Mater. Res. 28: 2584–2589.

Jiao, F. and P.G. Bruce. 2007. Mesoporous crystalline β-MnO_2-a reversible positive electrode for rechargeable lithium batteries. Adv. Mater. 19: 657–660.

Jung, Y.S., S. Lee, D. Ahn, A.C. Dillon and S.H. Lee. 2009. Electrochemical reactivity of ball-milled MoO_{3-y} as anode materials for lithium-ion batteries. J. Power Sources 188: 286–291.

Kanamura, K. and M. Kotobuki. 2009. New Material for Next Generation Rechargeable Batteries for Future Society. CMC Publishing Co., Ltd., Tokyo.

Kang, N., J.H. Park, J. Choi, J. Jin, J. Chun, I.G. Jung et al. 2012. Nanoparticulate iron oxide tubes from microporous organic nanotubes as stable anode materials for lithium ion batteries. Angew. Chem. Int. Ed. 51: 6626–6630.

Kasavajjula, U., C. Wang and A.J. Appleby. 2007. Nano- and bulk-silicon-based insertion anodes for lithium-ion secondary cells. J. Power Sources 163: 1003–1039.

Kaskhedikar, N.A. and J. Maier. 2009. Lithium storage in carbon nanostructures. Adv. Mater. 21: 2664–2680.

Ke, F.S., L. Huang, H.H. Juang, H.B. Wei, F.Z. Yang and S.G. Sunm. 2007. Fabrication and properties of three-dimensional macroporous Sn–Ni alloy electrodes of high preferential (110) orientation for lithium ion batteries. Electrochem. Comm. 9: 228–232.

Kepler, K.D., J.T. Vaughey and M.M. Thackeray. 1999. $Li_xCu_6Sn_5(0<x<13)$: an intermetallic insertion electrode for rechargeable lithium batteries. Electrochem. Solid-State Lett. 2: 307–309.

Kim, B.C., K. Takada, N. Ohta, Y. Seino, L.Q. Zhang, H. Wada et al. 2005a. All solid state Li-ion secondary battery with FeS anode. Solid State Ionics. 176: 2383–2387.

Kim, C., K.S. Yang, M. Kojima, K. Yoshida, Y.J. Kim, Y.A. Kim et al. 2006. Fabrication of electrospinning-derived carbon nanofiber webs for the anode material of lithium-ion secondary batteries. Adv. Funct. Mater. 16: 2393–2397.

Kim, H., S. Grugeon, G. Gachot, M. Armand, L. Sannier and S. Laruelle. 2014. Ethylene bis-carbonates as telltales of SEI and electrolyte health, role of carbonate type and new additives. Electrochim. Acta 136: 157–165.

Kim, I.T., A. Magasinski, K. Jacob, G. Yushin and R. Tannenbaum. 2013. Synthesis and electrochemical performance of reduced graphene oxide/maghemite composite anode for lithium ion batteries. Carbon 52: 56–64.

Kim, J.H., S.T. Myung and Y.K. Sun. 2004. Molten salt synthesis of $LiNi_{0.5}Mn_{1.5}O_4$ spinel for 5 V class cathode material of Li-ion secondary battery. Electrochimica Acta 49: 219–227.

Kim, M.G. and J. Cho. 2009. Reversible and high-capacity nanostructured electrode materials for Li-ion batteries. Adv. Funct. Mater. 19: 1497–1514.

Kim, S., Y.J. Jung and S.J. Park. 2005b. Effects of imidazolium salts on discharge performance of rechargeable lithium–sulfur cells containing organic solvent electrolytes. J. Power Sources 152: 272–277.

Kim, Y.U., B.W. Cho and H.J. Sohn. 2005c. The reaction mechanism of lithium insertion in vanadium tetraphosphide: a possible anode material in lithium-ion batteries. J. Electrochem. Soc. 152: A1475–A1478.

Kim, Y., H. Hwang, K. Lawler, S.W. Martin and J. Cho. 2008. Electrochemical behavior of Ge and GeX_2 (X = O, S) glasses: improved reversibility of the reaction of Li with Ge in a sulfide medium. Electrochim. Acta 53: 5058–5064.

Koo, B., H. Xiong, M.D. Slater, V.B. Prakapenka, M. Balasubramanian, P. Podsiadlo et al. 2012. Hollow iron oxide nanoparticles for application in lithium ion batteries. Nano Lett. 12: 2429–2435.

Kotobuki, M., Y. Isshiki, H. Munakata and K. Kanamura. 2010. All-solid-state lithium battery with a three-dimensionally ordered $Li_{1.5}Al_{0.5}Ti_{1.5}(PO_4)_3$ electrode. Electrochim. Acta 55: 6892–6896.

Kotobuki, M., Y. Mizuno, H. Munakata and K. Kanamura. 2011a. Improved performance of hydrothermally synthesized $LiMnPO_4$ by Mg doping. Electrochemistry 79: 467–469.

Kotobuki, M., K. Kanamura, Y. Sato and T. Yoshida. 2011b. Fabrication of all-solid-state lithium battery with lithium metal anode using Al_2O_3-added $Li_7La_3Zr_2O_{12}$ solid electrolyte. J. Power Sources 196: 7750–7754.

Kotobuki, M., N. Okada and K. Kanmura. 2011c. Design of a micro-pattern structure for a three dimensionally macroporous Sn–Ni alloy anode with high areal capacity. Chem. Comm. 47: 6144–6146.

Kotobuki, M. 2013a. Hydrothermal synthesis of carbon-coated $LiCoPO_4$ cathode material from various Co sources. International Journal of Energy and Environmental Engineering 4: 25–31.

Kotobuki, M. 2013b. Improved performance of hydrothermally synthesized $LiMnPO_4$ by ball-milling as a positive electrode for Li ion battery. ISRN Electrochemistry 1–5.

Kottegoda, I.R.M., Y. Kodama, H. Ikuta, Y. Uchimoto and M. Wakihara. 2002. Enhancement of rate capability in graphite anode by surface modification with zirconia. Electrochem. Solid State Lett. 5(12): A275–A278.

Krumeich, F., H.J. Muhr, M. Niederberger, F. Bieri, B. Schnyder and R. Nesper. 1999. Morphology and topochemical reactions of novel vanadium oxide nanotubes. J. Am. Chem. Soc. 121: 8324–8331.

Kumer, B. and Y.G. Scanlon. 1994. Polymer ceramic composite electrolytes. J. Power Sources 52: 261–268.

Kumta, P.N., D. Gallet, A. Waghray, G.E. Blomgren and M.P. Setter. 1998. Synthesis of $LiCoO_2$ powders for lithium-ion batteries from precursors derived by rotary evaporation. J. Power Sources 72: 91–98.

Kunduraci, M. and G.G. Amatucci. 2006. Synthesis and characterization of nanostructured 4.7 V $Li_xMn_{1.5}Ni_{0.5}O_4$ spinels for high-power lithium-ion batteries. J. Electrochem. Soc. 153: A1345–A1352.

Lai, C.H., M.Y. Lu and L.J. Chen. 2012. Metal sulfide nanostructures: synthesis, properties and applications in energy conversion and storage. J. Mater. Chem. 22: 19–30.

Landi, B.J., M.J. Ganter, C.D. Cress, R.A. DiLeo and R.P. Raffaelle. 2009. Carbon nanotubes for lithium ion batteries. Energy Environ. Sci. 2: 638–654.

Larcher, D., L.Y. Beaulieu, D.D. MacNeil and J.R. Dahn. 2000a. *In situ* X-ray study of the electrochemical reaction of Li with η'-Cu_6Sn_5. J. Electrochem. Soc. 147: 1658–1662.

Larcher, D., L.Y. Beaulieu, O. Mao, A.E. George and J.R. Dahn. 2000b. Study of the reaction of lithium with isostructural A_2B and various Al_xB alloys. J. Electrochem. Soc. 147: 1703–1708.

Larcher, D., G. Sudant, J.B. Leriche, Y. Chabre and J.M. Tarascon. 2002. The electrochemical reduction of Co_3O_4 in a lithium cell. J. Electrochem. Soc. 149: A234–A241.

Le, D.B., S. Passerini, J. Guo, J. Ressler, B.B. Owens and W.H. Smyrl. 1996. High surface area V_2O_5 aerogel intercalation electrodes. J. Electrochem. Soc. 143: 2099–2104.

Lee, S.H., Y.H. Kim, R. Deshpande, P.A. Parilla, E. Whitney, D.T. Gillaspie et al. 2008. Reversible lithium-ion insertion in molybdenum oxide nanoparticles. Adv. Mater. 20: 3627–3632.

Lee, Y.S., J.H. Lee, J.A. Choi, W.Y. Yoon and D.W. Kim. 2013. Cycling characteristics of lithium powder polymer batteries assembled with composite gel polymer electrolytes and lithium powder anode. Adv. Funct. Mater. 23: 1019–1027.

Li, C.C., Q.H. Li, L.B. Chen and T.H. Wang. 2011. Topochemical synthesis of cobalt oxide nanowire arrays for high performance binderless lithium ion batteries. J. Mater. Chem. 21: 11867–11872.

Li, C.C. and Y.W. Wang. 2013a. Importance of binder compositions to the dispersion and electrochemical properties of water-based $LiCoO_2$ cathodes. J. Power Sources 227: 204–210.

Li, H., X.J. Huang and L.Q. Chen. 1999. Anodes based on oxide materials for lithium rechargeable batteries. Solid State Ionics 123: 189–197.

Li, H., G. Richter and J. Maier. 2003. Reversible formation and decomposition of LiF clusters using transition metal fluorides as precursors and their application in rechargeable Li batteries. Adv. Mater. 15: 736–739.

Li, H. and H. Zhou. 2012. Enhancing the performances of Li-ion batteries by carbon-coating: present and future. Chem. Commun. 48: 1201–1217.

Li, H., Z. Wang, L. Chen and X. Huang. 2009. Research on advanced materials for Li-ion batteries. Adv. Mater. 21: 4593–4607.

Li, J.G., T. Ikegami, J.H. Lee and T. Mori. 2000. Well-sinterable $Y_3Al_5O_{12}$ powder from carbonate precursor. J. Mater. Res. 15: 1514–1523.

Li, L., Y. Peng and H. Yang. 2013a. Phase structure changes of MnP anode material during electrochemical lithiation and delithiation process. Electrochim. Acta 95: 230–236.

Li, T. and P.B. Balbuena. 2000. Theoretical studies of the reduction of ethylene carbonate. Chem. Phys. Lett. 317(3-5): 421–429.

Li, W., Y. Xing, X. Xing, Y. Li, G. Yang and L. Xu. 2013b. PVDF-based composite microporous gel polymer electrolytes containing a novelsingle ionic conductor $SiO_2(Li^+)$. Electrochimica Acta 112: 183–190.

Li, X. and C. Wang. 2013b. Engineering nanostructured anodes *via* electrostatic spray deposition for high performance lithium ion battery application. J. Mater. Chem. A. 1: 165–182.

Li, X., T. Li, Y. Zhang, X. Zhang, H. Li and J. Huang. 2014. Graphene nanoribbon-wrapping $LiFePO_4$ by electrostatic absorbing with improved electrochemical performance for rechargeable lithium batteries. Electrochim. Acta 139: 69–75.

Li, Y.G., B. Tan and Y.Y. Wu. 2008. Mesoporous Co_3O_4 nanowire arrays for lithium ion batteries with high capacity and rate capability. Nano Lett. 8: 265–270.

Lian, P., X. Zhu, S. Liang, Z. Li, W. Yang and H. Wang. 2010. Large reversible capacity of high quality graphene sheets as an anode material for lithium-ion batteries. Electrochim. Acta 55: 3909–3914.

Liang, M. and L. Zhi. 2009. Graphene-based electrode materials for rechargeable lithium batteries. J. Mater. Chem. 19: 5871–5878.

Lightfoot, P., M.A. Metha and P.G. Bruce. 1993. Crystal structure of the polymer electrolyte poly(ethylene oxide)$_3$: $LiCF_3$. Science 262: 883–885.

Liu, J., Y. Li, H. Fan, Z. Zhu, J. Jiang, R. Ding et al. 2009. Iron oxide-based nanotube arrays derived from sacrificial template-accelerated hydrolysis: large-area design and reversible lithium storage. Chem. Mater. 22: 212–217.

Liu, J. 2013. Addressing the grand challenges in energy storage. Adv. Funct. Mater. 23: 924–928.

Liu, S., N. Imanishi, T. Zhang, A. Hirano, Y. Takeda, O. Yamamoto et al. 2010. Effect of nano-silica filler in polymer electrolyte on li dendrite formation in li/poly(ethylene oxide)–Li $(CF_3SO_2)_2N$/Li. J. Power Sources 195: 6847–6853.

Liu, T.F., L. Zhao, J.S. Zhu , B. Wang , C.F. Guo and D.L. Wang. 2014. The composite electrode of $LiFePO_4$ cathode materials modified with exfoliated graphene from expanded graphite for high power Li-ion batteries. J. Mater. Chem. A. 2: 2822–2829.

Luo, W.B., S.L. Chou, Y.C. Zhai and H.K. Liu. 2014. Self-assembled graphene and $LiFePO_4$ composites with superior high rate capability for lithium ion batteries. J. Mater. Chem. A. 2: 4927–4931.

Luo, Z., D. Fan, X. Liu, H. Mao, C. Yao and Z. Deng. 2009. High performance silicon carbon composite anode materials for lithium ion batteries. J. Power Sources 189: 16–21.

Lux, S.F., T. Placke, C. Engelhardt, S. Nowak, P. Bieker, K.-E. Wirth et al. 2012. Enhanced electrochemical performance of graphite anodes for lithium-ion batteries by dry coating with hydrophobic fumed silica. J. Electrochem. Soc. 159: A1849–A1855.

Lv, R., L. Zou, X. Gui, F. Kang, Y. Zhu, H. Zhu et al. 2008. High-yield bamboo-shaped carbon nanotubes from cresol for electrochemical application. Chem. Commun. 2046–2048.

Ma, X.H., X.Y. Feng, C. Song, B.K. Zou, C.X. Ding, Y. Yu et al. 2013. Facile synthesis of flower-like and yarn-like α-Fe_2O_3 spherical clusters as anode materials for lithium-ion batteries. Electrochim. Acta 93: 131–136.

Mahmoud, A., J.M. Amarilla, K. Lasri and I. Saadoune. 2013. Influence of the synthesis method on the electrochemical properties of the $Li_4Ti_5O_{12}$ spinel in Li-half and Li-ion full-cells. A systematic comparison. Electrochim. Acta 93: 163–172.

Mao, O., R.A. Dunlap and J.R. Dahn. 1999a. Mechanically alloyed Sn-Fe(-C) powders as anode materials for Li-ion batteries: III. Sn_2Fe: $SnFe_3C$C: SnFe/inactive composites. J. Electrochem. Soc. 146: 423–427.

Mao, O., R.A. Dunlap and J.R. Dahn. 1999b. Mechanically alloyed Sn-Fe(-C) powders as anode materials for Li-ion batteries: I. The Sn_2Fe-C system. J. Electrochem. Soc. 146: 405–413.

Mao, O., R.L. Turner, I.A. Courtney, B.D. Fredericksen, M.I. Buckett, L.J. Krause et al. 1999c. Active/inactive nanocomposites as anodes for Li-ion batteries. Electrochem. Solid-State Lett. 2: 3–5.

Marcus, Y. 1985. Ion solvation. Wiley, New York.

Marmorstein, D., T.H. Yu, K.A. Striebel, F.R. McLarnon, J. Hou and E.J. Cairns. 2000. Electrochemical performance of lithium/sulfur cells with three different polymer electrolytes. J. Power Sources 89: 219–226.

Marom, R., O. Haik, D. Aurbach and I.C. Halalay. 2010. Revisiting $LiClO_4$ as an electrolyte for rechargeable lithium-ion batteries. J. Electrochem. Soc. 157(8): A972–983.

Marom, R., S.F. Amalraj, N. Leifer, D. Jacob and D. Aurbach. 2011. A review of advanced and practical lithium battery materials. J. Mater. Chem. 21(27): 9938–9954.

Martha, S.K., O. Haik, V. Borgel, E. Zinigrad, I. Exnar, T. Drezen et al. 2011. $Li_4Ti_5O_{12}/LiMnPO_4$ lithium-ion battery systems for load leveling application. J. Electrochem. Soc. 158: A790–A797.

Martos, M., J. Morales and L. Sanchez. 2003. Lead-based systems as suitable anode materials for Li-ion batteries. Electrochim. Acta 48: 615–621.

Meunier, V., J. Kephart, C. Roland and J. Bernholc. 2002. *Ab initio* investigations of lithium diffusion in carbon nanotube systems. Phys. Rev. Lett. 88: 075506.

Mikhaylik, Y.V. and J.R. Aldridge. 2004. Polysulfide shuttle study in the Li/S battery system. J. Electrochem. Soc. 151: A1969–A1976.

Mitra, S., P. Poizot, A. Finke and J.M. Tarascon. 2006. Growth and electrochemical characterization versus lithium of Fe_3O_4 electrodes made by electrodeposition. Adv. Funct. Mater. 16: 2281–2287.

Moretti, A., G.T. Kim, D. Bresser, K. Renger, E. Paillard, R. Marassi et al. 2013. Investigation of different binding agents for nanocrystalline anatase TiO_2 anodes and its application in a novel, green lithium-ion battery. J. Power Sources 221: 419–426.

Moshkovich, M., Y. Gofer and D. Aurbach. 2001. Investigation of the electrochemical windows of aprotic alkali metal (Li, Na, K) salt solutions. J. Electrochem. Soc. 148(4): E155–E167.

Muhr, H.J., F. Krumeich, U.P. Schoholzer, F. Bieri, M. Niederberger, L.J. Gauckler et al. 2000. Vanadium oxide nanotubes—a new flexible vanadate nanophase. Adv. Mater. 12: 231–234.

Mukaibo, H., T. Sumi, T. Yokoshima, T. Momma and T. Osaka. 2003. Electrodeposited Sn-Ni alloy film as a high capacity anode material for lithium-ion secondary batteries. Electrochem. Solid-State Lett. 6: A218–A220.

Mukaibo, H., T. Momma and T. Osaka. 2005a. Changes of electro-deposited Sn–Ni alloy thin film for lithium ion battery anodes during charge discharge cycling. J. Power Sources 146: 457–463.

Mukaibo, H., T. Momma, M. Mohamedi and T. Osaka. 2005b. Structural and morphological modifications of a nanosized 62 atom percent Sn-Ni thin film anode during reaction with lithium. J. Electrochem. Soc. 152: A560–A565.

Mukherjee, R., R. Krishnan, T.-M. Lu and N. Koratkar. 2012. Nanostructured electrodes for high-power lithium ion batteries. Nano Energy 1: 518–533.

Muraliganth, T., A. Vadivel Murugan and A. Manthiram. 2009. Facile synthesis of carbon-decorated single-crystalline Fe_3O_4 nanowires and their application as high performance anode in lithium ion batteries. Chem. Commun. 47: 7360–7362.

Nagaura, T. and K. Tozawa. 1990. Lithium ion rechargeable battery. Prog. Batteries Solar Cells 9: 209–217.

Naji, A., J. Ghanbaja, B. Humbert, P. Willmann and D. Billaud. 1996. Electroreduction of graphite in $LiClO_4$-ethylene carbonate electrolyte. Characterization of the passivating layer by transmission electron microscopy and fourier-transform infrared spectroscopy. J. Power Sources 63(1): 33–39.

Nishidate, K. and M. Hasegawa. 2005. Energetics of lithium ion adsorption on defective carbon nanotubes. Phys. Rev. B 71: 245418.

Nishikawa, K., K. Dokko, K. Kinoshita, S.W. Woo and K. Kanamura. 2009. Three-dimensionally ordered macroporous Ni–Sn anode for lithium batteries. J. Power Sources 189: 726–729.

Novak, P., D. Goers and M.E. Spahr 2010. Carbons for electrochemical energy storage and conversion systems. CRC Press, Florida.

O'Hare, D. 1991. Inorganic Materials. Wiley, New York.

Ohzuku, T., A. Ueda, N. Nagayama, Y. Iwakoshi and H. Komori. 1993. Comparative study of $LiCoO_2$, $LiNi_{1/2}Co_{1/2}O_2$ and $LiNiO_2$ for 4 volt secondary lithium cells. Electrochim. Acta 38: 1159–1167.

Ohzuku, T. and A. Ueda. 1994. Why transition metal (di)oxides are the most attractive materials for batteries. Solid State Ionics 69: 201–211.

Oktaviano, H.S., K. Yamad and K. Wak. 2012. Nano-drilled multiwalled carbon nanotubes: characterizations and application for LIB anode materials. J. Mater. Chem. 22: 25167–25173.

Orsini, F., A.d. Pasquier, B. Beaudouin, J.M. Tarascon, M. Trentin, N. Langenhuizen et al. 1999. *In situ* SEM study of the interfaces in plastic lithium cells. J. Power Sources 81-82: 918–921.

Ostrovskii, D., F. Ronci, B. Scrosati and P. Jacobsson. 2001. Reactivity of lithium battery electrode materials toward non-aqueous electrolytes: spontaneous reactions at the electrode–electrolyte interface investigated by FTIR. J. Power Sources 103(1): 10–17.

Park, M.H., M.G. Kim, J. Joo, K. Kim, J. Kim, S. Ahn et al. 2009. Arrays of sealed silicon nanotubes as anodes for lithium ion batteries. Nano Lett. 9: 3844–3847.

Park, T.H., J.S. Yeo, M.H. Seo, J. Miyawaki, I. Mochida and S.H. Yoon. 2013. Enhancing the rate performance of graphite anodes through addition of natural graphite/carbon nanofibers in lithium-ion batteries. Electrochim. Acta 93: 236–240.

Peng, C., B. Chen, Y. Qin, S. Yang, C. Li, Y. Zuo et al. 2012. Facile ultrasonic synthesis of CoO quantum dot/graphene nanosheet composites with high lithium storage capacity. ACS Nano 6: 1074–1081.

Pereira, N., L. Dupont, J.M. Tarascon, L.C. Klein and G.G. Amatucci. 2003a. Electrochemistry of Cu_3Nnectr lithium: a complex system with parallel processes. J. Electrochem. Soc. 159: A1273–A1280.

Pereira, N., M. Balasubramanian, L. Dupont, J. McBreen, L.C. Klein and G.G. Amatucci. 2003b. Kinetic characterization of ptRu fuel cell anode catalysts made by spontaneous Pt deposition on Ru nanoparticles. J. Electrochem. Soc. 150: A1118–A1128.

Persson, K., V.A. Sethuraman, L.J. Hardwick, Y. Hinuma, Y.S. Meng, A. van der Ven et al. 2010. Lithium diffusion in graphitic carbon. J. Phys. Chem. Lett. 1: 1176–1180.

Poizot, P., S. Laruelle, S. Grugeon, L. Dupont and J.M. Tarascon. 2000. Nano-sized transition-metal oxides as negative-electrode materials for lithium-ion batteries. Nature 407: 496–498.

Prakash, A.S., P. Manikandan, K. Ramesha, M. Sathiya, J.M. Tarascon and A.K. Shukla. 2010. Solution-combustion synthesized nanocrystalline $Li_4Ti_5O_{12}$ as high-rate performance Li-ion battery anode. Chem. Mater. 22: 2857–2863.

Pralong, V., D.C.S. Souza, K.T. Leung and L. Nazar. 2002. Reversible lithium uptake by CoP_3 at low potential: role of the anion. Electrochem. Commun. 4: 516–520.

Pralong, V., J.B. Leriche, B. Beaudoin, E. Naudin, M. Morcrette and J.M. Tarascon. 2004. Electrochemical study of nanometer Co_3O_4, Co, $CoSb_3$ and Sb thin films toward lithium. Solid State Ionics 166: 295–305.

Prosini, P.P., M. Carewska, S. Loreti, C. Minarini and S. Passerini. 2000. Lithium iron oxide as alternative anode for li-ion batteries. Int. J. Inorg. Mater. 2: 365–370.

Reimers, J.N. and J.R. Dahn. 1992. Electrochemical and *in situ* X-ray diffraction studies of lithium intercalation in Li_xCoO_2. J. Electrochem. Soc. 139: 2091–2097.

Riley, L.A., S.H. Lee, L. Gedvilias and A.C. Dillon. 2010. Optimization of MoO_3 nanoparticles as negative-electrode material in high-energy lithium ion batteries. J. Power Sources. 195: 588–592.

Saint, J., M. Morcrette, D. Larcher, L. Laffont, S. Beattie, J.P. Peres et al. 2007. Towards a fundamental understanding of the improved electrochemical performance of silicon–carbon composites. Adv. Funct. Mater. 17: 1765–1774.

Schauerman, M., M.J. Ganter, G. Gaustad, C.W. Babbitt, R.P. Raffaelle and B.J. Landi. 2012. Recycling single-wall carbon nanotube anodes from lithium ion batteries. J. Mater. Chem. 22: 12008–12015.

Schoellhorn, R. 1987. Chemical Physics of Intercalation. NATO Series B, New York.

Scrosati, B. and J. Garche. 2010. Lithium batteries: status, prospects and future. J. Power Sources 195: 2419–2430.

Shaju, K.M., F. Jiao, A. Debart and P.G. Bruce. 2007. Mesoporous and nanowire Co_3O_4 as negative electrodes for rechargeable lithium batteries. Phys. Chem. Chem. Phys. 9: 1837–1842.

Shen, L., E. Uchaker, X. Zhang and G. Cao. 2012. Hydrogenated $Li_4Ti_5O_{12}$ nanowire arrays for high rate lithium ion batteries. Adv. Mater. 24: 6502–6506.

Shi, Y., S.L. Chou, J.Z. Wang, D. Wexler, H.J. Li, H.K. Liu et al. 2012. Graphene wrapped $LiFePO_4$/C composites as cathode materials for Li-ion batteries with enhanced rate capability. J. Mater. Chem. 22: 16465–16470.

Shim, J., K.A. Striebel and E.J. Cairns. 2002. The lithium/sulfur rechargeable cell: effects of electrode composition and solvent on cell performance. J. Electrochem. Soc. 149: A1321–A1325.

Shukla, A.K. and T.P. Kumar. 2008. Materials for next-generation lithium batteries. Curr. Sci. 94: 314–331.

Si, Q., K. Hanai, N. Imanishi, M. Kubo, A. Hirano, Y. Takeda et al. 2009. Highly reversible carbon–nano-silicon composite anodes for lithium rechargeable batteries. J. Power Sources 189: 761–765.

Silva, D.C.C., O. Crosnier, G. Ouvrard, J. Greedan, A. SafaSefat and L. Nazar. 2003. Reversible lithium uptake by FeP_2. Electrochem. Solid-State Lett. 6: A162–A165.

Sohn, H., Z. Chen, Y.S. Jung, Q. Xiao, M. Cai, H. Wang et al. 2013. Robust lithium-ion anodes based on nanocomposites of iron oxide–carbon–silicate. J. Mater. Chem. A 1: 4539–4545.

Song, T., J. Xia, J.-H. Lee, D.H. Lee, M.-S. Kwon, J.-M. Choi et al. 2010. Arrays of sealed silicon nanotubes as anodes for lithium ion batteries. Nano Lett. 10: 1710–1716.

Stan, M.C., R. Klöpsch, A. Bhaskar, J. Li, S. Passerini and M. Winter. 2013. Cu_3P binary phosphide: synthesis *via* a wet mechanochemical method and electrochemical behavior as negative electrode material for lithium-ion batteries. Adv. Energy Mater. 3: 231–238.

Sun, Q. and Z.W. Fu. 2007. An anode material of CrN for lithium-ion batteries. Electrochem. Solid-State Lett. 10: A189–A193.

Sun, Q. and Z.W. Fu. 2008a. $Cr_{1-x}Fe_xN$ $(0 \le x \le 1) \le 1) \le$. a transition-metal nitrides as anode materials for lithium-ion batteries. Electrochem. Solid-State Lett. 11: A233–A237.

Sun, Q. and Z.W. Fu. 2008b. Vanadium nitride as a novel thin film anode material for rechargeable lithium batteries. Electrochim. Acta 54: 403–409.

Sun, Q., X.Q. Zhang, F. Han, W.C. Li and A.H. Lu. 2012. Controlled hydrothermal synthesis of 1D nanocarbons by surfactant-templated assembly for use as anodes for rechargeable lithium-ion batteries. J. Mater. Chem. 22: 17049–17054.

Sun, X.G. and S. Dai. 2010. Electrochemical investigations of ionic liquids with vinylene carbonate for applications in rechargeable lithium ion batteries. Electrochim Acta 55(15): 4618–4626.

Sun, Y., X. Hu, W. Luo and Y. Huang. 2012. Ultrathin CoO/graphene hybrid nanosheets: a highly stable anode material for lithium-ion batteries. J. Phys. Chem. C 116: 20794–20799.

Taberna, P.L., S. Mitra, P. Poizot, P. Simon and J.M. Tarascon. 2006. High rate capabilities Fe_3O_4-based Cu nano-architectured electrodes for lithium-ion battery applications. Nat. Mater. 5: 567–573.

Takada, K., K. Iwamoto and S. Kondo. 1999. Lithium iron sulfide as an electrode material in a solid state lithium battery. Solid State Ionics 117: 273–276.
Tamura, N., R. Ohshita, M. Fujimoto, S. Fujitani, M. Kamino and I. Yonezu. 2002. Study on the anode behavior of Sn and Sn–Cu alloy thin-film electrodes. J. Power Sources 107: 48–55.
Tarascon, J.M. and M. Armand. 2001. Issues and challenges facing rechargeable lithium batteries. Nature 414: 359–367.
Thackeray, M.M. 1995. Structural considerations of layered and spinel lithiated oxides for lithium ion batteries. J. Electrochem. Soc. 142: 2558–2563.
Tominaga, Y. and K. Yamazaki. 2014. Fast li-ion conduction in poly(ethylene carbonate)-based electrolytes and composites filled with TiO_2 nanoparticles. Chem. Commun. 50: 4448–4450.
Tsubouchi, S., Y. Domi, T. Doi, M. Ochida, H. Nakagawa and T. Yamanaka. 2012. Spectroscopic characterization of surface films formed on edge plane graphite in ethylene carbonate-based electrolytes containing film-forming additives. J. Electrochem. Soc. 159(11): A1786–A1790.
Ueda, A., M. Nagao, A. Inoue, A. Hayashi, Y. Seino, T. Ota et al. 2013. Electrochemical performance of all-solid-state lithium batteries with Sn_4P_3 negative electrode. J. Power Sources 244: 597–600.
Uysal, M., T. Cetinkaya, A. Alp and H. Akbulut. 2015. Active and inactive buffering effect on the electrochemical behavior of Sn–Ni/MWCNT composite anodes prepared by pulse electrodeposition for lithium-ion batteries. J. Alloy Comp. 645: 235–242.
Villevieille, C., F. Robert, P.L. Taberna, L. Bazin, P. Simon and L. Monconduit. 2008. The good reactivity of lithium with nanostructured copper phosphide. J. Mater. Chem. 18: 5956–5960.
Vinayan, B.P. and S. Ramaprabhu. 2013. Facile synthesis of SnO_2 nanoparticles dispersed nitrogen doped graphene anode material for ultrahigh capacity lithium ion battery applications. J. Mater. Chem. A. 1: 3865–3871.
Vincent, C.A. and B.C.H. Scrosati. 1997. Modern Batteries. An Introduction to Electrochemical Power Sources 2nd edn. Arnold, London.
Wagemaker, M. and F.M. Mulder. 2013. Properties and promises of nanosized insertion materials for Li-ion batteries. Acc. Chem. Res. 46: 1206–1215.
Wan, H., Z. Chen, M. Yuan, J. Wang and J. Zhang. 2015. Highly ordered nanoporous Sn-Ni alloy film anode with excellent lithium storage performance. Mater. Lett. 138: 139–142.
Wang, B., J.S. Chen, H.B. Wu, Z. Wang and X.W. Lou. 2011. Quasiemulsion-templated formation of α-Fe_2O_3 hollow spheres with enhanced lithium storage properties. J. Am. Chem. Soc. 133: 17146–17148.
Wang, B., X. Li, X. Zhang, B. Luo, M. Jin, M. Liang et al. 2013a. Adaptable silicon–carbon nanocables sandwiched between reduced graphene oxide sheets as lithium ion battery anodes. ACS Nano 7: 1437–1445.
Wang, F., C. Lu, Y. Qin, C. Liang, M. Zhao, S. Yang et al. 2013b. Solid state coalescence growth and electrochemical performance of plate-like Co_3O_4 mesocrystals as anode materials for lithium-ion batteries. J. Power Sources 235: 67–73.
Wang, G.X., L. Sun, D.H. Bradhurst, S.X. Dou and H.K. Liu. 2000. Lithium storage properties of nanocrystalline eta-Cu_6Sn_5 alloys prepared by ball-milling. J. Alloys Compd. 299: L12–L15.
Wang, H., M. Yoshio, T. Abe and Z. Ogumi. 2002a. Characterization of carbon-coated natural graphite as a lithium-ion battery anode material. J. Electrochem. Soc. 149: A499–A503.
Wang, K., Y. Wang, C. Wang and Y. Xia. 2014. Graphene oxide assisted solvothermal synthesis of $LiMnPO_4$ nanoplates cathode materials for lithium ion batteries. Electrochim. Acta 146: 8–14.
Wang, X., Q. Li, J. Xie, Z. Jin, J. Wang, L. Jinyong et al. 2009. Fabrication of ultralong and electrically uniform single-walled carbon nanotubes on clean substrates. Nano Letters 9: 3137–3141.

Wang, Y., S. Nakamura, M. Ue and P.B. Balbuena. 2001. Theoretical studies to understand surface chemistry on carbon anodes for lithium-ion batteries: reduction mechanisms of ethylene carbonate. J. Am. Chem. Soc. 123(47): 11708–11718.

Wang, Y. and P.B. Balbuena. 2002b. Theoretical insights into the reductive decompositions of propylene carbonate and vinylene carbonate: density functional theory studies. J. Phys. Chem. B. 106(17): 4486–4495.

Wang, Y., S. Nakamura, K. Tasaki and P.B. Balbuena. 2002. Theoretical studies to understand surface chemistry on carbon anodes for lithium-ion batteries: how does vinylene carbonate play its role as an electrolyte additive? J. Am. Chem. Soc. 124(16): 4408–4421.

Wang, Y. and P.B. Balbuena. 2004. Lithium Ion Batteries: Solid-Electrolyte Interphase. Imperial College Press, London.

Wang, Y., Z.W. Fu, X.L. Yue and Q.Z. Qin. 2004. Electrochemical reactivity mechanism of Ni_3N with lithium. J. Electrochem. Soc. 151: E162–E167.

Wang, Y. and G. Cao. 2008. Developments in nanostructured cathode materials for high-performance lithium-ion batteries. Adv. Mater. 20: 2251–2269.

Wang, Z., X. Huang and L. Chen. 2003. Performance improvement of surface-modified $LiCoO_2$ cathode materials: an infrared absorption and X-ray photoelectron spectroscopic investigation. J. Electrochem. Soc. 150(2): A199–A208.

Wang, Z., L. Zhou and X.W. Lou. 2012. Metal oxide hollow nanostructures for lithium-ion batteries. Adv. Mater. 24: 1903–1911.

Wang, C.S, G.T. Wu, X.B. Zhang, Z.F. Qi and W.Z. Li. 1998. Characterization of crystal quality by crystal originated particle delineation and the impact on the silicon wafer surface. J. Electrochem. Soc. 145: 275–284.

Whitacre, J.F., W.C. West, E. Brandon and B.V. Ratnakumar. 2001. Crystallographically oriented thin-film nanocrystalline cathode layers prepared without exceeding 300°C. J. Electrochem. Soc. 148: A1078–1084.

Whittingham, M.S. 1978. Chemistry of intercalation compounds: metal guests in chalcogenide hosts. Prog. Solid State Chem. 12: 41–99.

Wi, S., J. Kim, S. Nam, J. Kang, S. Lee, H. Woo et al. 2014. Enhanced rate capability of $LiMn_{0.9}Mg_{0.1}PO_4$ nanoplates by reduced graphene oxide/carbon double coating for Li-ion batteries. Curr. Appl. Phy. 14: 725–730.

Winter, M., J.O. Besenhard, M.E. Spahr and P. Novak. 1998. Insertion electrode materials for rechargeable lithium batteries. Adv. Mater. 10: 725–763.

Winter, M. and J.O. Besenhard. 1999. Electrochemical lithiation of tin and tin-based intermetallics and composites. Electrochim. Acta 45: 31–50.

Woo, S.G., J.H. Jung, H. Kim, M.G. Kim, C.K. Lee, H.J. Sohn et al. 2006. Electrochemical characteristics of Ti–P composites prepared by mechanochemical synthesis. J. Electrochem. Soc. 153: A1979–A1983.

Woo, S.W., N. Okada, M. Kotobuki, K. Sasajima, H. Munakata, K. Kajihara et al. 2010. Highly patterned cylindrical Ni–Sn alloys with 3-dimensionally ordered macroporous structure as anodes for lithium batteries. Electrochimica Acta 55: 8030–8035.

Wu, C., P. Yin, X. Zhu, C. OuYang and Y. Xie. 2006. Synthesis of hematite (α-Fe_2O_3) nanorods: diameter-size and shape effects on their applications in magnetism, lithium ion battery, and gas sensors. J. Phys. Chem. B. 110: 17806–17812.

Wu, H., G. Chan, J.W. Choi, I. Ryu, Y. Yao, M.T. McDowell et al. 2012. Stable cycling of double-walled silicon nanotube battery anodes through solid–electrolyte interphase control. Nat. Nano 7: 310–315.

Wu, Z.S., W. Ren, L. Wen, L. Gao, J. Zhao, Z. Chen et al. 2010. Graphene anchored with Co_3O_4 nanoparticles as anode of lithium ion batteries with enhanced reversible capacity and cyclic performance. ACS Nano 4: 3187–3194.

Xiao, W., X. Li, H. Guo, Z. Wang, Y. Zhang and Z. Zhang. 2012. Preparation of core–shell structural single ionic conductor $SiO_2@Li^+$ and its application in PVDF–HFP-based composite polymer electrolyte. Electrochimica Acta 85: 612–621.
Xing, L.D., C.Y. Wang, M.Q. Xu, W.S. Li and Z.P. Cai. 2009. Theoretical study on reduction mechanism of 1,3-benzodioxol-2-one for the formation of solid electrolyte interface on anode of lithium ion battery. J. Power Sources 189(1): 689–692.
Xu, C., Y. Zeng, X. Rui, N. Xiao, J. Zhu, W. Zhang et al. 2012a. Controlled soft-template synthesis of ultrathin C@FeS nanosheets with high-Li-storage performance. ACS Nano 6: 4713–4721.
Xu, C.C., L. Li, F.Y. Qiu, C.H. An, Y.A. Xu, Y. Wang et al. 2014. Graphene oxide assisted facile hydrothermal synthesis of $LiMn_{0.6}Fe_{0.4}PO_4$ nanoparticles as cathode material for lithium ion battery. J. Energy Chem. 23: 397–402.
Xu, J.S. and Y.J. Zhu. 2012. Monodisperse Fe_3O_4 and γ-Fe_2O_3 magnetic mesoporous microspheres as anode materials for lithium-ion batteries. ACS Appl. Mater. Interfaces 4: 4752–4757.
Xu, X., R. Cao, S. Jeong and J. Cho. 2012b. Spindle-like mesoporous α-Fe_2O_3 anode material prepared from MOF template for high-rate lithium batteries. Nano Lett. 12: 4988–4991.
Yamin, H. and E. Peled. 1983. Electrochemistry of a nonaqueous lithium/sulfur cell. J. Power Sources 9: 281–287.
Yang, J., Y. Takeda, N. Imanishi, C. Capiglia, J.Y. Xie and O. Yamamoto. 2002. SiO*x*-based anodes for secondary lithium batteries. Solid State Ionics 152-153: 125–129.
Yang, J., X. Zhou, J. Li, Y. Zou and J. Tang. 2012a. Study of nano-porous hard carbons as anode materials for lithium ion batteries. Mater. Chem. Phys. 135: 445–450.
Yang, J.L., J.J. Wang, D.N. Wang, X.F. Li, D.S. Geng, G.X. Liang et al. 2012b. 3D porous $LiFePO_4$/graphene hybrid cathodes with enhanced performance for Li-ion batteries. J. Power Sources 208: 340–344.
Yao, W., Z. Zhang, J. Gao, J. Li, J. Xu and Z. Wang. 2009. Vinyl ethylene sulfite as a new additive in propylene carbonate-based electrolyte for lithium ion batteries. Energy Environ. Sci. 2(10): 1102–1108.
Yashio, M., H. Wang, K. Fukuda, T. Umeno, T. Abe and Z. Ogumi. 2004. Improvement of natural graphite as a lithium-ion battery anode material, from raw flake to carbon-coated sphere. J. Mater. Chem. 14: 1754–1758.
Yu, Y., C. Cui, W. Qian, Q. Xie, C. Zheng, C. Kong et al. 2013. Carbon nanotube production and application in energy storage. Asia-Pacific J. Chem. Eng. 8: 234–245.
Yuan, W., D. Xie, Z. Dong, Q. Su, J. Zhang, G. Du et al. 2013. Preparation of porous Co_3O_4 polyhedral architectures and its application as anode material in lithium-ion battery. Mater. Lett. 97: 129–132.
Yun, Y.S., J.H. Kim, S.Y. Lee, E.G. Shim and D.W. Kim. 2011. Cycling performance and thermal stability of lithium polymer cells assembled with ionic liquid-containing gel polymer electrolytes. J. Power Sources 196(16): 6750–6755.
Zhang, H. and P.V. Braun. 2012. Three-dimensional metal scaffold supported bicontinuous silicon battery anodes. Nano Lett. 12: 2778–2783.
Zhang, H.L., S.H. Liu, F. Li, S. Bai, C. Liu, J. Tan et al. 2006. Electrochemical performance of pyrolytic carbon-coated natural graphite spheres. Carbon 44: 2212–2218.
Zhang, L., P. Hu, X. Zhao, R. Tian, R. Zou and D. Xia. 2011. Controllable synthesis of core–shell Co@CoO nanocomposites with a superior performance as an anode material for lithium-ion batteries. J. Mater. Chem. 21: 18279–18283.
Zhao, J., A. Buldum, J. Han and J.P. Lu. 2000. First-principles study of Li-intercalated carbon nanotube ropes. Phys. Rev. Lett. 85: 1706–1709.
Zhong, K., X. Xia, B. Zhang, H. Li, Z. Wang and L. Chen. 2010. MnO powder as anode active materials for lithium ion batteries. J. Power Sources 195: 3300–3308.
Zhou, H., S. Zhu, M. Hibino, I. Honma and M. Ichihara. 2003. Lithium storage in ordered mesoporous carbon (CMK-3) with high reversible specific energy capacity and good cycling performance. Adv. Mater. 15: 2107–2111.

Zhou, J., H. Song, B. Fu, B. Wu and X. Chen. 2010. Synthesis and high-rate capability of quadrangular carbon nanotubes with one open end as anode materials for lithium-ion batteries. J. Mater. Chem. 20: 2794–2800.

Zhu, G.N., L. Chen, Y.G. Wang, C.X. Wang, R.C. Che and Y.Y. Xia. 2013a. Binary $Li_4Ti_5O_{12}$-$Li_2Ti_3O_7$ nanocomposite as an anode material for Li-ion batteries. Adv. Funct. Mater. 23: 640–647.

Zhu, X., W. Wu, Z. Liu, L. Li, J. Hu, H. Dai et al. 2013b. A reduced graphene oxide–nanoporous magnetic oxide iron hybrid as an improved anode material for lithium ion batteries. Electrochim. Acta 95: 24–28.

Zhuo, K., M.G. Jeong and C.H. Chung. 2013. Highly porous dendritic Ni–Sn anodes for lithium-ion batteries. J. Power Sources 244: 601–605.

Zong, J. and X.J. Liu. 2014. Graphene nanoplates structured $LiMnPO_4$/C composite for lithium-ion battery. Electrochim. Acta 116: 9–18.

Nanocomposite Polymer Electrolytes for Electric Double Layer Capacitors (EDLCs) Application

Chiam-Wen Liew

1. Introduction

1.1 Polymer electrolytes

Liquid electrolytes have been widely used in the commercial electrochemical devices. However, these liquid electrolytes are harmful to the environment and consumers because of the solvent leakage. Liquid electrolytes also possess several disadvantages, such as electrolytic degradation, lithium dendrite growth formation, poor long-term stability, narrow operating temperature range, difficulty in handling and manufacturing, short life span and high probability of internal circuit shorting (Ramesh et al. 2011, Yang et al. 2008, Gray 1997, Stephan et al. 2006). Therefore, polymer electrolytes are introduced to substitute liquid electrolytes. Polymer electrolytes have gained much attention because of their attractive features. The advantages of polymer electrolytes are excellent safety performances, negligible vapor pressure, high automation potential, high energy density, high polymer

Department of Physical Science, Faculty of Applied Science, Tunku Abdul Rahman University College, 53300 Setapak, Kuala Lumpur, Malaysia.
Email: liewchiamwen85@gmail.com

flexibility, low volatility, superior electrochemical, structural, thermal, photochemical and chemical stabilities, and low electronic conductivity (Adebahr et al. 2003, Armand 1986, Gray 1991, Nicotera et al. 2002, Ramesh and Liew 2013). Other interesting characteristics are inherent viscoelasticity, suppression of lithium dendrite growth, excellent mechanical strength, light in weight, ease of handling and manufacturing, wide operating temperature range, cost effective and no complicated technology requirement (Baskaran et al. 2007, Imrie and Ingram 2000, Rajendran et al. 2004).

Polymer electrolytes have a wide application range, ranging from small scale production of commercial secondary lithium ion batteries to advanced high energy electrochemical devices. These electrochemical devices are fuel cells, chemical sensors, solid state reference electrode systems, electrochromic windows (ECWs), supercapacitors, thermoelectric generators, analogue memory devices and solar cells (Armand 1986, Gray 1991, Rajendran et al. 2004). These electrochemical devices are mainly used as power sources in portable electronic and personal communication devices such as laptops, mobile phones, MP3 players, personal digital assistants (PDAs), digital cameras, power tools, toys, digital clocks and video recorders. These devices can also be applied in fuel cell vehicles, hybrid electrical vehicles (EV) and start-light–ignition (SLI) as traction power source for electricity (Gray 1991, Ahmad et al. 2005). Polymer electrolytes are sub-divided into three main groups that are solid polymer electrolytes, gel polymer electrolytes and composite polymer electrolytes (nanocomposite polymer electrolytes).

1.2 *Solid polymer electrolytes (SPEs)*

Solid polymer electrolytes (SPEs) were first prepared by Wright, a polymer chemist from Sheffield in the year 1975 to overcome the drawbacks of liquid electrolytes. SPEs are truly interdisciplinary materials and used as flexible ion transporting medium in vital applications such as energy storage and electrochemical displays (Wright 1975). Solid polymer electrolytes (SPEs) are prepared by dissolving low lattice energy metal salts in polymer network. The charge carriers in salt are initially decoupled due to the low lattice energy of salt and then transported in the polymer complexes. The donor atom (also known as solvating group) of polymer forms the covalent bonding with the charge carriers which are dissociated from the salt for ion transport mechanism and leads to the formation of the coordination when positive charge on the cation interacts with the negative charge on the solvating group *via* electrostatic interactions. The charge carriers are dissociated and transported within the electrolytes. The ionic conduction in the polymer electrolytes arises from this ion dissociation from the coordination. The ion hopping mechanism in the polymer network could generate the electricity of the electrochemical devices (Bruce and Vincent 1993).

Superior safety performance is the key of SPEs because of its solvent free condition. The high elastic relaxation properties under stress, easy handling and processing, excellent mechanical strength and high flexibility of SPEs allow the fabrication of all solid-state electrochemical cells (Gray 1997, Ibrahim et al. 2012). SPEs exhibit excellent electrode-electrolyte interfacial contact over crystalline or glassy electrolytes. The interfacial contact can be maintained well under stress throughout the charging and discharging processes after prolonged time (Gray 1997, Gray 1991). The build up of the internal pressure in liquid electrolytes may cause the explosion during charge and discharge processes in the electrochemical cell. However, this problem is omitted in SPEs (Liew et al. 2013). The ionic transportation mechanism in the polymer electrolytes depends on the local relaxation processes in the polymer chains which has similar properties as liquid electrolytes (Gray 1997). As a result, SPEs are good replacements for liquid electrolytes. Wright and his groups invented the first generation SPEs, which were crystalline poly(ethylene oxide) (PEO)-based polymer electrolytes in the year 1975 (Fenton et al. 1973, Quartarone et al. 1998). Different types of alkali metal salts (sodium and potassium salts) were reported in this literature (Wright 1975, Fenton et al. 1973). However, the ionic conductivity of polymer electrolytes was still relatively low ($\sim 10^{-8}$–10^{-7} S cm^{-1}) due to their high crystallinity and rapid recrystallization (Wright 1975, Fenton et al. 1973).

Several ways have been proposed to reduce the degree of crystallinity and inhibit the recrystallization of the polymer electrolytes. These approaches are polymer blending of different types of polymers, polymer modifications, utilization of semi-crystalline or amorphous polymer and addition of additives like plasticizers, inorganic fillers and ionic liquids. Structural modifications onto the ethylene oxide group in PEO polymer chains such as cross-linking, random, block or comb polymerization, radical polymerization, cationic polymerization, epoxides copolymerization have been implemented to reduce and hinder the crystallization process (Quartarone et al. 1998). Bouridah et al. prepared cross-linked PEO and aliphatic isocyanate grafted poly(dimethylsiloxane) (PDMS) in year 1985. The ionic conductivity of polymer electrolytes reached $\sim 10^{-5}$ Scm^{-1} upon addition of 10 wt.% of lithium perchlorate ($LiClO_4$) into these cross-linked polymer complexes (Bouridah et al. 1985). Another attempt of crosslinking the triol type of PEO with poly(propylene oxide) (PPO) was prepared by Watanabe et al. (1986). The ionic conductivity of cross-linked polymer electrolytes was five times higher than PEO polymer electrolytes without cross-linking process (Watanabe et al. 1986). Copolymerization is also another attempt to reduce the percentage of crystallinity and inhibit the recrystallization. Polyacrylonitrile-polyethylene oxide (PAN-PEO) copolymer was synthesized by Yuan et al. (2005). The ionic conductivity of polymer electrolytes is increased through the copolymerization. The ionic conductivity of 6.79×10^{-4} Scm^{-1} was obtained with an [EO]/[Li] ratio of about 10 (Yuan et al. 2005). In general, SPEs possess low ionic conductivity that delays application in the electrochemical device even though they are safe.

1.3 Gel polymer electrolytes (GPEs)

The ionic conductivity of SPEs is still very low. Therefore, researchers came out with a brilliant idea to replace SPEs with gel polymer electrolytes (GPEs). There are two additives that can be used to produce GPEs (also known as gelionic solid polymer electrolytes):

- Plasticizers (Osinska et al. 2009, Rajendran et al. 2008)
- Ionic liquids (Pandey and Hashmi 2009, Jain et al. 2005, Vioux et al. 2010)

The liquid electrolytes would be entrapped in the polymer matrix of GPEs. The immobilization of liquid electrolytes in polymer complexes shows unique feature by providing liquid-like degree of freedom which is comparable to those conventional liquid electrolytes at the atomic level (Ramesh et al. 2012, Han et al. 2002). The cohesive property of solids and diffusive property of liquids make GPEs as ideal candidates to replace SPEs. GPEs exhibit many superb characteristics, for example, low interfacial resistance, low reactivity towards the electrode materials, improved safety performance, easy fabrication into desired shape and size and high ionic conductivity with a small portion of plasticizers (Pandey and Hashmi 2009, Ahmad et al. 2008, Zhang et al. 2011). In addition, GPEs possess improved electrochemical properties, wider operating temperature range and cheaper in comparison to liquid electrolytes (Ahmad et al. 2005, Zhang et al. 2011, Stephan et al. 2002).

1.3.1 Plasticizers

Numerous types of common plasticizers are used to prepare GPEs, for instance propylene carbonate (PC), ethylene carbonate (EC), dimethyl carbonate (DMC), diethyl carbonate (DEC), N,N-dimethyl formamide (DMF), N,N-dimethylacetamide (DMAc), γ-butyrolactone, dioctyl phthalate (DOP), dibutyl phthalate (DBP), dimethyl phthalate (DMP), diocthyl adipate (DOA) and poly(ethylene glycol) (PEG) (Ning et al. 2009, Pradhan et al. 2005, Suthanthiraraj et al. 2009). The advantages of adding plasticizers in the polymer electrolytes are summarized as below.

- enhance the salt solvating power (Rajendran et al. 2004, Ramesh and Arof 2001)
- increase the ion mobility (Rajendran et al. 2004, Ramesh and Arof 2001)
- provide a better contact between polymer electrolytes and electrodes due to its sticky behaviour (Rajendran et al. 2004, Ramesh and Arof 2001)
- superior miscibility with polymer (Ramesh and Chao 2011)
- high dielectric constant (Ramesh and Chao 2011)
- ease of processability (Ramesh and Chao 2011)
- reduce viscosity of polymer electrolytes (Ramesh and Chao 2011)
- improve the ionic conductivity (Ganesan et al. 2008)
- decrease glass transition temperature (T_g) of polymer electrolytes (Ganesan et al. 2008)
- enhance amorphousness of polymer electrolytes (Ganesan et al. 2008)

Plasticizers can cause the transition from the glassy state to rubbery region at progressively lower temperature when the T_g is decreased. Plasticizers also improve the chains flexibility of the polymer matrix and favors ionic migration in the electrolytes. Therefore, plasticized-GPEs illustrate a remarkable increase in ionic conductivity. Moreover, plasticizer can promote the salt dissociation and increase the number of charge carriers (Sukeshini et al. 1998).

The effect of adding plasticizers into polymer electrolytes have been widely prepared and studied in recent years. Sukeshini prepared PEO-lithium trifluoromethanesulfonate ($LiCF_3SO_3$) polymer electrolytes. The ionic conductivity of polymer electrolytes was increased almost three orders of magnitude from 7.13×10^{-7} S cm^{-1} to 6.03×10^{-4} S cm^{-1} by adding DBP (Sukeshini et al. 1998). Apart from ionic conductivity, cationic transport number in the polymer electrolytes was enhanced. These findings can be observed in Bhide and Hariharan (2007) when PEG which acts as plasticizer were added into PEO-sodium metaphosphate ($NaPO_3$) complexes (Bhide and Hariharan 2007). PEG had also been studied by Kuila and co-workers. The reported ionic conductivity of plasticizer-free PEO-sodium perchlorate ($NaClO_4$) system was 1.05×10^{-6} S cm^{-1}. However, the maximum conductivity of 2.60×10^{-4} S cm^{-1} was achieved at 300 K with addition of 30 wt.% of PEG as plasticizer. The results imply that the addition of plasticizer reduces the degree of crystallinity and energy barrier for ion transport (Kuila et al. 2007). Three different classes of ester-based plasticizers, namely DOP, DBP and DMP were investigated by Michael et al. (1997). Among all these three plasticizers, DOP showed the best performance as it is a thermally stable compound (Michael et al. 2007). Similar work was also conducted by Rajendran et al. (2004). EC-plasticized polymer electrolytes containing PVA/PMMA–$LiBF_4$ portrayed the highest ionic conductivity of 1.29 mS cm^{-1} because of higher dielectric constant of EC (ε = 85.1). Plasticization can improve the ionic conductivity of polymer electrolytes extensively but it has some limitations, such as low safety performances, poor electrical, electrochemical, mechanical and thermal stabilities, slow evaporation of solvent, high vapor pressure and narrow electrochemical window as well as low flash point (Pandey and Hashmi 2009, Kim et al. 2006, Raghavan et al. 2010).

1.3.2 Ionic liquids

A new material, that is ionic liquid (IL) has been discovered to replace plasticizer. ILs are known as non-volatile molten salts that remain in their liquid state at ambient temperature with a low melting temperature, $T_m < 100°C$ (Osinska et al. 2009, Quartarone and Mustarelli 2011). These ILs consist of a bulky and asymmetric organic cation and a highly delocalized-charge inorganic anion. There are several types of ionic liquids available for polymer electrolytes preparation. Organic cations are 1,3-dialkylimidazolium, 1,3-dialkylpyridinium, tetraalkylammonium, trialkylsulphonium, tetraalkylphosphonium, *N*-methyl-*N*-alkylpyrrolidinium, *N,N*-dialkylpyrrolidinium, *N*-alkylthiazolium, *N,N*-dialkyltriazolium, *N,N*-dialkyloxazolium, *N,N*-dialkylpyrazolium and guanidinium (Jain et al. 2005, Ye et al. 2013). In contrast, the inorganic anions can be acetate (CH_3COO^-), triflate (Tf^-), hexaflurophosphate (PF_6^-), tetrafluoroborate (BF_4^-), bis(trifluoromethylsulfonyl imide) ($TFSI^-$), bis (perfluoroethyl sulfonyl) imide [$N(C_2F_5SO_2)^{2-}$], nitrate (NO_3^-) and halides (Cl^-, Br^- and I^-). The combination of cation with anion in ILs alter their physical properties such as melting point, viscosity, hydrophobicity, dielectric constant, miscibility with water and other solvents, polarity and density as well as dissolution ability (Jain et al. 2005, Vioux et al. 2010).

Ionic liquids have emerged as promising candidates because of their unique and fascinating physicochemical properties. These beneficial properties are:

- wide electrochemical potential window (up to 6V) (Cheng et al. 2007, Patel et al. 2011, Pandey and Hashmi 2013, Ye et al. 2013)
- wide decomposition temperature range (Cheng et al. 2007, Patel et al. 2011, Pandey and Hashmi 2013, Ye et al. 2013)
- negligible vapor pressure (Cheng et al. 2007, Patel et al. 2011, Pandey and Hashmi 2013, Ye et al. 2013)
- non-toxic (Cheng et al. 2007, Patel et al. 2011, Pandey and Hashmi 2013, Ye et al. 2013)

- non-volatile (Cheng et al. 2007, Patel et al. 2011, Pandey and Hashmi 2013, Ye et al. 2013)
- non-flammable (Cheng et al. 2007, Patel et al. 2011, Pandey and Hashmi 2013, Ye et al. 2013)
- environmental friendly (Cheng et al. 2007, Patel et al. 2011, Pandey and Hashmi 2013, Ye et al. 2013)
- excellent chemical, thermal and electrochemical stabilities (Cheng et al. 2007, Patel et al. 2011, Pandey and Hashmi 2013, Ye et al. 2013)
- high ionic conductivity due to high ion concentration (Cheng et al. 2007, Patel et al. 2011, Pandey and Hashmi 2013, Ye et al. 2013)
- good oxidative stability (Cheng et al. 2007, Patel et al. 2011, Pandey and Hashmi 2013, Ye et al. 2013)
- superior ion mobility (Cheng et al. 2007, Patel et al. 2011, Pandey and Hashmi 2013, Ye et al. 2013)
- high cohesive energy density (Cheng et al. 2007, Patel et al. 2011, Pandey and Hashmi 2013, Ye et al. 2013)
- high ability to dissolve a wide range of organic, inorganic and organometallic compounds (Reiter et al. 2006, Vioux et al. 2010)
- excellent safety performance (Reiter et al. 2006, Vioux et al. 2010)
- excellent contact between electrolyte and electrode (Reiter et al. 2006)
- strong plasticizing effect (Singh et al. 2009)

Strong plasticizing effect of ionic liquids is the main contributor to the highly conductive polymer electrolytes by softening the polymer backbone. As a result, the ion dissociation and transport are favored. Same as plasticizers, ionic liquids also help in reducing the degree of crystallinity. The bulky cations paired with anions could be detached easily from this ionic compound due to its poor packing efficiencies. Therefore, there is more mobile charge carriers migrated within the electrolytes. The immobilization of ionic liquids within polymer matrices is a unique characteristic. The ionic liquid-based polymer electrolytes can behave as liquid electrolytes, while it can maintain itself in the solid state.

Ionic liquids-based polymer electrolytes have received attention from many researchers. Sirisopanaporn et al. had prepared flexible poly(vinylidenefluoride-co-hexafluoropropylene) (PVdF-co-HFP) based copolymer polymer electrolytes using N-butyl-N-ethylpyrrolidinium N,N-bis(trifluoromethane)sulfonimide-lithium N,N-bis(trifluoromethane) sulfonamide (Py_{24}TFSI-LiTFSI). The ionic conductivity of polymer electrolytes was increased from 0.34 to 0.94 mS cm^{-1}. Most conducting polymer electrolyte exhibited excellent thermal and interfacial stabilities as they can be operated up to 110°C without any degradation and any leakage within 4 months of storage time (Sirisopanaporn et al. 2009). PVdF-co-HFP copolymer electrolytes were also investigated by Sekhon and co-workers with addition of 2,3-dimethyl-1-octylimidazolium trifluromethanesulfonylimide (DMOImTFSI). This proton conductor reached the highest ionic conductivity of 2.74 $mScm^{-1}$ was achieved at 130°C, along with good mechanical stability (Sekhon et al. 2006). The effect of ionic liquids on biodegradable polymer was also discussed in Ning et al. (2009). The ionic conductivity of $10^{-1.6}$ Scm^{-1} was achieved upon addition of 30 wt.% of 1-ally-3-methylimidazoliumchloride (AmImCl) into corn starch-based electrolytes (Ning et al. 2009). Biopolymer electrolytes comprised of corn starch, lithium hexafluorophosphate ($LiPF_6$) and 1-butyl-3-methylimidazolium trifluoromethanesulfonate (BmImTf). This biopolymer electrolyte showed the maximum ionic conductivity of 3.21×10^{-4} Scm^{-1} at room temperature (Liew and Ramesh 2013). Doping of ionic liquids is a feasible way to improve the ionic conductivity greatly without degrading the polymer electrolytes.

1.4 Composite polymer electrolytes and nanocomposite polymer electrolytes (CPEs and NCPEs)

Composite polymer electrolytes (CPEs) have witnesses an upsurge in attention, recently, due to their potential to solve the drawbacks of plasticizers which are related to the safety performances and mechanical integrity. Dispersion of little amount of organic or inorganic fillers into the polymer solution will produce CPEs (Osinska et al. 2009). Nanocomposite Polymer Electrolytes (NCPEs) are the designation of the polymer electrolytes with addition of nanometer grain size fillers. Both CPEs and NCPEs exhibit several advantages:

- superior physical and mechanical properties of polymer electrolytes (Gray 1997)
- good interfacial contact at electrode-electrolyte region (Gray 1997)
- improved ion transport (Gray 1997)
- high flexibility (Gray 1997)
- high ionic conductivity (Gray 1997)
- excellent thermodynamic stability towards lithium and other alkali metals (Gray 1997)
- superior interfacial properties towards lithium metal anode (Krawiec et al. 1995)
- improved electrochemical properties (Jian-hua et al. 2008)

Fillers are classified into two main classes that are organic fillers and inorganic fillers. Some examples of organic fillers are graphite fiber, aromatic polyamide and cellulosic rigid rods (whiskers). On the other hand, the common inorganic fillers used in polymer electrolytes are titania (TiO_2), zirconia (ZrO_2), fumed silica (SiO_2), alumina (Al_2O_3) and manganese oxide (MnO_2). Inorganic fillers are divided into two types: active fillers and passive fillers. Active fillers are the fillers that contribute to the charge carrier concentration in transport mechanism, for example, lithium-nitrogen (Li_2N), lithium containing ceramics like lithium-alumina ($LiAl_2O_3$), lithium aluminate ($LiAlO_2$), titanium carbides (TiC_x), titanium carbonitrides (TiC_xN_y) and titanium nitrides (TiN_x) (Ishkov and Sagalakov 2005). On the contrary, the passive fillers are not involved in the ionic conduction process (Giffin et al. 2012). There are several types of inactive fillers which include inert metal oxide (e.g., TiO_2, ZrO_2, Al_2O_3, MnO_2), treated silica (SiO_2), molecular sieves and zeolites (e.g., aluminosilicate molecular sieves and ordered mesoporous silica, OMS), rare earth oxide (e.g., $SrBi_4Ti_4O_{15}$, $La_{0.55}Li_{0.35}TiO_3$ fibers), ferroelectric materials (e.g., barium titanate, $BaTiO_3$), solid superacid (e.g., sulphates and phosphates, including SO_4^{2-}/ZrO_2, SO_4^{2-}/Fe_2O_3, and SO_4^{2-}/TiO_2), mica and nano-clay (e.g., montmorillonite, MMT), carbon (e.g., carbon nanotubes, CNTs, fly ash) and heteropolyacid (e.g., silicotungstic acid (SiWA), phosphotungstic acid (PWA), molybdophosphoric acid, phosphomolibdicacid (PMoA)) and biodegradable ceramics (e.g., calcium carbonate, calcium aluminates) (Jung et al. 2009, Noto et al. 2012, Samir et al. 2005, Zapata et al. 2012). An invention has been made to combine organic and inorganic phases to produce organic-inorganic hybrid fillers. Zhang et al. had synthesized poly(cyclotri-phosphazene-co-4,40-sulfonyldiphenol) (PZS) microspheres organic-inorganic hybrid fillers. They found out that the ionic conductivity and lithium ion transference number increased upon addition of PZS fillers into PEO-based electrolytes (Zhang et al. 2010).

Fillers are great materials as mechanical stiffener and they possess many advantages in the development of polymer electrolytes. Fillers can reduce glass transition temperature (T_g) and the crystallinity of polymer membrane (Kim et al. 2002, Saikia et al. 2009). Fillers can also enhance the thermal stability of polymer matrix and improve the morphological properties of polymer electrolytes (Kim et al. 2002, Saikia et al. 2009). Apart from that, fillers increase cationic diffusivity by altering the transport properties, improve the physical properties of polymer matrix and reduce the water retention of polymer electrolytes (Samir et al. 2005, Hammami et al. 2013, Jian-hua et al. 2008). The effect of fillers on the electrochemical performances of devices ought to be focused. Dispersion of fillers can improve interfacial stability and reduce interfacial resistance between electrode and electrolyte, reduce the capacity fading, widen the electrochemical stability of polymer electrolytes because of the excellent membrane stability as well as enhance the long-term electrochemical stability of polymer electrolytes and electrochemical devices (Jian-hua et al. 2008, Kim et al. 2002, Lue et al. 2008, Polu and Kumar 2013, Raghava et al. 2008, Saikia et al. 2009, Yang et al. 2012, Zhang et al. 2010).

Fillers have been widely used and studied in the development of polymer electrolytes. Silica has been widely used as filler for use in the polymer electrolytes. Ketabi and Lian had prepared polymer-ionic liquid electrolyte containing PEO and 1-ethyl-3-methylimidazolium hydrogensulfate ($EMIHSO_4$). The ionic conductivity of 2.15 mS cm^{-1} at room temperature, which is more than 2-fold increase over the electrolyte without filler upon dispersion of amorphous SiO_2 nano-sized filler was reported (Ketabi and Lian 2013). In order to prevent the recrystallization of PEO-based polymer electrolytes and improve the ionic conductivity, special modification and treatment on the conventional fillers have to be employed. Fan et al. (2003) reported that ionic conductivity of the polymer electrolytes containing silane-modified SiO_2 was much higher than that of unmodified SiO_2 (Fan et al. 2003). Xi et al. (2005) had prepared NCPEs based on PEO-$LiClO_4$ using solid acid sulphated-zirconia (SO_4^{2-}-ZrO_2, abbreviated as SZ) as filler. The prepared NCPEs had higher ionic conductivity than pristine PEO-$LiClO_4$ polymer electrolytes (Xi et al. 2005). The ionic conductivity of SZ-treated NCPEs increased by two orders of magnitude, from 1.5×10^{-7} S cm^{-1} to 4.0×10^{-5} S cm^{-1}. The effect of nano-scaled TiO_2 has also been studied in two copolymers, PVdF-co-HFP and poly (ethylene oxide-co-ethylene carbonate) (P(EO-co-EC)) by Jeon and co-workers. Based on the findings, the ionic conductivity of NCPEs was slightly higher than the polymer electrolyte without the addition of TiO_2 nanoparticles at ambient temperature as reported in Jeon et al. (2006). New nano-sized organic-inorganic hybrid materials were introduced by Wang et al. (2009). In their work, they synthesized high surface area nano-scaled zinc aluminate ($ZnAl_2O_4$) with a mesoporous network. Addition of 8 wt.% of $ZnAl_2O_4$ into PEO-$LiClO_4$, the highest ionic conductivity of 2.23×10^{-6} Scm^{-1} was achieved at ambient temperature. These nano-sized organic-inorganic hybrid fillers reduced the crystallinity of polymer membrane and increased the lithium ion transference number (Wang et al. 2009). The fabricated battery containing PEG-magnesium acetate [$Mg(CH_3COO)_2$]-alumina showed the current density of 13.91 $\mu A/cm^2$, discharge capacity of 1.721 mA h, power density of 13.14 mW/kg and energy density of 1.84 W h/kg with an open circuit voltage (OCV) of 1.85 V (Polu and Kumar 2013). Dispersion of filler is an alternative method to improve the ionic conductivity of polymer electrolytes.

2. Supercapacitors

Supercapacitor (also known as ultracapacitor or electrochemical capacitor) is an energy storage-based electrochemical device which is mainly used as a power source. The energy storage of a supercapacitor arises from the ion accumulation at the electrode-electrolyte boundary through rapid and reversible adsorption and/or desorption of charges carriers (Frackowiak 2007, Pandey et al. 2011). Supercapacitors consist of one pair of electrodes and an electrolyte. The electrode can be derived from many materials, such as carbon, metal oxide and conducting polymers. On the other hand, the electrolyte can be liquid electrolyte, solid polymer electrolyte, gel polymer electrolyte or composite polymer electrolyte, as discussed above. However, there are two basic requirements for choosing the suitable electrolytes to be used in supercapacitors which are conductive and high ionic mobility. The ion accessibility from electrolyte to the electrode becomes an important parameter that governs the capacitance of supercapacitors.

2.1 Types of supercapacitors

Supercapacitors fall broadly into three main types: pseudocapacitors, electric double layer capacitors (EDLCs) and hybrid capacitors which is the combination of pseudocapacitors and EDLCs.

2.1.1 Pseudocapacitors

Pseudocapacitors (also recognized as redox capacitors) are the capacitors that are involved in fast Faradaic processes. These Faradaic processes can be intercalation, under-potential deposition and redox reaction occurring at the surface of electrode at an appropriate applied potential (Choudhury et al. 2009). The term "pseudo" originated from the double layer capacitance from the quick Faradaic charge transfer reactions (Frackowiak and Béguin 2001). The active materials used as pseudocapacitive electrodes are chemically modified carbon materials like reduced grapheme oxide (rGO), nobel metal oxide and electroactive conducting polymers (Choi et al. 2012). Examples of transition metal oxide include ruthenium dioxide (RuO_2), iron (III) oxide (Fe_2O_3), nickel (II) oxide (NiO), TiO_2, cobalt (II, III) oxide (Co_3O_4), tin dioxide (SnO_2), iridium (IV) oxide (IrO_2) and manganese (IV) oxide (MnO_2). Polypyrrole (PPy) and poly(thiophene) derivatives such as poly(3,4-ethylenedioxythiophene) (PEDOT) and poly(aniline) (PANI) are the common conducting polymers used in pseudocapacitors fabrication (Hashmi and Upadhyaya 2002, Peng et al. 2008, Yu et al. 2015).

There are many treatments to prepare the pseudocapacitive electrodes. These methods include chemical vapour deposition (CVD), thermal exfoliation, nitrogen (N_2) plasma treatment, electrochemical polarization, ultrasonication and *in situ* reduction, self-assembly chemical reduction, *in situ* polymerization, electric deposition, oxidative polymerization, dip and dry deposition, sol-gel process, hydrothermal process electrophoretic deposition and chemical bath deposition (Choi et al. 2012, Frackowiak and Béguin 2001). RuO_2 is a capacitive material as reported in Trasatti et al. (1971). However, high cost of this metal oxide limits the practical application. Other metal oxide materials are synthesized and used for pseudocapacitance application, but the specific capacitance is lower (between 20 F g^{-1} and 200 F g^{-1}) with poor electrical conductivity (Peng et al. 2008, Wang et al. 2012a). Conducting polymers are suitable for pseudocapacitive electrodes because of their good electrical conductivity, high pseudocapacitance and low cost compared to metal oxide (Peng et al. 2008). Although conducting polymers show large pseudocapacitance, they exhibit poor mechanical stability due to the repeated intercalation and ion depletion during charging and discharging processes (Peng et al. 2008). Even though pseudocapacitors show high capacitance, these capacitors also exhibit some limitations such as shorter lifespan, poor electrochemical stability, low rate capability, relatively expensive due to high raw material cost, high response time because it takes longer time to move electrons during the redox reaction and difficulty in processing which limits their practical applications (Choi et al. 2012, Shao et al. 2012).

2.1.2 Electric double layer capacitors (EDLCs)

In contrast, EDLCs are another type of supercapacitors which do not involve any electrochemical Faradic reaction over the operation range. The energy storage in these non-Faradaic capacitors arises from the formation of Helmholtz layer (or well-known as double layer) through the electrostatic interaction. The double layer is produced due to the charge accumulation at the electrode-electrolyte interface without any chemical reaction. In other words, the capacitive behaviour of an EDLC depends on the ability to form an electrical double layer at the polarizable electrode-electrolyte boundary between the high specific area of porous carbon-based electrodes and an organic electrolyte (Ingram et al. 1998). Porous carbonaceous materials are the common electrode materials for an EDLC. High power density (up to 10 kW kg^{-1}), low cost, long durability (> 100000 cycles) high dynamic of charge propagation (short term pulse), fast energy storage, higher ability to be charged and discharged continuously without degrading, short charging time, environmental friendly and maintenance-free during long life operation are the main features of an EDLC (Choudhury et al. 2009, Endo et al. 2001, Frackowiak and Béguin 2001, Lewandowski et al. 2001, Yu et al. 2012). Solid state EDLC which was prepared and fabricated using PAA-based polymer electrolytes in this research work is presented.

2.1.3 Hybrid capacitors

Hybrid capacitors are the combinations of pseudocapacitor and EDLC which comprise both capacitors' features. There are many intensive ways to hybridize the electrodes. The easiest way is to add the electrochemical active materials into the carbon-based electrode which produces composite hybrid capacitor (Deng et al. 2013). Asymmetric hybrid capacitor is another type of hybrid capacitors. These asymmetric capacitors consist of a pseudocapacitive metal oxide/hydroxide electrode and a capacitive carbon electrode such as activated carbon (AC)//$Ni(OH)_2$ and AC//MnO_2 (Jiang et al. 2013). Lithium insertion electrode with a capacitive carbon electrode like $Li_4Ti_5O_{12}$//AC is another brand new hybrid capacitor. Asymmetric supercapacitors have complementary properties which takes advantage of the best properties of each component to eliminate their drawbacks and to get synergic effect (Amitha et al. 2009). However, the depletion of electrolyte in these hybrid capacitors has been identified as a major challenge (Jiang et al. 2013). So, researchers come out with a brilliant idea which is the production of battery-like hybrid capacitor. Lithium cations (Li^+) will be intercalated to a cathode compound, such as $LiMn_2O_4$ as cathode, while activated carbon will be used as anode material (Jiang et al. 2013).

2.2 *Advantages of supercapacitors*

Supercapacitors have emerged as the new type of electrochemical devices to replace lithium ion batteries and conventional electrolytic capacitors. Supercapacitors deliver higher power density, faster charge-discharge rate and longer cycle life than lithium ion secondary batteries (Choudhury et al. 2009, Wang et al. 2012b, Wu et al. 2013). Supercapacitors also possess several advantages over lithium batteries. For instance, supercapacitors are safer against short circuit than batteries in terms of the possibility of self-ignition. They do not contain any hazardous or toxic or flammable materials (Choi et al. 2012). Apart from that, supercapacitors show higher energy density than conventional solid state and electrolytic dielectric capacitors due to the large surface area of the electrode materials (Choudhury et al. 2009, Wang et al. 2012, Wu et al. 2013).

2.3 *Applications of supercapacitors*

These tailor-made supercapacitors are suitable for everyone. There is high demand of supercapacitors in the market. These devices can not only be used as power buffer and power saving units, but also for energy recovery (Simon and Gogotsi 2010). High dynamic ability of charge propagation in supercapacitors is very useful as hybrid power sources for digital telecommunication systems (e.g., wireless communication, GPS systems and two-way pagers), uninterruptible power supply (UPS), and automotive applications such as automobiles equipped with advanced automotive subsystems, fuel cell vehicles, metro trains, tramways, buses and hybrid electrical vehicles (mainly for engine start, acceleration and braking energy recovery) (Mitra et al. 2001, Hastak et al. 2012, Korenblit et al. 2010, Kwon et al. 2014, Lei et al. 2011, Pandey et al. 2010, Simon and Gogotsi 2010). These are widely used as complementary power sources in pulsed-light generators, medical electronics, electrical utilities, transportation, military defense systems, heavy-load starting assists for diesel locomotives and aerospace applications (Choudhury et al. 2009, Hastak et al. 2012, Pandey et al. 2010). These capacitors can also work as power sources in portable electronic devices such as personal digital assistants (PDAs), digital cameras, power tools, toys, digital clocks, video recorder and mobile phones (Endo et al. 2001, Simon and Gogotsi 2010). Besides that, supercapacitors can sustain low drain-rate memories and work as semiconductor memory back-up systems in microprocessors (Choudhury et al. 2009, Endo et al. 2001). They can be used for load levelling in stop and go traffic in automotive applications to solve the problem of electrical power steering (Simon and Gogotsi 2010). Supercapacitors are also applied in emergency doors on the Airbus A380 in recent years (Simon and Gogotsi 2010). Moreover, supercapacitors can be particularly important for levelling the subsecond disturbances in power lines which cost billions of dollars to the world economy because of their excellent charge-discharge rate without altering their energy storage characteristics (Korenblit et al. 2010).

2.4 *Electrodes materials*

The common electrode material for EDLC is carbon. These carbon-based materials are activated carbon powder, carbon black, carbon nanotubes (CNTs), graphite, carbon fibre, carbon fabrics, carbon composites, carbon aerogel and carbon monoliths (Frackowiak and Béguin 2001, Hashmi et al. 1997, Lei et al. 2013, Wang et al. 2012). Carbon is abundant in this world and we can obtain it from the renewable natural resources. The agriculture and biomass materials such as coal, pitch coke, wood, rice husk, coconut husk, walnut shell, waste tyre, plant bark, fruit peel, fruit shell, fungi, vegetable waste and chicken waste can act as precursors to produce activated carbon (Elmouwahidi et al. 2012, Gupta and Gupta 2015, Peng et al. 2008, Peng et al. 2013, Wang et al. 2012). Activation pre-treatment on the existing carbon materials is needed to produce porous activated carbon. Several extensive methods such as laser ablation, electrical arc, chemical-vapor decomposition (CVD), nanocasting, chemical or physical activation and heat treatment under nitrogen atmosphere, carbon dioxide or steam flux can be used to activate the porous carbon for different applications (Lv et al. 2012, Peng et al. 2008).

In order to design a high performance supercapacitor, the choice of electrode should be taken into account. The basic criteria of the electrodes are:

- high specific capacitance
- large rate capability
- superior cycle stability
- low toxicity
- cost effective (Hashmi and Upadhyaya 2002)

Carbonaceous materials are the most preferable materials because of their low cost, abundant, non-toxic, low environmental impact, excellent cycle stability, good conductivity, large specific area and easy processability (Peng et al. 2008). Other advantages are well polarizable, excellent electrochemical properties due to its amphoteric behavior, easy accessibility of ions, high power capability, chemically stable in different solution (from acidic to basic), maintenance-free, wide operating temperature with excellent performances, superior physicochemical stability and well-developed pore-size distribution (Fang and Binder 2006, Frackowiak and Béguin 2001, Elmouwahidi et al. 2012, Lei et al. 2013, Wu et al. 2013).

2.4.1 Activated carbon (AC)

Porous carbon electrode is normally composed of activated carbon powder, conductive additives such as carbon black and/or carbon nanotubes and a polymer binder with a metal current collector. The materials used for the electrode preparation are activated carbon, CNTs and carbon black in this present work. Activated carbon (AC) is a predominant electrode material used in EDLCs because of its attractive properties such as large specific surface area (1000–2500 m^2g^{-1}), high porosity, chemical inertness, good mechanical stability, low mass density, excellent thermal stability and inexpensiveness (Frackowiak 2007, Kwon et al. 2014, Lei et al. 2011, Ruan et al. 2014).

2.4.2 *Carbon nanotubes (CNTs)*

There is a variety of carbon allotropes, such as buckyminsterfullerene, carbon nanotubes, graphene, amorphous carbon, lonsdaleite and nanodiamond (Wang et al. 2012). High microporosity (pore dimension: < 2 nm) of activated carbon could limit the accessibility of charge carriers into the micropores of carbon due to its low mesoporosity (Lu et al. 2011). Therefore, larger charge carriers have difficulty for the diffusion into the smaller pores of carbon (Frackowiak 2007, Kumar et al. 2012). Hence, CNTs are introduced to mix with AC to allow the electrolyte accessibility as CNTs have mesoporous structure (pore size: 2–50 nm) which enhances ion accumulation properties through its unique entanglement network onto the bigger pores of carbon (An et al. 2002). Carbon nanotubes-based electrode offers several advantages, for example, good electrical conductivity, large specific surface area (100–1315 m^2g^{-1}), superior mechanical stability, excellent electrical properties, high dimensional ratio, low mass density, superior charge-discharge capability and better chemical stability with well-defined hollow core shape (Amitha et al. 2009, An et al. 2002, Chang et al. 2013, Emmenegger et al. 2003, Lu et al. 2011, Peng et al. 2008, Portet et al. 2005). CNTs have been recognized as potential materials also owing to their superior chemical stability, high flexibility, low resistivity, narrow distribution of mesopores (or high mesoporosity) and high mechanical properties as well as good adsorption characteristic (Amitha et al. 2009, Emmenegger et al. 2003, Kim et al. 2012, Li et al. 2012, Lu et al. 2011, Wang et al. 2005). Combining high surface area of AC with the high mesoporosity of CNTs is a desirable way to attain a balanced surface area and mesoporosity on the surface of carbon electrode and ultimately improve the capacitive performances of an EDLC.

3. Description of Electrolytes

3.1 Poly(acrylic acid) (PAA)

PAA is a non-toxic, hydrophilic and biocompatible superabsorbent polymer with three dimensional (3-D) network (Boonsin et al. 2012, Shan et al. 2009, Tang et al. 2011). PAA is chosen as host polymer in this research due to its fascinating behaviors:

- excellent stability in acidic and basic media (Gu and Wang 2011, Shaikh et al. 2011a, Yin et al. 2010)
- high ionic conductivity (Gu and Wang 2011, Shaikh et al. 2011a, Yin et al. 2010)

- strong adhesive properties (Gu and Wang 2011, Shaikh et al. 2011b, Yin et al. 2010)
- superior selectivity and permeability (Gu and Wang 2011, Shaikh et al. 2011b, Yin et al. 2010)
- high ability to associate with a variety of multivalent metal ions in solution (Gu and Wang 2011, Shaikh et al. 2011b, Yin et al. 2010)
- suppress the crystallization (Dasenbrock et al. 1998, Huang et al. 1998, Shaikh et al. 2011b)
- form stable complexes with metal (Dasenbrock et al. 1998, Huang et al. 1998, Shaikh et al. 2011b)

The main reason for selecting PAA as host polymer is because of its high charge density. PAA has carboxylic (–COOH) functional group on the polymer backbone. This functional group favors the bond formation, for example, ionic, covalent, hydrogen and coordination which can be used to form complexation with the nanoparticles (Ohya et al. 1994, Shaikh et al. 2011b). This is very useful in this research as nano-sized fillers are incorporated in the polymer electrolytes preparation. The ionization of carboxylic group in PAA mainly depends on pH and ionic strength (Elliott et al. 2004, Moscoso-Londoño et al. 2013). The charge density of PAA is low in acidic media due to the poor degree of dissociation. The ions can just be dissociated well in the solution at higher pH value (pH > 5). This is the reason why water is used as solvent in this polymer electrolyte preparation.

3.2 *Lithium bis(trifluoromethanesulfonyl)imide (LiTFSI)*

Lithium bis(trifluoromethane)sulfonimide (LiTFSI) consists of lithium cation and bis(trifluromethanesulfonyl) imide anion. The imide anion is stabilized by two trifluoromethanesulfonyl (triflic) groups. The imide anion is very stable due to the delocalization of the formal negative charge. The occurrence of this delocalization is a result of combination of the inductive effect by the electron withdrawing group and the conjugated structure. In other words, the strong electron-withdrawing behavior of triflic groups, and the conjugation between triflic group and the lone electron pair on the nitrogen favors the delocalization process of this negative charge.

The extensive delocalized electrons in $TFSI^-$ anion can promote the ion dissociation and thereby increase the ionic conductivity by weakening the interactions between alkali metals and nitrogen with the polyether oxygen (Ramesh and Lu 2008). The attempt of using LiTFSI in this work is because of its non-corrosive behavior towards electrodes, wide electrochemical stability, excellent thermal stability and superior thermal properties. Apart

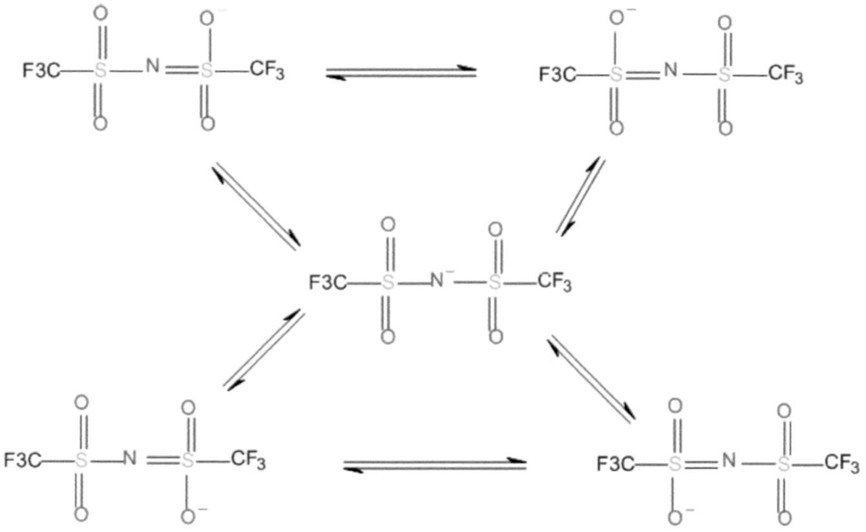

Figure 1. Resonance structures of bis(trifluoromethane)sulfonimide (TFSI) anions.

from that, this salt can dissociate very well even in low dielectric solvents. It is a new designed metal salt to replace the poor conducting lithium triflate (LiTf), the hazardous lithium perchlorate ($LiClO_4$), the thermally unstable $LiBF_4$ and lithium hexafluorophosphate ($LiPF_6$), and the toxic lithium hexafluoroarsenate ($LiAsF_6$) (Kang 2004).

3.3 *Barium titanate (IV) ($BaTiO_3$)*

Barium titanate (IV) ($BaTiO_3$) is one type of crystalline ferroelectric materials with perovskite-type (ABO_3) structure (He et al. 2009, Sun et al. 1999). $BaTiO_3$ can be prepared from decomposition of ethylene diamine modified titanium (IV) isopropoxide and barium hydroxide [$Ba(OH)_2$] in aqueous solution (Li et al. 2008). This filler has been widely used in advanced electronic devices and energy power system such as multilayer ceramic capacitors, chemical sensors and nonvolatile memories backup system due to its superb dielectric ($\varepsilon = 10^2$ to 10^5) and ferroelectric characteristics (He et al. 2009, Sun et al. 1999). $BaTiO_3$ was chosen because of its attractive properties such as bimetallic oxide structure, large surface-to-volume ratio, high stability, reduced interfacial resistance between electrolyte and electrode, and significant Lewis acid character (Forsyth et al. 2002, He et al. 2009, Li et al. 2001).

4. Methodology

4.1 Materials

Poly(acrylic acid) (PAA), lithium bis(trifluoromethanesulfonyl)imide (LiTFSI) and nano-sized barium titanate ($BaTiO_3$) were used as polymer, salt and fillers, respectively to prepare NCPEs. PAA (Sigma-Aldrich, USA, molecular weight of 3000000 $gmol^{-1}$), LiTFSI (Sigma-Aldrich, USA), $BaTiO_3$ (dielectric constant of 150, cubic crystalline phase with particle size < 100 nm from Aldrich, USA) were used as received.

4.2 Sample preparation

PAA-based NCPEs were prepared by solution casting method. PAA was initially dissolved in distilled water. Appropriate amount of LiTFSI was subsequently mixed in PAA solution. The weight ratio of PVA:LiTFSI was kept at 70:30 as this ratio achieved the maximum ionic conductivity in the preliminary step. Different mass ratio of nano-filler was thus doped into the PAA-LiTFSI aqueous solution to prepare NCPEs. The resulting solution was placed and heated in a sonicator for 1 hour at 70°C to make sure well dispersion of filler into the solution. The solution was thus stirred thoroughly and heated at 80°C for several hours. The solution was eventually cast in a glass Petri dish and dried in an oven overnight at 60°C to obtain a free-standing polymer electrolyte film. The appearance of the polymer electrolyte is displayed below.

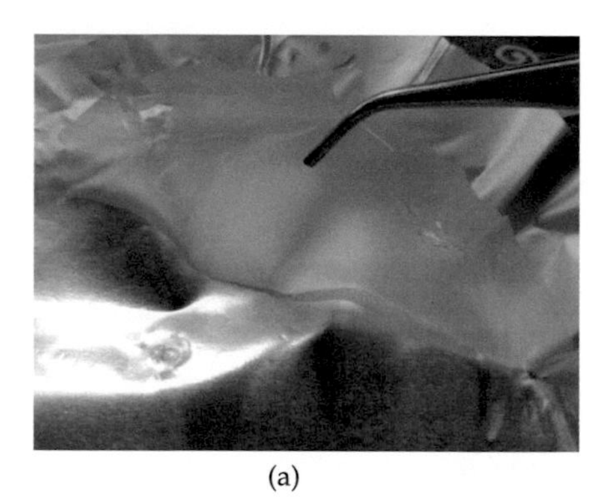

(a)

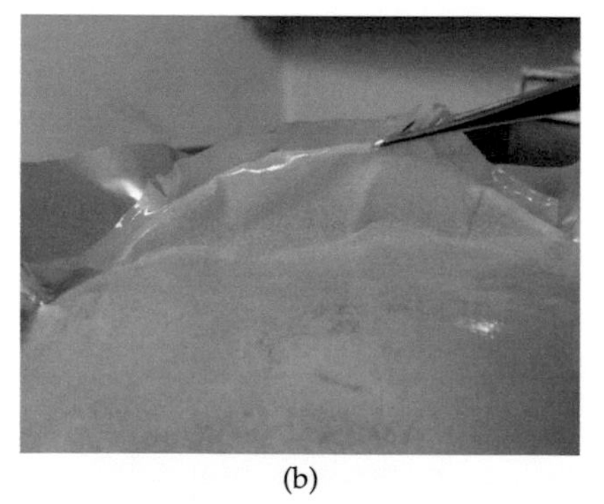

(b)

Figure 2. The image of (a) filler-free polymer electrolytes and (b) the most conducting nano-composite polymer electrolyte.

4.3 *Sample characterization*

4.3.1 *Differential scanning calorimetry (DSC)*

DSC analysis was performed using TA Instrument Universal Analyzer 200 which consists of a DSC Standard Cell FC as main unit and Universal V4.7A software. The whole analysis was analyzed under nitrogen atmosphere at a flow rate of 60 ml min^{-1}. Samples weighing 3–5 mg were hermetically sealed in an aluminium Tzero pan. A tiny hole was punched on top of the pan to eliminate the water and moisture which would be released during the heating process. In contrast, an empty aluminium pan was hermetically sealed as reference cell. The samples were heated from 25°C to 110°C at a heating rate of 20°C min^{-1} to remove any trace amount of water and moisture as a preliminary step. The heating process was maintained at 110°C for 2 minutes to ensure the complete evaporation. After that, the samples were cooled rapidly to –100°C at a cooling rate of 20°C min^{-1}. Beyond this preliminary step, the samples were heated to 250°C and maintained for 2 minutes. The samples were cooled to –100°C thereafter at the pre-set heating rate. This heat-cool process is repeated for 3 times to reach the equilibrium state. Glass transition temperature (T_g) was evaluated using the final scan with the provided software.

4.3.2 *Ambient temperature-ionic conductivity and temperature dependent-ionic conductivity studies*

Freshly prepared NCPEs were subjected to AC-impedance spectroscopy for ionic conductivity determination. A digital micrometer screw gauge was used to measure the thickness of the samples. The impedance of the polymer electrolytes was measured using the HIOKI 3532–50 LCR HiTESTER impedance analyzer over the frequency range between 50 Hz and 5 MHz at ambient temperature. Continuous heating was then applied onto the samples from room temperature to 120°C to examine the temperature dependent conductivity study. The measurement was taken by sandwiching the NCPE between two stainless steel (SS) blocking electrodes at a signal level of 10 mV. The filler-free and the highest conducting NCPE of each system were subjected to the linear sweep voltammetry (LSV) study and EDLC fabrication. The bulk ionic conductivity of polymer electrolytes was determined by using the equation below.

$$\sigma = \frac{l}{R_b A} \qquad \text{(Equation 1)}$$

where l is the thickness (cm), R_b is bulk resistance (Ω) and A is the known surface area (cm^2) of the polymer electrolytes. The semicircle fitting of the impedance plot was accomplished to obtain R_b value. R_b of the thin electrolytes film was calculated from extrapolation of the semicircular region on Z real axis (Z').

4.3.3 Linear sweep voltammetry (LSV)

CHI600D electrochemical analyzer was used to evaluate LSV responses of the samples. These cells were analyzed at a scan rate of 10 mVs^{-1} by placing the polymer electrolyte between the two SS electrodes in the potential range of ±3V.

4.4 Electrode preparation

Activated carbon-based EDLC electrodes were prepared by dip coating technique. The preparation of carbon slurry was performed by mixing 80 wt% activated carbon (Kuraray Chemical Co. Ltd., Japan), having particle size between 5–20 µm, and surface area between 1800–2000 m^2g^{-1}, with 5 wt.% carbon black (Super P), 5 wt.% multi-walled carbon nanotubes (CNTs) (Aldrich, USA) with outer diameter between 7–15 nm and length, L ranging from 0.5 to 10 µm and 10 wt.% poly(vinylidene fluoride) (PVdF) binder (molecular weight of 534000 $gmol^{-1}$ from Aldrich) and dissolving them in 1-methyl-2-pyrrolidone (Purity ≥ 99.5% from Merck, Germany). Activated carbon was initially treated with sodium hydroxide (NaOH) and sulphuric acid (H_2SO_4) to increase the porosity of carbon. This slurry was stirred thoroughly for several hours at ambient temperature. The obtained carbon slurry was then applied on an aluminium electrode. The coated electrodes were subsequently dried in an oven at 110°C for drying purposes. After that, the electrolyte (liquid form) was spin-coated on the surface of the electrodes for 10 times with spinning time of 60 s at 2000 rpm. The spin coating technique is applied to ensure that carbon pores are filled with the electrolytes. In addition, it helps to enhance the interfacial contact between electrolyte and electrode.

4.5 Electrical double layer capacitors (EDLCs) fabrication

EDLC cell was assembled in the configuration of electrode/polymer electrolyte/electrode. The EDLC cell configuration was eventually placed in a cell kit for further electrochemical analyses.

4.6 Electrical double layer capacitors (EDLCs) characterization

The fabricated EDLC cell was subsequently subjected to cyclic voltammetry (CV) and galvanostatic charge-discharge (GCD) measurement.

4.6.1 Cyclic voltammetry (CV)

The CV study of EDLC was investigated using CHI600D electrochemical analyzer. The cell was rested for 2 seconds prior to the measurement to achieve an equilibrium state. The EDLC cell was then evaluated at 10 mVs^{-1} scan rate in the potential range between 0 and 1 V at intervals of 0.001 V. The specific capacitance (C_{sp}) of EDLC was computed using the following equation:

$$C_{sp} = \frac{i}{Sm} \ (\mathrm{F\ g^{-1}}) \qquad \text{(Equation 2)}$$

$$C_{sp} = \frac{i}{SA} \ (\mathrm{Fcm^{-2}}) \qquad \text{(Equation 3)}$$

where i is the average anodic-cathodic current (A), S is the potential scan rate (Vs^{-1}), m refers to the average mass of active materials and A represents surface area of the electrodes, that is 1 cm^{-2}. In this work, the average mass of the electrode materials was around 0.01 g.

4.6.2 Galvanostatic charge-discharge analysis (GCD)

The charge-discharge study was carried out using a Neware battery cycler. EDLC was charged and discharged at the current of 1 mA. EDLC was allowed to rest for 10 minutes before commencement of the measurements. The specific discharge capacitance (C_{sp}) was obtained from charge-discharge curves, according to the following relation:

$$C_{sp} = \frac{I}{m\left({}^{dV}/_{dt}\right)} \qquad \text{(Equation 4)}$$

where I is the applied current (A), m is the average mass of electrode materials (including the binder and carbon black), dV represents the potential change of a discharging process excluding the internal resistance drop occurring at the beginning of the cell discharge and dt is the time interval of discharging process. The dV/dt is determined from the slope of the discharge curve. The mass of the electrode used in this study was 0.01 g. Energy density, E (W h kg^{-1}), power density, P (kW kg^{-1}) and Coulombic efficiency, η (%) were assessed using the equations below:

$$E = \frac{C_{sp} \times (dV)^2}{2} \times \frac{1000}{3600} \qquad \text{(Equation 5)}$$

$$P = \frac{I \times dV}{2 \times m} \times 1000 \qquad \text{(Equation 6)}$$

$$\eta = \frac{t_d}{t_c} \times 100 \qquad \text{(Equation 7)}$$

where t_d and t_c are the discharging and charging times, respectively.

5. Results and Discussion

5.1 Differential scanning calorimetry (DSC)

Figure 3 depicts the DSC curve of PAA, filler-free polymer electrolytes and nanocomposite polymer electrolytes.

PAA shows glass transition temperature (T_g) at 131°C. The T_g of polymer electrolyte is expected to be lower upon addition of LiTFSI as this salt has plasticizing effect. This can be seen in Fig. 3 where the obtained T_g of polymer electrolyte is 122°C by dispersing LiTFSI into PAA. The T_g of polymer electrolytes is further decreased with the addition of nano-scaled $BaTiO_3$. The T_g of polymer electrolytes is reduced to 110°C and 103°C upon addition of 2 wt.% and 8 wt.% of $BaTiO_3$, respectively. This observation denotes that fillers can act as solid plasticizer which provides the plasticizing effect to the polymer matrix. Fillers can ease the ion conduction mechanism with the aid of plasticizing effect by softening the polymer chains in the

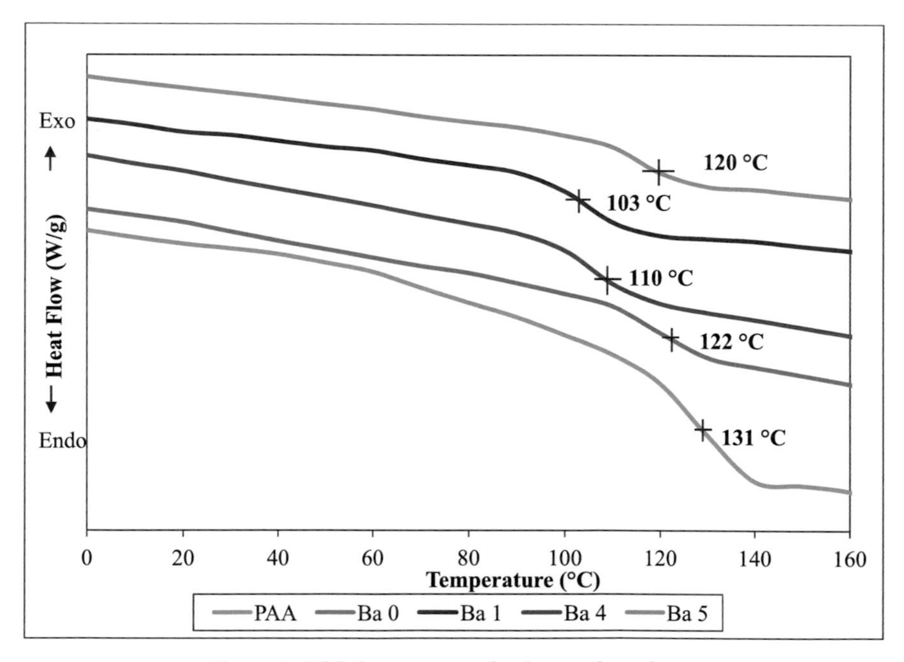

Figure 3. DSC thermogram of polymer electrolytes.

matrix. This plasticizing effect can produce highly flexible polymer chains. Consequently, the ions (also known as charge carriers) are dissociated and can migrate easily. However, the T_g of polymer electrolytes is increased with higher dosage of $BaTiO_3$. We opine that it is because of the excess of the filler particles. Fillers tend to agglomerate when the nano-sized $BaTiO_3$ particles are in excessive form. This fillers agglomeration is prone to form entanglement between the polymer chains. Hence, higher temperature is needed to break all the entangled chains. That is why we observed higher T_g in the polymer electrolytes containing high mass loading of $BaTiO_3$.

5.2 *Ambient temperature-ionic conductivity studies*

Figure 4 illustrates the ionic conductivity of the polymer electrolytes with doping of different mass fraction of $BaTiO_3$ at room temperature. The ionic conductivity of polymer electrolytes is increased at two orders of magnitude from 1.04×10^{-6} S cm^{-1} to 1.30×10^{-4} S cm^{-1} upon addition of 2 wt.% of nano-sized $BaTiO_3$. The ionic conductivity of polymer electrolytes is increased up to a maximum level of 5×10^{-4} S cm^{-1} with increasing weight percent of fillers further.

The enhancement of ionic conductivity of nanocomposite polymer electrolytes is related to effective-medium-theory (EMT) model. This EMT approach implies the existence of a conductive space charge layer at the interface between the electrolyte and filler. The highly conductive layer in

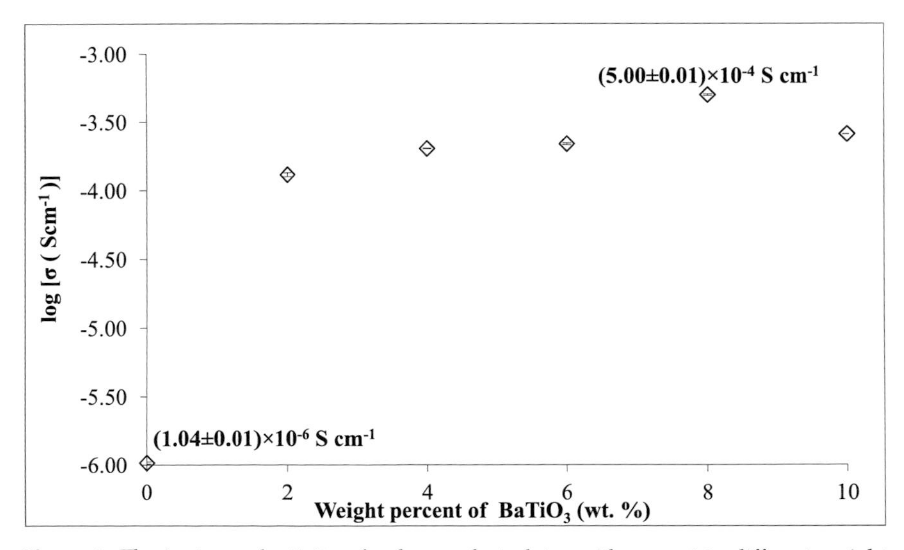

Figure 4. The ionic conductivity of polymer electrolytes with respect to different weight percent of $BaTiO_3$ at ambient temperature.

the neighborhood of grain boundaries arising from the Lewis acid-base interaction is produced when filler grains are dispersed in the electrolyte (Sun et al. 2000). The permanent dipole moments in the tetragonal phase of nanoferroelectric materials form a strong Lewis acid-base reaction and thus extend the conducting pathway through its stronger Lewis acid-base interaction. We suggest that the electrostatic forces originating from the dipole moments of the grains can weaken the Li^+-O^- in the polymer matrix (Sun et al. 2000). As a result, the charge carriers can be transported easily in the vicinity of the grain boundary through this conducting trail.

Moreover, fillers can act as solid plasticizer to improve the ionic conductivity of ion conductors. The decrease in ionic conductivity is also attributed to the lower T_g of the polymer electrolytes upon the addition of $BaTiO_3$. We imply that the fillers can soften the polymer membrane by decreasing the T_g. Therefore, the polymer chains become more flexible and hence promote the charge carrier dissociation from the native coordination bonding. This ion dissociation can produce more mobile ions which are the carriers for the transportation and conduction mechanism. It is expected that the ionic conductivity of nanocomposite polymer electrolytes is higher than filler-free polymer electrolyte. The ionic conductivity of polymer electrolytes is decreased when we add the filler further into the polymer electrolyte system. This is due to the agglomeration of the filler particles which inhibits the ion conducting pathway. In addition, the particles tend to interact with the polymer chains at high mass fraction of fillers. So, this interaction will form entanglement of polymer chains. The ion dissociation will be harder in the entangled chains. Apart from that, the charge carriers will have hurdle in transportation in the intertwined polymer chains as the entangled polymer chains can impede the ion conduction process.

5.3 *Temperature dependent-ionic conductivity studies*

Figure 5 portrays the temperature dependent-ionic conductivity plot from room temperature to 120°C. The temperature dependent ionic conductivity plot was fitted with Arrhenius equation. The ionic conductivity is expressed as follows in this thermally activated principle:

$$\sigma = A \exp\left(\frac{-E_a}{kT}\right) \qquad \text{(Equation 8)}$$

where A is a constant which is proportional to the amount of charge carriers, E_a is activation energy, k is Boltzmann constant, that is 8.6173×10^{-5} eV K^{-1} and T represents the absolute temperature in K.

There is no abrupt change in ionic conductivity with respect to the temperature and the regression value of all the plots is close to unity. Therefore, all these well-fitted plots reflect that the polymer electrolytes obey Arrhenius theory which indicates the ionic hopping mechanism for the ion transport in the electrolytes. We suggest that dipole moments of the grains can weaken the Li^+-O^- bonding. The charge carriers (refer to lithium cations) are initially dissociated from coordination bond with TFSI anions due to the strong electron delocalization. Thus, the charge carriers interact with the electron withdrawing group (O^-) in PAA. As a result, the charge carriers are detached from the interactive bond due to the disturbance from the dipole moment of the nano-sized particles. The charge carriers could transport to an empty vacant site to generate the ionic hopping mechanism in the electrolyte.

Activation energy (E_a) for the ionic hopping mechanism was then evaluated. The E_a of sample Ba 0 is 0.16 eV. The E_a value is reduced to 0.15 eV, 0.12 eV, 0.11 eV and 0.08 eV upon addition of 2 wt.%, 4 wt.%, 6 wt.% and 8 wt.% $BaTiO_3$ respectively. The most conducting polymer electrolyte shows the lowest E_a among all the samples. This explains why sample Ba 4 has the highest ionic conductivity. The lowest E_a value denotes that sample Ba 4 requires lesser energy for ion hopping mechanism. So, the ionic transport in the electrolyte is favorable in sample Ba 4. However, the E_a value is increased to 0.11 eV by adding 10 wt.% of $BaTiO_3$. This is correlated to the agglomeration of the nano-sized fillers which hinders the ionic transport in the electrolyte. Therefore, more energy is needed to break and reform the coordination bond.

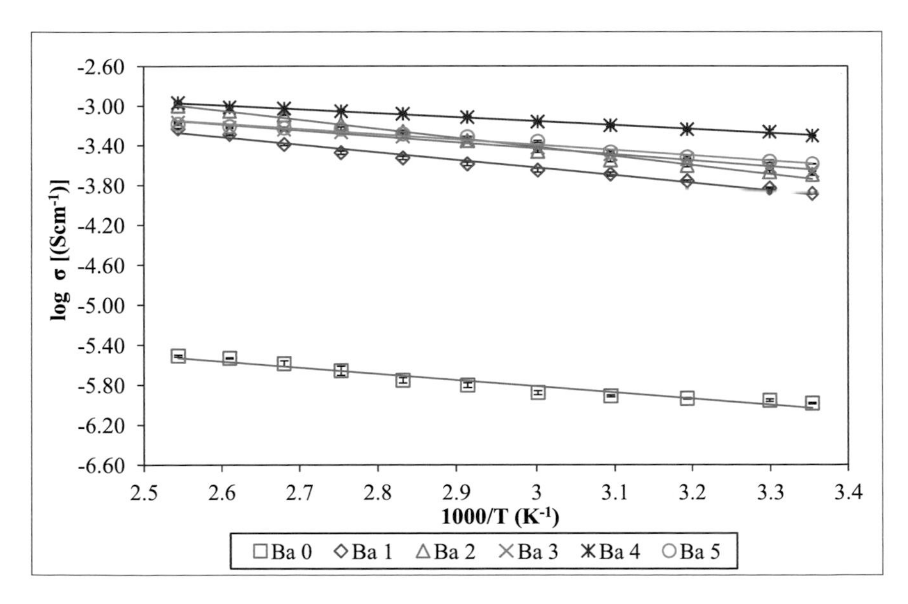

Figure 5. Arrhenius theory-fitted temperature dependent ionic conductivity plot of the polymer electrolytes from room temperature to 120°C.

5.4 Linear sweep voltammetry (LSV)

Figure 6 depict the LSV range of filler-free polymer electrolyte and the most conducting filler-doped polymer electrolyte. Polymer electrolyte without addition of filler shows LSV range from –2.2 V to 2.2 V. Upon incorporation of filler, the electrochemical potential window of polymer electrolyte is

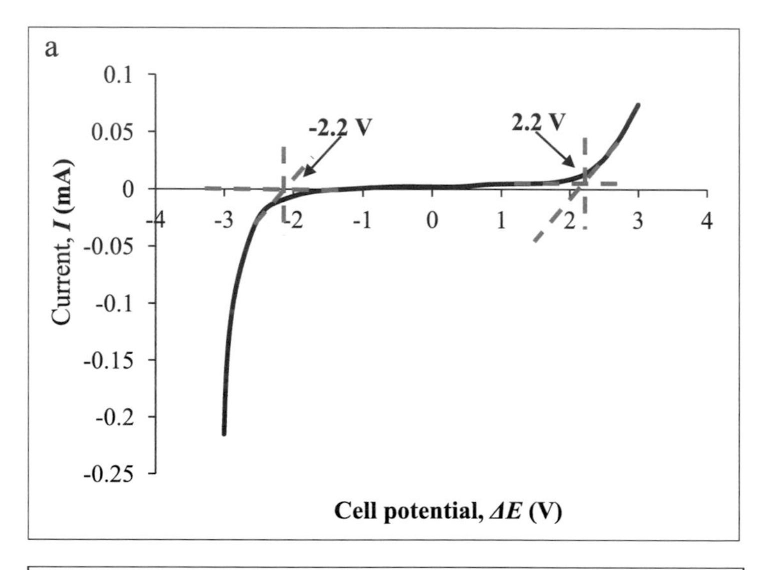

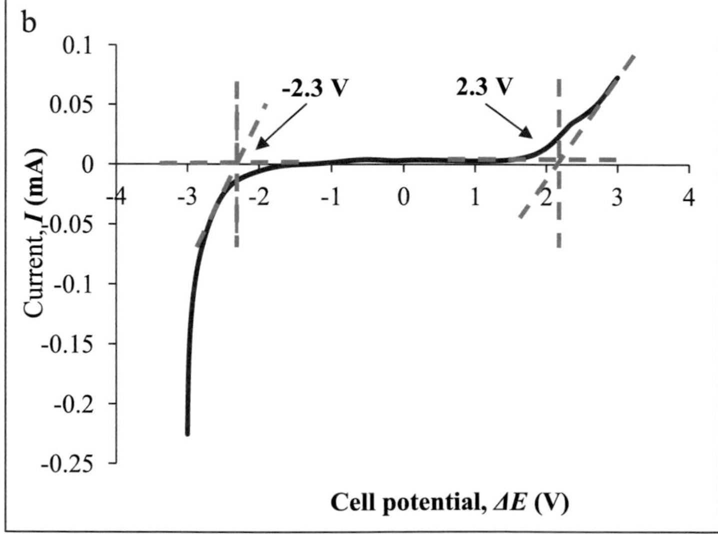

Figure 6. (a) LSV range of $BaTiO_3$-free polymer electrolyte. (b) LSV range of the most conducting sample $BaTiO_3$-added polymer electrolyte.

expanded to 4.6 V, ranging from –2.3 V to 2.3 V. The effect of addition of filler on the potential window of polymer electrolytes is insignificant in this current work.

5.5 *Cyclic voltammetry (CV)*

Figure 7 portray the EDLCs containing filler-free polymer electrolyte and the most conducting filler-doped polymer electrolyte. The cyclic voltammetry of sample Ba 0-based EDLC shows non-ideal rectangular shape with specific capacitance of 13.46 F g^{-1} (or equivalent to 15 mF cm^{-2}).

The CV shape approaching to the ideal rectangular shape is observed for the EDLC based on polymer electrolyte with the most conducting polymer electrolyte. The specific capacitance of EDLC is almost doubled up which is around 25 F g^{-1} (or equivalent to 29 mF cm^{-2}) upon dispersion of 8 wt.% of $BaTiO_3$. The enhancement of specific capacitance in the EDLC is attributed to the higher ionic conductivity of the electrolyte. Capacitance of an EDLC cell arises from the charge accumulation at the boundary of electrode and electrolyte. The working principle of ion conductivity of a polymer electrolyte is the dissociation and reattachment of the charge carriers (or ions) in the electrolyte. Therefore, ions can move freely in the electrolyte. Rapid ion transportation in the most conducting nanocomposite polymer electrolyte also promotes the ion adsorption at the electrode-electrolyte interface which gives rise to higher capacitance of EDLC. The increase in the capacitance of the EDLC with the most conducting polymer electrolyte might be related to lower interfacial resistance between the electrolyte and electrode. The nanocomposite polymer electrolyte has sticky behavior due to the presence of plasticizing effect of the filler. As a result, the contact at the electrode-electrolyte is intimate. Therefore, the charge carriers can be built up easily at the interface between electrode and electrolyte.

5.6 *Galvanostatic charge-discharge analyses (GCD)*

The EDLC cell was charged and discharged galvanostatically in order to measure the electrochemical stability of the cell. The finding of the GCD performance is presented in Fig. 8.

Symmetrical ideal shape of the GCD curve is observed in Fig. 8. The cell is charged at 0.19 V instead of 0 V. This implies the occurrence of the internal resistance (also known as ohmic loss). The ohmic loss arises from the electrode and electrolyte layers, such as charge transfer resistance and bulk resistance of polymer electrolyte (Arof et al. 2012, Mitra et al. 2001). The discharge specific capacitance of EDLC for the first cycle of charge-

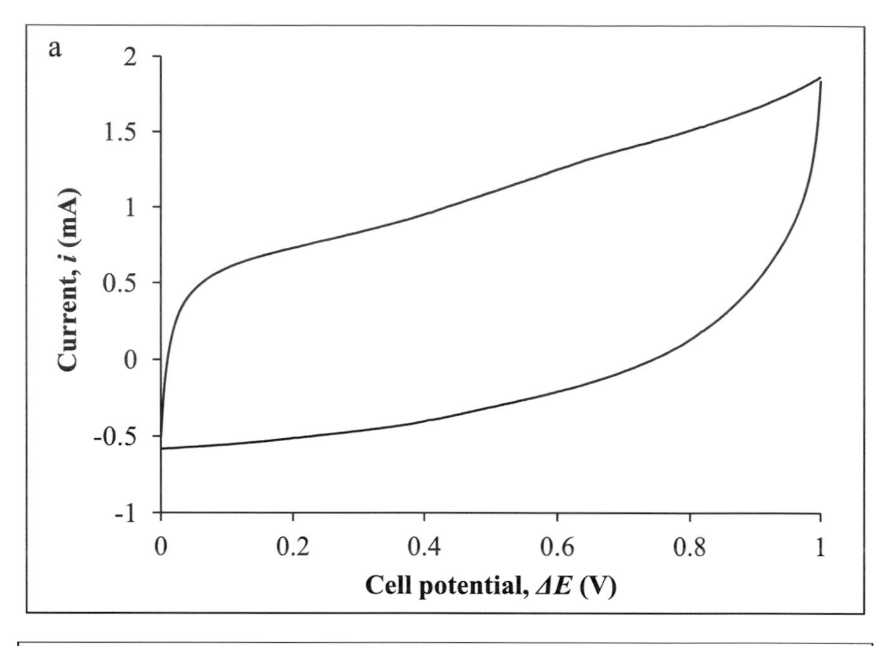

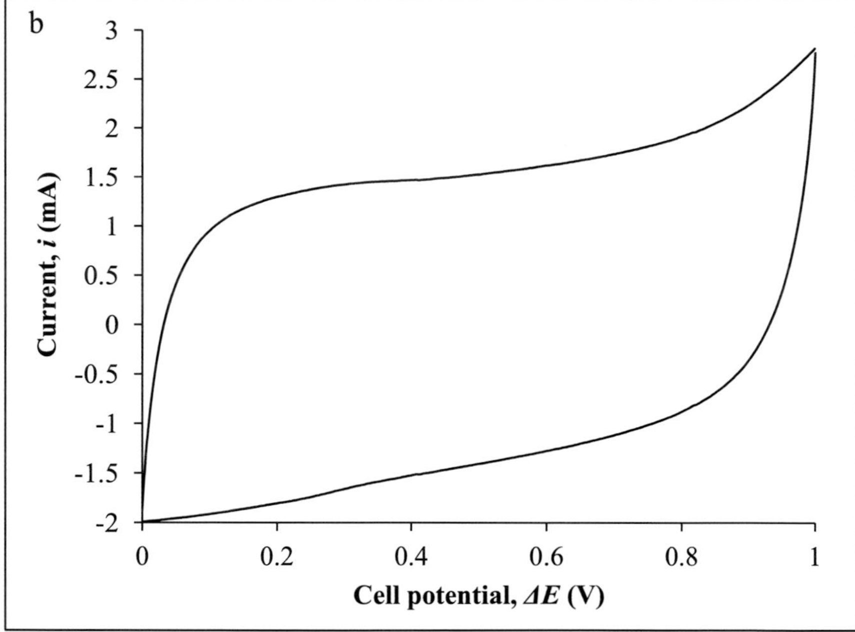

Figure 7.(a) Cyclic voltammetry of Ba 0-based EDLC. (b) Cyclic voltammetry of Ba 4-based EDLC.

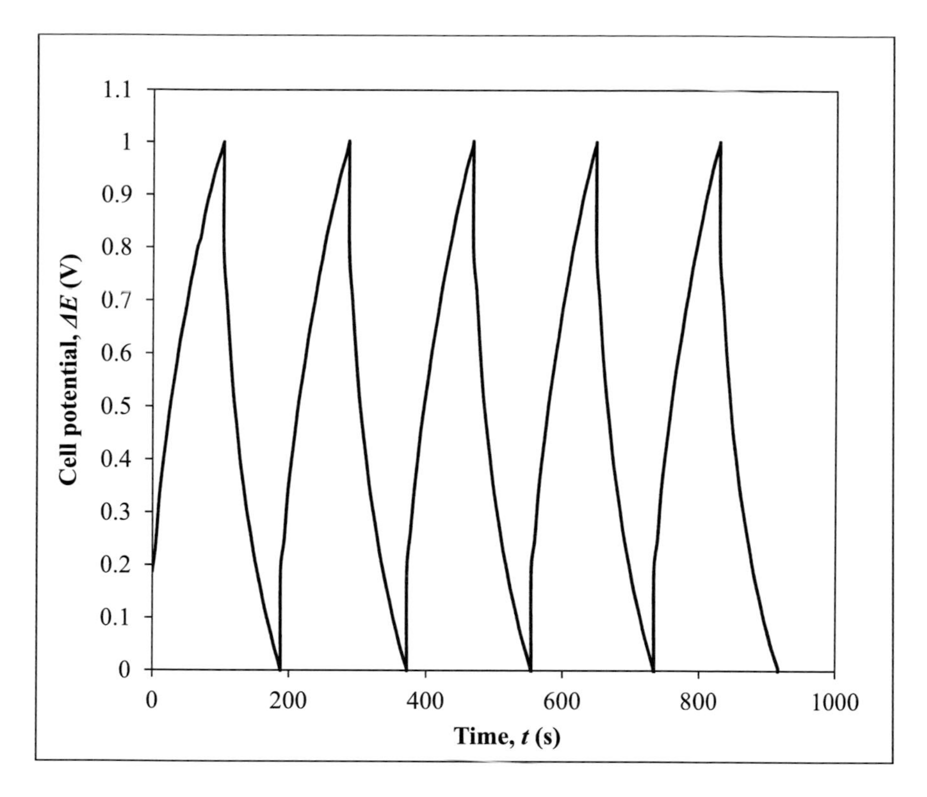

Figure 8. Galvanostatic charge-discharge performance of EDLC cell using the most conducting polymer electrolyte over first five cycles of charge and discharge processes.

discharge is 25.54 F g^{-1}. The cell also shows an energy density of 2.23 $Whkg^{-1}$ and power density of 67.91 Wkg^{-1} with Coulombic efficiency of 86%.

The reliability of the fabricated cell was also studied by charging and discharging for 20000 cycles. Figures 9 and 10 depict the electrochemical stability of the EDLC over 20,000 cycles.

It is great to observe that the Coulombic efficiency of EDLC remains at its level at above 90%. This indicates that the cell used almost the same time for charging and discharging processes. Beyond the 1st cycle of charging and discharging, the specific capacitance, energy density and power density are increased up to a maximum level. All these parameters are expected to decrease after the cell undergoes charging and discharging processes. However, different observation is attained in GCD curve. The cell shows a moderate increase in all these parameters from 1000 cycles to 3000 cycles. The EDLC achieves the highest specific discharge capacitance of 34.22 F g^{-1}, energy density of 3.32 $Whkg^{-1}$ and power density of 71.47 Wkg^{-1} upon charging and discharging for 3000 cycles. The result is then accompanied

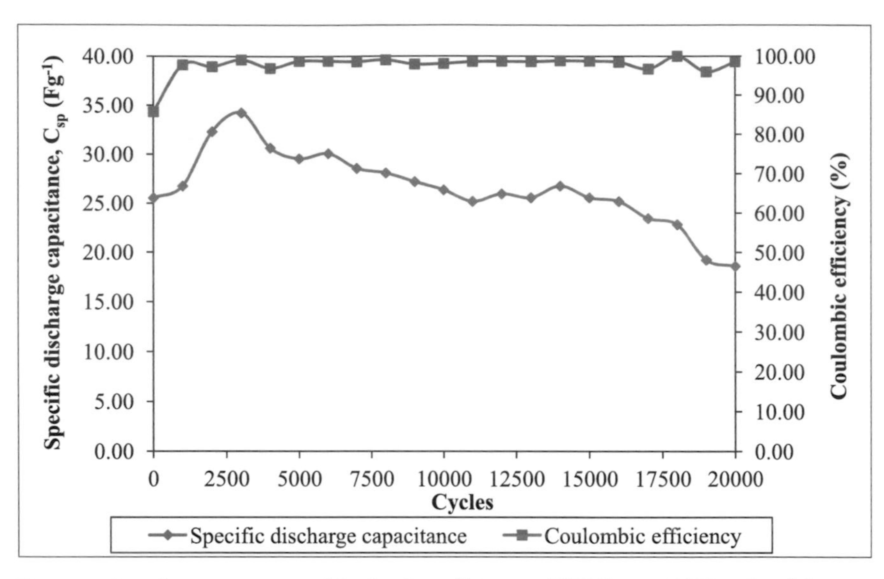

Figure 9. Specific capacitance and Coulombic efficiency of EDLC over 20000 cycles of charging and discharging processes.

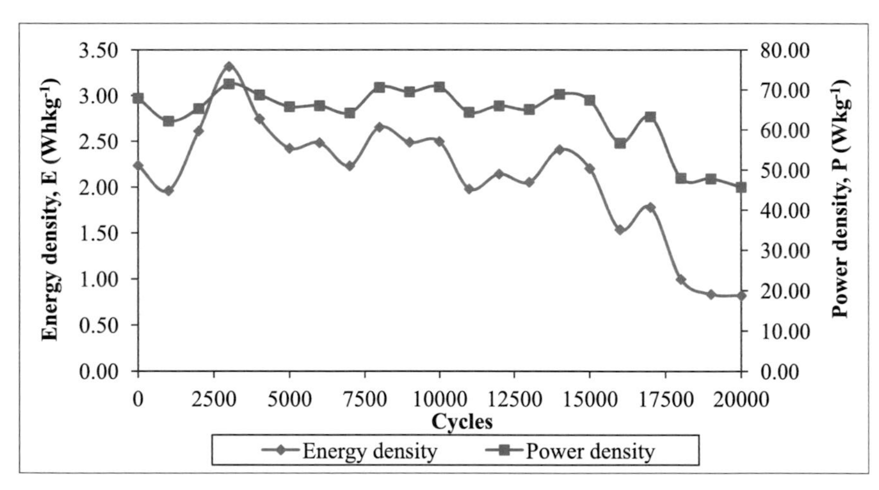

Figure 10. Energy density and power density of EDLC over 20000 cycles of charging and discharging processes.

by a gradual decrease thereafter. Even though there's fluctuation in specific capacitance, energy density and power density as illustrated within the 20,000 cycles in Figs. 9 and 10; however, there is no abrupt change in all these parameters. Long cycle life is the main feature of an EDLC compared with pseudocapacitor and hybrid capacitors. Hence, $BaTiO_3$-doped polymer electrolyte is a great material to be used in EDLC application as it can exhibit a long cycle life.

6. Conclusions

Polymer electrolytes comprising of PAA and LiTFSI were prepared using solution casting technique. Ferroelectric material, $BaTiO_3$ is a promising candidate as filler for increasing the ionic conductivity of polymer electrolytes. The ionic conductivity was enhanced by two orders of magnitude, that is from 1.04 µS cm^{-1} to 0.5 mS cm^{-1} upon addition of 8 wt.% of $BaTiO_3$. The polymer electrolytes follow the Arrhenius theory which is associated with ionic hopping mechanism. $BaTiO_3$-added polymer electrolytes illustrate good thermal and electrochemical properties. Doping of fillers can reduce the T_g of polymer electrolytes as fillers act as solid plasticizer. The most conducting polymer electrolyte can be operated up to 4.6 V. Electric double layer capacitors were assembled using polymer electrolytes (filler-free polymer electrolyte and the most conducting electrolyte) and carbon-based electrodes. The EDLC containing filler-free polymer electrolyte showed a capacitance value of 13.46 Fg^{-1}. The capacitance was then increased extensively to 25 Fg^{-1} with the addition of filler into the electrolyte as shown in the CV plot. The cyclability test of EDLC was also scrutinized to investigate the electrochemical stability of the cell over 20000 cycles of charging and discharging processes. This nanocomposite polymer electrolyte is a good candidate to be used as an electrolyte in EDLC device because the specific capacitance of the cell decreased gradually upon 20000 charging cycles. The cell can still remain stable at high number of cycles.

7. Future Aspects

Incorporation of fillers can improve the ionic conductivity of polymer electrolytes; however, the enhancement of ionic conductivity is still restricted. Ionic liquid, a new additive, can be added into the polymer electrolytes to boost the ionic conductivity effectively. Ionic liquid is a type of molten salt that can increase the ionic conductivity greatly without affecting the thermal and electrochemical properties of the electrolyte. Strong plasticizing effect is the main feature of this environmental friendly material. The ionic liquids can be integrated in the polymer network like liquid electrolytes. This is why the ionic liquid-added polymer electrolytes can perform well, in terms of ionic conductivity, thermal behavior and electrochemical properties. In addition, liquid crystal is the newest type of material to be used in the polymer electrolyte preparation. The aligned liquid crystal can provide a superior charge carriers pathway inside the polymer electrolytes. Nevertheless, the ionic conduction mechanism in the liquid crystal-added polymer electrolytes has not been fully investigated and understood. A more extensive research is needed to be done to make liquid crystal a new highly ion conductive organic material for use in the electrochemical devices.

Apart from electrolyte, electrode materials can be investigated for future aspects. Growing high quality carbon materials from these low-valued carbon sources opens an effective way to convert the waste carbon sources into a high value-added product. The conversion from solid wastes to carbon sources is an alternative way to reduce the waste problem and protect the environment. The use of biomass materials to produce porous carbon is an interesting topic when costs, wide availability and environmental concern are considered. Carbon can be derived from residues and wastes such as dried leaves, fruit peels, fruit shells, plants and hair to produce sustainable and renewable energy systems. Several raw materials such as rice husk, coconut husk, waste tyre, plant bark, vegetable waste and chicken waste can be used for the synthesis of activated carbon. The synthesis of carbon from waste is an alternative way to solve the environmental crisis and produce reliable, stable, and sustainable large-scale use of renewable energy.

8. Acknowledgement

Author gratefully acknowledges the contribution from Professor Dr. Ramesh T. Subramaniam from Centre for Ionics University of Malaya, University of Malaya, Malaysia in supporting this work.

References

Adebahr, J., N. Byrne, M. Forsyth, D.R. MacFarlane and P. Jacobsson. 2003. Enhancement of ion dynamics in PMMA-based gels with addition of TiO_2 nano-particles. Electrochimi. Acta 48: 2099–2103.

Ahmad, S., S. Ahmad and S.A. Agnihotry. 2005. Nanocomposite electrolytes with fumed silica in poly(methyl methacrylate): Thermal, rheological and conductivity studies. J. Power Sources 140: 151–156.

Ahmad, S., M. Deepa and S.A. Agnihotry. 2008. Effect of salts on the fumed silica-based composite polymer electrolytes. Sol. Energy Mater. Sol. Cells 92: 184–189.

Amitha, F.E., A.L.M. Reddy and S. Ramaprabhu. 2009. A non-aqueous electrolyte-based asymmetric supercapacitor with polymer and metal oxide/multiwalled carbon nanotube electrodes. J. Nanopart. Res. 11: 725–729.

An, K.H., K.K. Jeon, J.K. Heo, S.C. Lim, D.J. Bae and Y.H. Leez. 2002. High–capacitance supercapacitor using a nanocomposite electrode of single–walled carbon nanotube and polypyrrole. J. Electrochem. Soc. 149: 1058–1062.

Armand, M.B. 1986. Polymer electrolytes. Ann. Rev. Mater. Sci. 16: 245–261.

Arof, A.K., M.Z. Kufian, M.F. Syukur, M.F. Aziz, A.E. Abdelrahman and S.R. Majid. 2012. Electrical double layer capacitor using poly(methyl methacrylate)-C_4BO_8Li gel polymer electrolyte and carbonaceous material from shells of mata kucing (Dimocarpus longan) fruit. Electrochim Acta 74: 39–45.

Baskaran, R., S. Selvasekarapandian, N. Kuwata, J. Kawamura and T. Hattori. 2007. Structure, thermal and transport properties of PVAc-$LiClO_4$ solid polymer electrolytes. J. Phys. Chem. Solids 68: 407–412.

Bhide, A. and K. Hariharan. 2007. Ionic transport studies on $(PEO)_6$:$NaPO_3$ polymer electrolyte plasticized with PEG_{400}. Eur. Polym. J. 43: 4253–4270.

Boonsin, R., J. Sudchanham, N. Panusophon, P. Sae-Heng, C. Sae-Kung and P. Pakawatpanurut. 2012. Dye-sensitized solar cell with poly(acrylic acid-co-acrylonitrile)-based gel polymer electrolyte. Mater. Chem. Phys. 132: 993–998.

Bouridah, A., F. Dalard, D. Deroo, H. Cheradame and J.F. LeNest. 1985. Poly(dimethylsiloxane)-poly(ethylene oxide) based polyurethane networks used as electrolytes in lithium electrochemical solid state batteries. Solid State Ionics 15: 233–240.

Bruce, P.G. and C.A. Vincent. 1993. Polymer electrolytes. J. Chem. Soc. Faraday Transactions 89: 3187–3203.

Chang, H.-Y., H.-C. Chang and K.-Y. Lee. 2013. Characteristics of NiO coating on carbon nanotubes for electric double layer capacitor application. Vac. 87: 164–168.

Cheng, H., C. Zhu, B. Huang, M. Lu and Y. Yang. 2007. Synthesis and electrochemical characterization of PEO-based polymer electrolytes with room temperature ionic liquids. Electrochim. Acta 52: 5789–5794.

Choi, H.-J., S.-M. Jun, J.-M. Seo, D.W. Chang, M. Dai and J.-B. Baek. 2012. Graphene for energy conversion and storage in fuel cells and supercapacitors. Nano Energy 1: 534–551.

Choudhury, N.A., S. Sampath and A.K. Shukla. 2009. Hydrogel-polymer electrolytes for electrochemical capacitors: An overview. Energy Environ. Sci. 2: 55–67.

Deng, L., G. Zhang, L. Kang, Z. Lei, C. Liu and Z.-H. Liu. 2013. Graphene/VO_2 hybrid material for high performance electrochemical capacitor. Electrochim. Acta 112: 448–457.

Dasenbrock, C.O., T.H. Ridgway, C.J. Seliskar and W.R. Heineman. 1998. Evaluation of the electrochemical characteristics of a poly(vinyl alcohol)/poly(acrylic acid) polymer blend. Electrochimi. Acta 43: 3497–3502.

Elliott, J.E., M. Macdonald, J. Nie and C.N. Bowman. 2004. Structure and swelling of poly(acrylic acid) hydrogels: effect of pH, ionic strength, and dilution on the crosslinked polymer structure. Polym. 45: 1503–1510.

Elmouwahidi, A., Z. Zapata-Benabithe, F. Carrasco-Marín and C. Moreno-Castilla. 2012. Activated carbons from KOH-activation of argan (Argania spinosa) seed shells as supercapacitor electrodes. Bioresour. Technol. 111: 185–190.

Emmenegger, C., P. Mauron, P. Sudan, P. Wenger, V. Hermann, R. Gallay et al. 2003. Investigation of electrochemical double-layer (EDLC) capacitors electrodes based on carbon nanotubes and activated carbon materials. J. Power Sources 124: 321–329.

Endo, M., T. Takeda, Y.J. Kim, K. Koshiba and K. Ishii. 2001. High power electric double layer capacitor (EDLC's); from operating principle to pore size control in advanced activated carbons. Carbon Sci. 1: 117–128.

Fan, L., C.-W. Nan and S. Zhao. 2003. Effect of modified SiO_2 on the properties of PEO-based polymer electrolytes. Solid State Ionics 164: 81–86.

Fang, B. and L. Binder. 2006. A novel carbon electrode material for highly improved EDLC performance. J. Phys. Chem. B 110: 7877–7882.

Fenton, D.E., J.M. Parker and P.V. Wright. 1973. Complexes of alkali metal ions with poly(ethylene oxide). Polym. 14: 589.

Frackowiak, E. and F. Béguin. 2001. Carbon materials for the electrochemical storage of energy in capacitors. Carbon 39: 937–950.

Frackowiak, E. 2007. Carbon materials for supercapacitor application. Phys. Chem. Chem. Physics 9: 1774–1785.

Forsyth, M., D.R. MacFarlane, A. Best, J. Adebahr, P. Jacobsson and A.J. Hill. 2002. The effect of nano-particle TiO_2 fillers on structure and transport in polymer electrolytes. Solid State Ionics 147: 203–211.

Ganesan, S., B. Muthuraaman, V. Mathew, J. Madhavan, P. Maruthamuthu and S.A. Suthanthiraraj. 2008. Performance of a new polymer electrolyte incorporated with diphenylamine in nanocrystalline dye-sensitized solar cell. Sol. Energy Mater. Sol. Cells 92: 1718–1722.

Giffin, G.A., M. Piga, S. Lavina, M.A. Navarra, A. D'Epifanio, B. Scrosati et al. 2012. Characterization of sulfated-zirconia/Nafion composite membranes for proton exchange membrane fuel cells. J. Power Sources 198: 66–75.

Gray, F.M. 1991. Solid polymer electrolytes: Fundamentals of technological applications: Wiley-VCH, United Kingdom.

Gray, F.M. 1997. Polymer electrolytes. The Royal Society of Chemistry, United Kingdom.

Gu, X. and G. Wang. 2011. Interfacial morphology and friction properties of thin PEO and PEO/PAA blend films. Appl. Surf. Sci. 257: 1952–1959.

Gupta, H. and B. Gupta. 2015. Adsorption of polycyclic aromatic hydrocarbons on banana peel activated carbon. Desalin. Water Treat. 1–2.

Han, H.S., H.R. Kang, S.W. Kim and H.T. Kim. 2002. Phase separated polymer electrolyte based on poly(vinyl chloride)/poly(ethyl methacrylate) blend. J. Power Sources 112: 461–468.

Hammami, R., Z. Ahamed, K. Charradi, Z. Beji, I.B. Assaker, J.B. Naceur et al. 2013. Elaboration and characterization of hybrid polymer electrolytes Nafion-TiO_2 for PEMFCs. Int. J. Hydrogen Energy 38: 11583–11590.

Hashmi, S.A., R.J. Latham, R.G. Linford and W.S. Schlindwein. 1997. Studies on all solid state electric double layer capacitors using proton and lithium ion conducting polymer electrolytes. J. Chem. Soc. Faraday Trans. 93: 4177–4182.

Hashmi, S.A. and H.M. Upadhyaya. 2002. Polypyrrole and poly(3-methyl thiophene)-based solid state redox supercapcitors using ion conducting polymer electrolyte. Solid State Ionics 152-153: 883–889.

Hastak, R.S., P. Sivaraman, D.D. Potphode, K. Shashidhara and A.B. Samui. 2012. All solid supercapacitor based on activated carbon and poly[2,5-benzimidazole] for high temperature application. Electrochim. Acta 59: 296–303.

Ibrahim, S., S.M.M. Yasin, R. Ahmad and M.R. Johan. 2012. Conductivity, thermal and morphology studies of PEO based salted polymer electrolytes. Solid State Sci. 14: 1111–1116.

He, X., C. Hu, Y. Xi, B. Wan and C. Xia. 2009. Electroless deposition of $BaTiO_3$ nanocubes for electrochemical sensing. Sensor Actuat. B 137: 62–66.

Huang, J., Q. Guo, H. Ohya and J. Fang. 1998. The characteristics of crosslinked PAA composite membrane for separation of aqueous organic solutions by reverse osmosis. J. Membr. Sci. 144: 1–11.

Imrie, C.T. and M. Ingram. 2000. Bridging the gap between polymer electrolytes and inorganic glasses: side group liquid crystal polymer electrolytes. Mol. Cryst. Liq. Cryst. 347: 199–210.

Ingram, M.D., A.J. Pappin, F. Delalande, D. Poupard and G. Terzulli. 1998. Development of electrochemical capacitors incorporating processable polymer electrolytes. Electrochim. Acta 43: 1601–1605.

Ishkov, A.V. and A.M. Sagalakov. 2005. Effect of active fillers on properties of heat-resistant composites. Russ. J. Appl. Chem. 78: 1512–1516.

Jain, N., A. Kumar, S. Chauhan and S.M.S. Chauhan. 2005. Chemical and biochemical transformations in ionic liquids. Tetrahedron 61: 1015–1060.

Jeon, J.-D., M.-J. Kim and S.-Y. Kwak. 2006. Effect of addition of TiO_2 nanoparticles on mechanical properties and ionic conductivity of solvent-free polymer electrolytes based on porous P(VdF-HFP)/P(EO-EC) membranes. J. Power Sources 162: 1304–1311.
Jian-hua, T., G. Peng-fei, Zhang, Zhi-yuan, L. Wen-hui and S. Zhong-qiang. 2008. Preparation and performance evaluation of a Nafion-TiO_2 composite membrane for PEMFCs. Int. J. Hydrogen Energy 33: 5686–5690.
Jiang, J., G. Tan, S. Peng, D. Qian, J. Liu, D. Luo et al. 2013. Electrochemical performance of carbon-coated $Li_3V_2(PO_4)_3$ as a cathode material for asymmetric hybrid capacitors. Electrochim. Acta 107: 59–65.
Jung, S., D.W. Kim, S.D. Lee, M. Cheong, D.Q. Nguyen, B.W. Cho et al. 2009. Fillers for solid-state polymer electrolytes: Highlight. Bull. Korean Chem. Soc. 30: 2355–2361.
Ketabi, S. and K. Lian. 2013. Effect of SiO_2 on conductivity and structural properties of PEO-$EMIHSO_4$ polymer electrolyte and enabled solid electrochemical capacitors. Electrochim. Acta 103: 174–178.
Kim, K.M., N.-G. Park, K.S. Ryu and K.S. Chang. 2002. Characterization of poly (vinylidenefluoride-co-hexafluoropropylene)-based polymer electrolytes filled with TiO_2 nanoparticles. Polym. 43: 3951–3957.
Kim, K.S., S.Y. Park, S. Choi and H. Lee. 2006. Ionic liquid-polymer gel electrolytes based on morpholinium salt and PVdF(HFP) copolymer. J. Power Sources 155: 385–390.
Kim, B.C., S.J. Kim, J.K. Chung, J. Chen, S.Y. Park and G.G. Wallace. 2012. Charge storage in carbon nanotube-TiO_2 hybrid nanoparticles. Synth. Met. 162: 650–654.
Korenblit, Y., M. Rose, E. Kockrick, L. Borchardt, A. Kvit, S. Kaskel et al. 2010. High-rate electrochemical capacitors based on ordered mesoporous silicon carbide-derived carbon. ACS Nano 4: 1337–1344.
Krawiec, W., Jr., L.G. Scanlon, J.P. Fellner, R.A. Vaia, S. Vasudevan and E.P. Giannelis. 1995. Polymer nanocomposites: A new strategy for synthesizing solid electrolytes for rechargeable lithium batteries. J. Power Sources 54: 310–315.
Kumar, Y., G.P. Pandey and S.A. Hashmi. 2012. Gel polymer electrolyte based electrical double layer capacitors: Comparative study with multiwalled carbon nanotubes and activated carbon electrodes. J. Phys. Chem. C 116: 26118–26127.
Kwon, S.H., E. Lee, B.-S. Kim, S.-G. Kim, B.-J. Lee, M.-S. Kim et al. 2014. Activated carbon aerogel as electrode material for coin-type EDLC cell in organic electrolyte. Curr. Appl. Phys. 14: 603–607.
Kuila, T., H. Acharya, S.K. Srivastava, B.K. Samantaray and S. Kureti. 2007. Enhancing the ionic conductivity of PEO based plasticized composite polymer electrolyte by $LaMnO_3$ nanofiller. Mater. Sci. Eng. B 137: 217–224.
Lei, C., P. Wilson and C. Lekakou. 2011. Effect of poly(3,4-ethylenedioxythiophene) (PEDOT) in carbon-based composite electrodes for electrochemical supercapacitors. J. Power Sources 196: 7823–7827.
Lei, C., F. Markoulidis, Z. Ashitaka and C. Lekakou. 2013. Reduction of porous carbon/Al contact resistance for an electric double-layer capacitor (EDLC). Electrochim. Acta 92: 183–187.
Lewandowski, A., M. Zajder, E. Frąckowiak and F. Béguin. 2001. Supercapacitor based on activated carbon and polyethylene oxide-KOH-H_2O polymer electrolyte. Electrochim. Acta 46: 2777–2780.
Liew, C.-W., Y.S. Ong, J.Y. Lim, C.S. Lim, K.H. Teoh and S. Ramesh. 2013. Effect of ionic liquid on semi-crystalline poly(vinylidene fluoride-co-hexafluoropropylene) solid copolymer electrolytes. Inter. J. Electrochem. Sci. 8: 7779–7794.
Liew, C.-W. and S. Ramesh. 2013. Studies on ionic liquid-based corn starch biopolymer electrolytes coupling with high ionic transport number. Cellulose 20: 3227–3237.

Li, Q., H.Y. Sun, Y. Takedaa, N. Imanishi, J. Yang and O. Yamamoto. 2001. Interface properties between a lithium metal electrode and a poly(ethylene oxide) based composite polymer electrolyte. J. Power Sources 94: 201–205.

Li, J., J. Claude, L.E. Norena-Franco, S.I. Seok and Q. Wang. 2008. Electrical energy storage in ferroelectric polymer nanocomposites containing surface-functionalized $BaTiO_3$ nanoparticles. Chem. Mater. 20: 6304–6306.

Li, Z.J., T.X. Chang, G.Q. Yun and Y. Jia. 2012. Coating single walled carbon nanotube with SnO2 and its electrochemical properties. Powder Technol. 224: 306–310.

Lu, W., R. Hartman, L. Qu and L. Dai. 2011. Nanocomposite Electrodes for high-performance supercapacitors. J. Phys. Chem. Lett. 2: 655–660.

Lue, S.J., D.-T. Lee, J.-Y. Chen, C.-H. Chiu, C.-C. Hu, Y.C. Jean and J.-Y. Lai. 2008. Diffusivity enhancement of water vapor in poly(vinyl alcohol)-fumed silica nano-composite membranes: Correlation with polymer crystallinity and free-volume properties. J. Membr. Sci. 325: 831–839.

Lv, Y., L. Gan, M. Liu, W. Xiong, Z. Xu, D. Zhu et al. 2012. A self-template synthesis of hierarchical porous carbon foams based on banana peel for supercapacitor electrodes. J. Power Sources 209: 152–157.

Michael, M.S., M.M.E. Jacob, S.R.S. Prabaharan and S. Radhakrishana. 1997. Enhanced lithium ion transport in PEO-based solid polymer electrolytes employing a novel class of plasticizers. Solid State Ionics 98: 167–174.

Mitra, S., A.K. Shukla and S. Sampath. 2001. Electrochemical capacitors with plasticized gel-polymer electrolytes. J. Power Sources 101: 213–219.

Moscoso-Londoño, O., J.S. Gonzalez, D. Muraca, C.E. Hoppe, V.A. Alvarez, A. López-Quintela et al. 2013. Structural and magnetic behavior of ferrogels obtained by freezing thawing of polyvinyl alcohol/poly(acrylic acid) (PAA)-coated iron oxide nanoparticles. Eur. Polym. J. 49: 279–289.

Nicotera, I., G.A. Ranieri, M. Terenzi, A.V. Chadwick and M.I. Webster. 2002. A study of stability of plasticized PEO electrolytes. Solid State Ionics 146: 143–150.

Ning, W., Z. Xingxiang, L. Haihui and H. Benqiao. 2009. 1-Allyl-3-methylimidazolium chloride plasticized-corn starch as solid biopolymer electrolytes. Carbohydr. Polym. 76: 482–484.

Noto, V.D., M. Bettiol, F. Bassetto, N. Boaretto, E. Negro, S. Lavina et al. 2012. Hybrid inorganic-organic nanocomposite polymer electrolytes based on Nafion and fluorinated TiO_2 for PEMFCs. Inter. J. Hydrogen Energy 37: 6169–6181.

Ohya, H., M. Shibata, Y. Negish, Q.H. Guo and H.S. Choi. 1994. The effect of molecular weight cut-off of PAN ultrafiltration support layer on separation of water-ethanol mixtures through pervaporation with PAA-PAN composite membrane. J. Membr. Sci. 90: 91–100.

Osinska, M., M. Walkowiak, A. Zalewska and T. Jesionowski. 2009. Study of the role of ceramic filler in composite gel electrolytes based on microporous polymer membranes. J. Membrane Sci. 326: 582–588.

Pandey, G.P. and S.A. Hashmi. 2009. Experimental investigations of an ionic-liquid-based, magnesium ion conducting, polymer gel electrolyte. J. Power Sources 187: 627–634.

Pandey, G.P., Y. Kumar and S.A. Hashmi. 2010. Ionic liquid incorporated polymer electrolytes for supercapacitor application. Indian J. Chem. B 49: 743–751.

Pandey, G.P., Y. Kumar and S.A. Hashmi. 2011. Ionic liquid incorporated PEO based polymer electrolyte for electric double layer capacitors: A comparative study with lithium and magnesium systems. Solid State Ionics 190: 93–98.

Pandey, G.P. and S.A. Hashmi. 2013. Ionic liquid 1-ethyl-3-methylimidazolium tetracyanoborate-based gel polymer electrolyte for electrochemical capacitors. J. Mater. Chem. A 1: 3372–3378.

Patel, M., M. Gnanavel and A.J. Bhattacharyya. 2011. Utilizing an ionic liquid for synthesizing a soft matter polymer "gel" electrolyte for high rate capability lithium-ion batteries. J. Mater. Chem. 21: 17419–17424.

Peng, C., S. Zhang, D. Jewell and G.Z. Chen. 2008. Carbon nanotube and conducting polymer composites for supercapacitors. Prog. Nat. Sci. 18: 777–788.

Peng, C., X.-B. Yan, R.-T. Wang, J.-W. Lang, Y.-J. Ou and Q.-J. Xue. 2013. Promising activated carbons derived from waste tea-leaves and their application in high performance supercapacitors electrodes. Electrochim. Acta 87: 401–408.

Pradhan, D.K., B.K. Samantaray, R.N.P. Choudhary and A.K. Thakur. 2005. Effect of plasticizer on structure-property relationship in composite polymer electrolytes. J. Power Sources 139: 384–393.

Polu, A.R. and R. Kumar. 2013. Effect of Al_2O_3 ceramic filler on PEG-based composite polymer electrolytes for magnesium batteries. Adv. Mater. Lett. 4: 543–547.

Portet, C., P.L. Taberna, P. Simon and E. Flahaut. 2005. Influence of carbon nanotubes addition on carbon-carbon supercapacitor performances in organic electrolyte. J. Power Sources 139: 371–378.

Quartarone, E., P. Mustarelli and A. Magistris. 1998. PEO-based composite polymer electrolytes. Solid State Ionics 110: 1–14.

Quartarone, E. and P. Mustarelli. 2011. Electrolytes for solid-state lithium rechargeable batteries: Recent advances and perspectives. Chem. Soc. Rev. 40: 2525–2540.

Raghava, P., X. Zhao, J.-K. Kim, J. Manuel, G.S. Chauhan, J.-H. Ahn et al. 2008. Ionic conductivity and electrochemical properties of nanocomposite polymer electrolytes based on electrospun poly(vinylidenefluoride-co-hexafluoropropylene) with nano-sized ceramic fillers. Electrochim. Acta 54: 228–234.

Raghavan, P., X. Zhao, J. Manuel, G.S. Chauhan, J.H. Ahn, H.S. Ryub et al. 2010. Electrochemical performance of electrospun poly(vinylidene fluoride-co-hexafluoropropylene)-based nanocomposite polymer electrolytes incorporating ceramic fillers and room temperature ionic liquid. Electrochimi. Acta 55: 1347–1354.

Rajendran, S., M. Sivakumar and R. Subadevi. 2004. Investigations on the effect of the various plasticizers in PVA-PMMA solid polymer blend electrolytes. Mater. Lett. 58: 641–649.

Rajendran, S., M.R. Prabhu and M.U. Rani. 2008. Ionic conduction in poly(vinyl chloride)/ poly(ethyl methacrylate)-based polymer blend electrolytes complexed with different lithium salts. J. Power Sources 180: 880–883.

Ramesh, S. and A.K. Arof. 2001. Ionic conductivity studies of plasticized poly(vinyl chloride) polymer electrolytes. Mater. Sci. Eng. B 85: 11–15.

Ramesh, S. and S.C. Lu. 2008. Effect of nanosized silica in poly(methyl methacrylate)-lithium bis(trifluoromethanesulfonyl)imide based polymer electrolytes. J. Power Sources 185: 1439–1443.

Ramesh, S., C.-W. Liew, P.Y. Lau and M. Ezra. 2012. Characterization of high molecular weight poly (vinyl chloride)-lithium tetraborate electrolyte plasticized by propylene carbonate. *In*: Mohammad Luqman (eds.). Recent Advances in Plasticizers. In Tech., Rijeka, Crotia.

Ramesh, S. and L.Z. Chao. 2011. Investigation of dibutyl phthalate as plasticizer on poly(methyl methacrylate)-lithium tetraborate based polymer electrolytes. Ionics 17: 29–34.

Kang, X. 2004. Nonaqueous liquid electrolytes for lithium-based rechargeable batteries. Chem. Rev. 104: 4303–4417.

Ramesh, S., C.-W. Liew and K. Ramesh. 2011. Evaluation and investigation on the effect of ionic liquid onto PMMA-PVC gel polymer blend electrolytes. J. Non-Cryst. Solids 357: 2132–2138.

Ramesh, S. and C.-W. Liew. 2013. Development and investigation on PMMA-PVC blend-based solid polymer electrolytes with LiTFSI as dopant salt. Polym. Bull. 70: 1277–1288.

Reiter, J., J. Vondrak, J. Michalek and Z. Micka. 2006. Ternary polymer electrolytes with 1-methylimidazole based ionic liquids and aprotic solvents. Electrochim. Acta 52: 1398–1408.

Ruan, C., K. Ai and L. Lu. 2014. Biomass-derived carbon materials for high-performance supercapacitor electrodes. RSC Adv. 4: 30887–30895.

Saikia, D., Y.W. Chen-Yang, Y.T. Chen, Y.K. Li and S.I. Lin. 2009. ^{7}Li NMR spectroscopy and ion conduction mechanism of composite gel polymer electrolyte: A comparative study with variation of salt and plasticizer with filler. Electrochim. Acta 54: 1218–1227.
Samir, M.A.S.A., F. Alloin, J.-Y. Sanchez and A. Dufresne. 2005. Nanocomposite polymer electrolytes based on poly(oxyethylene) and cellulose whiskers. Polímeros: Ciência e Tecnologia 15: 109–113.
Sekhon, S.S., P. Krishnan, B. Singh, K. Yamada and C.S. Kim. 2006. Proton conducting membrane containing room temperature ionic liquid. Electrochim. Acta 52: 1639–1644.
Shaikh, J.S., R.C. Pawar, N.L. Tarwal, D.S. Patil and P.S. Patil. 2011a. Supercapacitor behavior of CuO-PAA hybrid films: Effect of PAA concentration. J. Alloy Compd. 509: 7168–7174.
Shaikh, J.S., R.C. Pawar, A.V. Moholkar, J.H. Kim and P.S. Patil. 2011b. CuO-PAA hybrid films: Chemical synthesis and supercapacitor behavior. Appl. Surf. Sci. 257: 4389–4397.
Shan, D., G. Cheng, D. Zhu, H. Xue, S. Cosnier and S. Ding. 2009. Direct electrochemistry of hemoglobin in poly(acrylonitrile-co-acrylic acid) and its catalysis to H_2O_2. Sensor Actuat. B 137: 259–265.
Shao, Y., Z. Yi, F. Lu, F. Deng and B. Li. 2012. Poorly crystalline $Ru_{0.4}Sn_{0.6}O_2$ nanocomposites coated on Ti substrate with high pseudocapacitance for electrochemical supercapacitors. Adv. Chem. Eng. Sci. 2: 118–122.
Simon, P. and Y. Gogotsi. 2010. Charge storage mechanism in nanoporous carbons and its consequence for electrical double layer capacitors. Philos. Trans. R. Soc. A 368: 3457–3467.
Singh, P.K., K.-W. Kim and H.-W. Rhee. 2009. Development and characterization of ionic liquid doped solid polymer electrolyte membranes for better efficiency. Synth. MET. 159: 1538–1541.
Sirisopanaporn, C., A. Fernicola and B. Scrosati. 2009. New, ionic liquid-based membranes for lithium battery application. J. Power Sources 186: 490–495.
Stephan, A.M. and K.S. Nahm. 2006. Review on composite polymer electrolytes for lithium batteries. Polym. 47: 5952–5964.
Sukeshini, A.M., A.R. Kulkarni and A. Sharma. 1998. PEO based solid polymer electrolyte plasticized by dibutyl phthalate. Solid State Ionics 113–115: 179–186.
Sun, H.Y., H.-J. Sohn, O. Yamamoto, Y. Takeda and N. Imanishi. 1999. Enhanced lithium-ion transport in PEO-based composite polymer electrolytes with ferroelectric $BaTiO_3$. J. Electrochem. Soc. 146: 1672–1676.
Sun, H.Y., Y. Takeda, N. Imanishi, O. Yamamoto and H.-J. Sohn. 2000. Ferroelectric materials as a ceramic filler in solid composite polyethylene oxide-based electrolytes. J. Electrochem. Soc. 147: 2462–2467.
Suthanthiraraj, S.A., D.J. Sheeba and B.J. Paul. 2009. Impact of ethylene carbonate on ion transport characteristics of PVdF-$AgCF_3SO_3$ polymer electrolyte system. Mater. Res. Bull. 44: 1534–1539.
Stephan, A.M., Y. Saito, N. Muniyandi, N.G. Renganathan, S. Kalyanasundaram and R.N. Elizabeth. 2002. Preparation and characterization of PVC/PMMA blend polymer electrolytes complexed with $LiN(CF_3SO_2)_2$. Solid State Ionics 148: 467–473.
Tang, Z., Q. Liu, Q. Tang, J. Wu, J. Wang, S. Chen et al. 2011. Preparation of PAA-g-CTAB/PANI polymer based gel-electrolyte and the application in quasi-solid-state dye-sensitized solar cells. Electrochimi. Acta 58: 52–57.
Trasatti, S. and G. Buzzanca. 1971. Ruthenium dioxide: a new interesting electrode material. Solids state structure and electrochemical behaviour. J. Electroanal. Chem. 29: A1–5.
Vioux, A., L. Viau, S. Volland and J.L. Bideau. 2010. Use of ionic liquid in sol-gel; ionogels and applications. Comptes Rendus Chimie 13: 242–255.
Wang, G.–X., B.–L. Zhang, Z.–L. Yu and M.–Z. Qu. 2005. Manganese oxide/MWNTs composite electrodes for supercapacitors. Solid State Ionics 176: 1169–1174.

Wang, L., W. Yang, J. Wang and D.G. Evans. 2009. New nanocomposite polymer electrolyte comprising nanosized $ZnAl_2O_4$ with a mesopore network and PEO-$LiClO_4$. Solid State Ionics 180: 392–397.
Wang, H., C.M.B. Holt, Z. Li, X. Tan, B.S. Amirkhiz, Z. Wu et al. 2012a. Graphene-nickel cobaltite nanocomposite asymmetrical supercapacitor with commercial level mass loading. Nano Res. 5: 605–617.
Wang, R., P. Wang, X. Yan, J. Lang, C. Peng and Q. Xue. 2012b. Promising Porous carbon derived from celtuce leaves with outstanding supercapacitance and CO_2 capture performance. ACS Appl. Mater. Interfaces 4: 5800–5806.
Watanabe, M., S. Nagano, K. Sanui and N. Ogata. 1986. Ion conduction mechanism in network polymers from poly(ethylene oxide) and poly(propylene oxide) containing lithium perchlorate. Solid State Ionics 18 and 19: 338–342.
Wright, V.P. 1975. Electrical conductivity in ionic complexes of poly(ethylene oxide). Brit. Polym. J. 7: 319–327.
Wu, Z.-S., K. Parvez, X. Feng and K. Müllen. 2013. Graphene-based in-plane micro-supercapacitors with high power and energy densities. Nat. Commun. 4: 1–8.
Xi, J., X. Qiu, S. Zheng and X. Tang. 2005. Nanocomposite polymer electrolyte comprising PEO/$LiClO_4$ and solid super acid: Effect of sulphated-zirconia on the crystallization kinetics of PEO. Polym. 46: 5702–5706.
Yang, C.-C., C.-T. Lin and S.-J. Chiu. 2008. Preparation of the PVA/HAP composite polymer membrane for alkaline DMFC application. Desalination 233: 137–146.
Yang, Y., X.-Y. Guo and X.-Z. Zhao. 2012. A novel composite polysaccharide/inorganic oxide electrolyte for high efficiency quasi-solid-state dye-sensitized solar cell. Procedia Engineering 36: 13–18.
Ye, Y.-S., J. Rick and B.-J. Hwang. 2013. Ionic liquid polymer electrolytes. J. Mater. Chem. A 1: 2719–2743.
Yin, Z., R. Cui, Y. Liu, L. Jiang and J.-J. Zhu. 2010. Ultrasensitive electrochemical immunoassay based on cadmium ion-functionalized PSA@PAA nanospheres. Biosens. Bioelectron. 25: 1319–1324.
Yuan, F., H.-Z. Chen, H.-Y. Yang, H.-Y. Li and M. Wang. 2005. PAN-PEO solid polymer electrolytes with high ionic conductivity. Mater. Chem. Phys. 89: 390–394.
Yu, H., J. Wu, L. Fan, Y. Lin, K. Xu and Z. Tang. 2012. A novel redox-mediated gel polymer electrolyte for high-performance supercapacitor. J. Power Sources 198: 402–407.
Yu, Z., L. Tetard, L. Zhai and J. Thomas. 2015. Supercapacitor electrode materials: nanostructures from 0 to 3 dimensions. Energy Environ. Sci. 8: 702–730.
Zapata, P., J.-H. Lee and J.C. Meredith. 2012. Composite proton exchange membranes from zirconium-based solid acids and PVDF/acrylic polyelectrolyte blends. J. Appl. Polym. Sci. 124: E241–E250.
Zhang, J., X. Huang, H. Wei, J. Fu, Y. Huang and X. Tang. 2010. Novel PEO-based solid composite polymer electrolytes with inorganic-organic hybrid polyphosphazene microspheres as fillers. J. Appl. Electrochem. 40: 1475–1481.
Zhang, P., L.C. Yang, L.L. Li, M.L. Ding, Y.P. Wua and R. Holze. 2011. Enhanced electrochemical and mechanical properties of P(VDF-HFP)-based composite polymer electrolytes with SiO_2 nanowires. J. Membrane Sci. 379: 80–85.

5

Nanostructured Electrodes for Emerging Solar Cells

Nadia Shahzad

1. Introduction

The significance of developing new sources of energy is obvious as the global energy needs are expected to be nearly doubled till the year 2050 along with the exponential increase in CO_2 emission. This shocking increase in worldwide energy consumption is due to the rapid global economic expansion, world population growth and ever increasing human dependence on energy appliances. On this planet, the existing sources of energy are generally divided into two wide categories: energy capital and energy income or renewable sources. The capital sources of energy are the sources which once used, cannot be recycled at any time scale less than millions of years whereas the renewable sources are more or less continuously refreshed and are considered to be available for millions of years by keeping their current level of potential. Thus, the renewable sources can provide sustainable energy services based on the routinely accessible native resources. Solar energy is the most dominant form of renewable energy, and can theoretically fulfill the world's energy demand many times.

U.S.-Pakistan Center for Advanced Studies in Energy, National University of Sciences and Technology, Sector H-12, Islamabad, Pakistan.
E-mail: nadia-shahzad@live.com

2. A Brief Overview of Emerging Solar Cells

Since the discovery of photovoltaic effect by Alexandre-Edmond Becquerel, a French physicist, in 1839 (Kalyanasundaram 2010, Mart´i and Luque 2004), the efforts to convert the sunlight directly into electricity have achieved notable progress. The most employed photovoltaic (PV) process to convert light energy into electrical power includes the absorption of light in semiconductors, the generation of electron-hole pairs, and the separation and collection of charge carriers. In the terrestrial applications, the most frequently used solar cells are the bulk-type single or multi-crystalline silicon solar cells. The typical cell structure of this solar cell consists of a thin n-type emitter layer (less than 1 μm) on a thick p-type substrate (about 300 μm) (Kalyanasundaram 2010). Despite the fact that the first silicon (Si) based solar cell was developed in 1950s, the extensive usage of these traditional devices is still limited due to the complexities in adjusting the band gap of Si crystals, high price linked with the fabrication processes and lower power conversion efficiencies. Although the solar cell module manufacturing cost has continuously decreased till now, a drastic reduction of the cell cost together with a radical enhancement of the photon conversion efficiency cannot be expected from traditional materials and solar cell structures. Moreover, a shortage of the high-purity silicon feedstock is predicted in the near future.

Nanoscience and nanotechnology have unlocked the state-of-the-art frontlines to encounter the above mentioned challenges by crafting new materials (zero, one and two-dimensional nanostructures), particularly carbon nanomaterials (CNMs) (Lang et al. 2015, Singh and Nalwa 2015, Wang et al. 2008, Habisreutinger et al. 2014, Yahya 2010), nanostructured semiconducting oxides (Srivastava 2013), 2D materials (Pang et al. 2015, Yang et al. 2015) and quantum dots (QDs) (Tian and Cao 2013) for the resourceful energy conversion and storage. In fact, nanostructured materials offer some unique features including controllable morphology, large surface-to-volume ratio, ease for large area processing, fast charge transport to surface and potential for low-cost fabrication on flexible substrates. Some examples of their applications are the thin film solar cells, the organic and carbon-based solar cells, and the so-called dye/QD/perovskite sensitized solar cells (Green 2003, L´opez et al. 2012, Lee et al. 2016, Park 2015).

The efficient PV process, in all mesoscopic solar cells, requires the nanocrystalline morphology of the semiconductor film. A whole range of nanostructures, from simple assemblies of nanoparticles (Saito and Fujihara 2008, Cicero et al. 2013, Bauer et al. 2002) to nanosheets (Qiu et al. 2010), tetrapods (Chen et al. 2010, Zhang et al. 2009), nanorods (Fujihara et al. 2007, Schlur et al. 2013), nanotubes (Chen and Xu 2009, Lamberti et al. 2013) and branched structures (Sacco et al. 2012), have been tested so far as transparent conductive electrodes (TCEs), electron transporting layers

(ETLs), photoactive layers, electro-catalyst for counter electrodes (CEs), fillers for quasi liquid (QL) electrolytes and the hole transport materials (HTMs). For example, in a dye-sensitized solar cell (DSC) (O'Regan and Grätzel 1991), the use of interpenetrating network junctions is essential to gain efficient performance of the device. A DSC is a photo-electro-chemical system, consisting of a photosensitized anode, an electrolyte and a counter electrode. In a typical DSC, a fluorine-doped tin oxide (FTO) coated glass plate is used as the transparent conducting substrate, on which a highly porous thin layer of titanium dioxide (TiO_2) sensitized with suitable dye is deposited. A conductive sheet (typically platinum film on FTO) covered with a layer of electrolyte (containing iodide/triiodide redox couple) is then used as the counter electrode (Grätzel 2001). A monolayer of dye adsorbed on a flat surface can absorb only a small percent of light because of its smaller optical cross section. Employing multi-layers of the sensitizer cannot suggest a feasible solution for this problem, as those molecules would be photoactive that are in direct contact with the oxide surface. Consequently, the sensitization of flat electrodes has yielded notoriously low conversion efficiencies. The use of nanocrystalline films with high specific surface area is highly recommended to support the adsorption of light absorber and collection of the photo-injected electrons resulting in a vivid enhancement of the performance of sensitized hetero-junction devices.

Another example is quantum dot-sensitized solar cell (QDSC), which utilizes quantum dots (QDs) as absorbing PV material (Tian and Cao 2013, Chen et al. 2010). The band gap of QDs (which is fixed for bulk materials) is tunable and dependent on the size of the dot. This property makes QDs a strong candidate for sensitized solar cells, where varieties of absorber materials are used to improve the efficiency by harvesting a wide range of the solar spectrum. Effective separation of photo-generated electron-hole pairs and their rapid transfer to electrodes are the main hurdle in the development of high performance QDSCs. Carbon based nanomaterials having appropriate band energies, for instance, fullerenes and SWCNTs, are being used in these devices as efficient electron transporting materials (Chen et al. 2010).

Another type of emerging solar cell is based on organic–inorganic metal halide perovskite absorbers which represent the state-of-the-art light harvesting device. Today, in the race of emerging solid state photovoltaics, perovskite solar cells (PSCs) are at the top, achieving record making efficiencies among all the third generation solar cells. It has only been 7 years from 2009, when Miyasaka and co-workers employed the perovskite material for the very first time as sensitizer in a photo-electrochemical cell based DSC architecture, to recent year where the efficiency of this evolving type of solar cells was boosted from 3.8% (Kojima et al. 2009) to 20% (Zhou et al. 2014, Yang et al. 2015). The representative device structure of this cell consists of a photoactive layer (methylammonium lead halide perovskite

[$CH_3NH_3(MA)PbX_3$, X = I^-, Br^-, or Cl^-]), electron-/hole-transporting layers and charge collecting electrodes. These cells exhibit potential to compete with commercially available solar cells, if the issues regarding their long-term stability can be resolved. Several researchers are working on this task and proposing novel ideas such as: compositional engineering of perovskite materials (Burschka et al. 2013), employing hydrophobic passivation over the organic HTM layer (Hwang et al. 2015) or carbon nanomaterial based hydrophobic polymeric composites as HTM or nanostructured inorganic HTMs to protect perovskite layer from moisture contact (Habisreutinger et al. 2014), and using metal oxide nanostructure based HTMs with excellent ambient stability (Kim et al. 2015). PSCs can be classified into three major types based on the solar cell structure.

Mesoscopic sensitized configuration is similar to a traditional DSC and it was used for the fabrication of the first perovskite based solar cell (Kojima et al. 2009), where perovskite ($CH_3NH_3PbI_3$-$MAPbI_3$) nanocrystals were used as the sensitizer (light absorber) in place of the dye in a device based on liquid electrolyte. The device achieved a photon conversion efficiency (PCE) of 3.8%. Later on, Im and co-workers (Im et al. 2011) sensitized the mesoporous TiO_2 films with $MAPbI_3$ quantum dots, leading to an improved liquid electrolyte based device with an efficiency of 6.5%. However, the life of such liquid electrolyte based PSCs was only a few minutes due to rapid degradation of the perovskite material in contact with the electrolyte. In this scenario, another group employed a solid-state hole transport material (spiro-MeOTAD) to fabricate the first high performance PSC which gave a record breaking PCE value of 9.7%. This state-of-the-art device with improved stability was based on nanoporous TiO_2 film sensitized by $MAPbI_3$ using one step process (Kim et al. 2012). Subsequently, Graetzel et al. reported a relatively cheaper p-type polymer, poly-triarylamine (PTAA), as a remarkable hole conducting material for perovskite-sensitized solar cells, achieving a PCE value up to 12% (Heo et al. 2013). Nevertheless the efficiency of PSCs, fabricated with the widely adopted one-step deposition, is limited as the precipitation of perovskite nanocrystals remains uncontrollable. Hence, the produced devices demonstrated a broad distribution in performance because of the wide variations in the perovskite crystal morphology. A sequential deposition method was introduced to resolve these issues. In this case, at first PbI_2 was deposited on the porous TiO_2 layer and was then transformed into the perovskite light absorber by reacting with methylammonium iodide (MAI). The resulting device, three-dimensional bulk-heterojunction structures (3D-BHJ) of TiO_2/perovskite was formed as the perovskite material fully penetrated into the pores of the TiO_2 film. The two-step sequential deposition approach has led to the modest PSCs reaching a PCE as high as 15% (Burschka et al. 2013). The device configurations are schematically shown in Fig. 1 for a typical mesoscopic sensitized PSC (part a) and a 3D-BHJ sensitized PSC (part b).

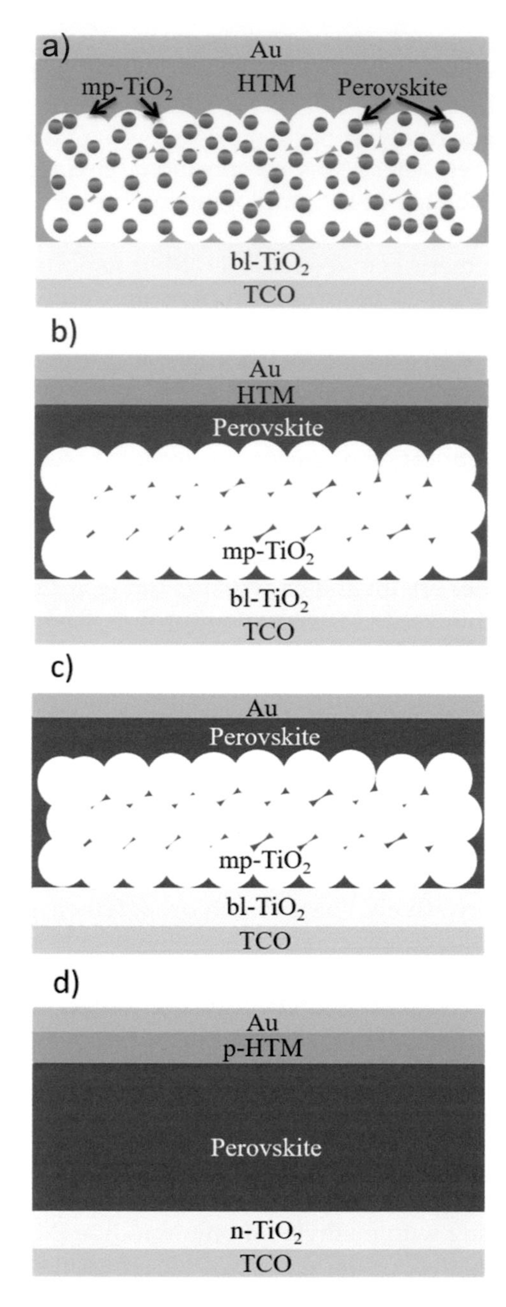

Figure 1. Schematic diagram of the perovskite solar cell, (a) traditionally sensitized cell configuration, (b) 3D-BHJ sensitized configuration, (c) HTM-free mesoscopic p-n configuration and (d) p-i-n configuration.

In the traditional sensitized cell (Fig. 1a), the photo-excited electrons are at first transferred into the mesoporous layer of TiO_2 (mp-TiO_2) and then diffused to the dense TiO_2 blocking layer (bl-TiO_2)/TCO. As the electron mobility in the perovskite (25 cm^2 V^{-1} s^{-1}) is higher than the one in the TiO_2 (7.5 cm^2 V^{-1} s^{-1}), therefore a considerable fraction of the electron transport could take place through the organo-lead halide perovskite phase itself (Ponseca et al. 2014). In the 3D-BHJ sensitized configuration (see Fig. 1b) a continuous film of perovskite is formed across the whole mesoporous layer. So, the structure of 3D-BHJ sensitized PSCs is a combination of the traditional mesoscopic sensitized configuration and a p-i-n architecture, as described below. Currently, PSCs utilizing the 3D-BHJ architecture are among the highest performing lab scale cells (Chen et al. 2016) which are composed of TCO/bl-TiO_2/mp-TiO_2/perovskite/HTM/Au, where perovskite could be methylammonium lead halide (Burschka et al. 2013) or formamidinium lead halide (Lee et al. 2014) or mixture of both (Jeon et al. 2015).

HTM-free mesoscopic p-n configuration has also been proposed as a well-functioning PSC architecture (see Fig. 1c) where perovskite can conduct holes in place of HTM. Since $MAPbI_3$, having ambipolar characteristics, is little more p-type than n-type (Yang and Zhang 2014), therefore, it is reasonable to fabricate p-n junction-like HTM-free solar cells. Etgar and co-workers (Etgar et al. 2012) reported a $MAPbI_3$/mp-TiO_2 heterojunction device (where mp = mesoporous) which demonstrated a PCE of 5.5% under standard illumination (AM 1.5 solar light with 1000 W/m^2 intensity), which was further improved to over 8% by optimizing the perovskite and mp-TiO_2 layer deposition procedures (Laban and Etgar 2013). Using a similar mesoporous p-n configuration, Zhou et al. have fabricated a compatible PSC introducing an insulating Al_2O_3 interfacial layer between the mp-TiO_2 film and the nanoporous gold back contact (Zhou et al. 2015). Recently, another research group (Liu et al. 2016) have developed a low-cost HTM free PSC achieving a PCE of about 12.5%, which was based on a controllable ETL composed of commercially available P25 nanoparticles. Generally, HTM free devices show poor IV-characteristics as compared to the ones having hole conducting layers.

PSCs based on *p-i-n configuration* have also achieved immense attention of researchers after assessment of the fact that the fabrication of efficient PSCs is possible by complete replacement of n-type conducting mp-TiO_2 with the insulating nanoporous Al_2O_3 (Lee et al. 2012). In this work, a mixed halide perovskite material ($MAPbI_2Cl$) served not only as a light absorber but also as an ETL. Such a device was named as "meso-superstructured solar cell" which operates in a similar mode to the standard p-i-n junction solar cell, apart from its 3D architecture (Yang and Zhang 2014). This fact suggests that a thin-film planar device architecture could be used to manufacture an efficient perovskite based solar cell. Actually, such a planar junction device based on extremely uniform perovskite thin film without any pinholes was successfully made by Liu et al. (Liu et al. 2013). This p-i-n configuration (shown in Fig. 1d) based device, comprising a compact layer of n-type TiO_2 as the ETL, $MAPbI_{3-x}Cl_x$ as the light absorber and spiro-MeOTAD as the hole conducting layer, achieved a maximum photonconversion efficiency of over 15%. This study implies that perovskites can work well in the simplified cell configuration without a complex nanoporous structure. Implementing the similar planar p-i-n structure, Liu and Kelly (Liu and Kelly 2014) have fabricated a highly efficient PSC composed of ITO/ZnO/$MAPbI_3$/spiro-OMeTAD/Ag, which exhibited a maximum efficiency of 15.7%. As ZnO nanoparticle based ETL of this device was deposited at room temperature using a wet chemical method, benefiting from the fact, a flexible PSC was also prepared with PCE exceeding 10%.

It is yet ambiguous whether PSCs based on different architectures will demonstrate the same high performance or if certain configurations will have an advantage over the others. As for all other novel light harvesting materials, reliable measurement procedures need to be established for a steadfast comparison of device performance parameters among diverse research laboratories around the globe. Normally PCE of a solar cell is determined by measuring the current–voltage (IV) curves under standard illumination conditions. Actually, from these IV curves the typical device performance parameters like: short-circuit current (J_{SC}), open-circuit voltage (V_{OC}) and the fill factor (FF) are attained which further define the device efficiency. Derivation of authentic device efficiencies is only possible, if the IV characteristics are obtained under quasi-steady state conditions. In literature, the stated values of device efficiency have been found increasing rapidly, whereas a little debate is evidenced regarding slow transient effects observed in perovskite-based solar cells that harshly affect the IV measurements (Unger et al. 2014, Ono et al. 2015, van Reenen et al. 2015). Device efficiency values can be both over and underestimated because of the hysteresis effects, raised by the slow transients, observed between IV-curves measured in different directions and scan rates (Unger et al. 2014). Therefore, it is important to report the directions and the scan rates, at which IV-curves were measured, as these factors may have a massive impact on the device performance parameters. It has been observed that hysteresis in IV measurements, performed on the device, appears in all of the PSC architectures. Initially, it was assumed that hysteresis originates from the perovskite/spiro-OMeTAD interface, but afterward experiments showed that hysteresis persisted even after omitting the spiro-OMeTAD from the devices and contacting the perovskite-absorber directly with gold contact (Unger et al. 2014). This arena is still an open question and needs to be investigated thoroughly.

3. Fabrication Techniques for Nanostructured Materials

The fabrication techniques to grow/deposit nanostructured materials can be classified into two major categories based on the phase of materials during the growth process; one is vapor phase growth and the other one is wet chemical synthesis. Vapor phase techniques can be further divided into physical and chemical vapor deposition. The most employed techniques are listed in the flowchart as shown in Fig. 2 and each technique is briefly described below.

Physical vapor deposition (PVD)

PVD is probably the most famous and vastly practiced method among the various routes of depositing thin films (i.e., the layers of material having thickness in the range from fractions of nanometer to several micrometers). It typically includes a range of clean and dry vacuum deposition techniques where layers are deposited on a solid surface by atom-by-atom or molecule-by-molecule growth. These techniques involve purely physical processes like thermal evaporation or plasma sputtering instead of chemical reaction. Typically, the material to be deposited (i.e., target) is positioned in an entropic surrounding, to facilitate the escape of particles from its surface. In front of the source, a cooler surface known as substrate is placed.

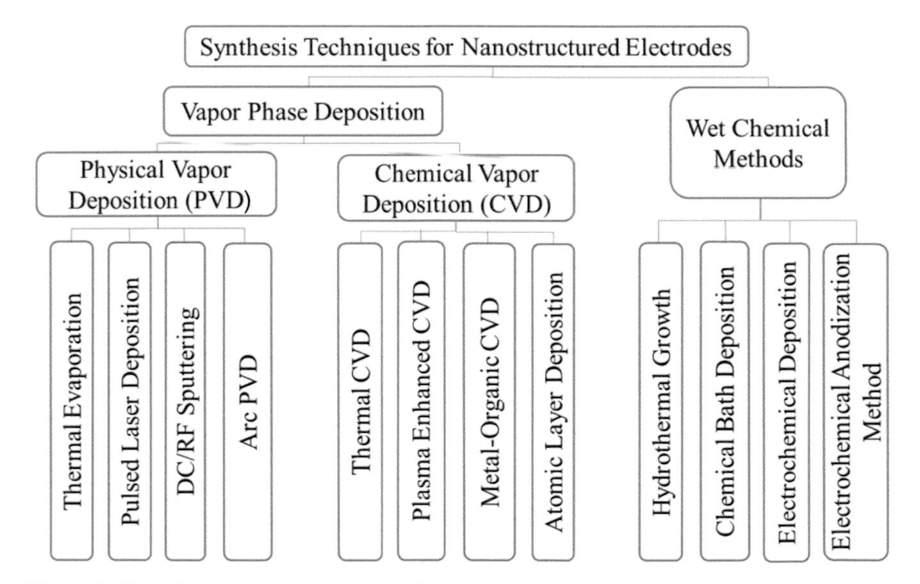

Figure 2. Flow chart representing the synthesis techniques to grow nanostructured materials.

The ejected/emitted particles then form a solid layer on the substrate, delivering their energy to its surface. PVD is performed in a vacuum chamber, so that the particles can move freely following a straight path and as a result, the deposited films are normally directional, rather than conformal. The following four stages can define the entire PVD process:

(I) Evaporation of the target material by bombardment of a high-energy source which extricates atoms from the surface, "vaporizing" them.

(II) Transporting the "vaporized" material from the target to the substrate.

(III) Reaction between atoms emitted from the target and the appropriate gas during the transport stage, which is an important step for the PVD of metal oxide nanostructured electrodes, whereby oxygen reacts with metallic atoms ejected from the target to form oxides.

(IV) Deposition of the thin film/coating on the surface of substrate. Based on the first stage, i.e., the target metal atom dislodgment process, various vapor phase deposition techniques are discussed as follows:

Thermal Evaporation. In this technique, a solid material is heated in a high vacuum chamber to a specified high temperature producing some vapor pressure which leads to a vapor cloud. Subsequently, this vaporized material, in the form of vapor stream, passes through the chamber and strikes the substrate creating a thin film or coating layer on its surface. In general configuration, the substrate is mounted in appropriate fittings facing down on the top of the upright crucible containing melted precursor material which is located at the bottom of the chamber. There are two primary heating ways to evaporate the source material: filament evaporation and electron beam evaporation. In the first method, to heat the precursor material, a simple electrical resistive heat element/filament is used which operates at low voltage, though very high current is needed (several hundred amps). Alternatively, electron beam evaporation (EBE) can be an added "high-tech" methodology to heat up a material but it involves relatively dangerous high voltage (about 10,000 volts). Therefore, EBE methods always contain extra safety sorts. In this process, a very high vacuum condition about ~10^{-4} torr is desired. As compared to filament evaporation, EBE, however, results in a high deposition rate (0.1 μm/min to 100 μm/min) (Srivastava 2013). The material utilization efficiency rate is very high in e-beam assisted evaporation and relatively low substrate temperature is required. In this methodology, helium carrier gas jet at high flow rate is combined with EBE to produce metal vapor and proficiently transport it to the substrate. To deposit metal oxide thin films, oxygen is added to the carrier gas.

The deposition of silver and gold contacts employing simple thermal evaporation has been performed by several researchers for PSC applications (Kim et al. 2015, Eames et al. 2015). Moreover, an evaporation system equipped with an e-beam source was used to evaporate TiO_2 pellets and to deposit thin film onto ITO substrates under nearly 10^{-5} Torr partial pressure of O_2, maintaining the stoichiometry of the film (Qiu et al. 2015). This compact film was successfully employed as ETL in the fabrication of a PSC, achieving cell efficiency beyond 14%. Recently, Huang et al. reported the thermal evaporation controlled growth of a new group of ZnO hierarchical nanostructures consisting of various spines with flowers blooming at polar sites (Huang et al. 2015). Tailoring the ZnO nanostructures by the combination of dislocations and polar cites during evaporation deposition implies a promising strategy to design nanostructured electrodes for solar cell applications. Flower-like ZnO structures have also been grown by Mehmood et al. and Khan using a simple evaporation deposition method for DSC photonode application (Mahmood et al. 2010, Khan 2010).

Pulsed laser deposition (PLD). PLD is a PVD technique in which high-energy laser pulses irradiate a solid target to produce the confined ablation of the target surface. Consequently, an expanding plume of plasma is created due to the ejection of the ablated material (Chrisey and Hubler 1994, Eason 2007). When the solid is ablated in vacuum, the resulting plasma plume is regarded as a nearly collisionless propagation regime of the ejected species. The spatial distribution of these species is predominantly forward directed (Chrisey and Hubler 1994). In these circumstances, the deposition regime, largely consisting of high energy atoms/ions (tens to hundreds of electron volts), impinges the substrate which leads to the atom-by-atom deposition of compact layers. If an inert background gas at a sufficiently high pressure (i.e., 7–7000 mTorr) is introduced into the PLD chamber, the dynamics of the expanding plasma plume will be extremely affected (Yahya 2010). In this state, increasing the gas pressure will result in enhanced spatial confinement of the plasma plume and in higher collision rate of the ablated species. Thus the film growth processes are affected because of the cluster nucleation resulting in a lower cluster kinetic energy. Moreover, PLD can also be carried out using a reactive gas such as oxygen which opens the way for deposition of oxidized species ablated from metallic targets as the composition of cluster may also be modified by the chemical interaction with the neighboring gas (Labis et al. 2015, Han et al. 2016). PLD in presence of reactive gasses, in conjunction with appropriate post-deposition heat treatments, provides an efficient tool for the synthesis of numerous oxide materials with tailored composition, improved crystal structure, and variety of morphologies. PLD deposited nanostructured films have not only found applications in DSCs (Labis et al. 2015, Han et al. 2016, Ghosh et al. 2011) but also in the recently introduced perovskite based solar cells (Liang et al. 2016, Park et al. 2015). Several nanostructured oxides produced by PLD have been reported such as nanoporous ZnO NPs (Han et al. 2016), 3D ZnO nanowall network (Labis et al. 2015), Nb_2O_5 nanoforest (Ghosh et al. 2011) to be employed as DSC photoanode and nanostructured p-type NiO electrode as HTM for efficient perovskite solar cell application (Park et al. 2015).

DC/RF Magnetron Sputtering. Sputtering is a mechanism in which high-energy particles collide with the target material dislodging atoms/molecules to be deposited on substrate surface as a film. There are three major sputtering techniques:

(I) Direct Current (DC) sputtering is the simplest sputtering procedure, which is used to produce the metallic thin films as well as the oxide layers. At first, metal particles are formed by DC sputtering under an argon atmosphere and then a post-oxidation process is responsible for the reaction between these metal particles and the supplied oxygen gas.

(II) Radio frequency (RF) Sputtering is relatively more flexible and suitable for all electrically conductive and non-conductive targets (polymers and silicon oxides) compared with DC Sputtering. Metal oxide deposition involves RF sputtering of a metallic target with different mixtures of argon and oxygen gases. Since the specified chamber pressure is required to be maintained in this method, therefore, oxygen should be persistently supplied to the process chamber in order to recompense the quantity consumed.

(III) Reactive sputtering is the most intricate among the three methods whereby a reactive gas is used to form a plasma in addition to the inert argon gas. Oxygen and nitrogen are the commonly used reactive gases. During the deposition process, the target atoms can react with the chemically activated reactive gas which results in the formation of the required compound. This reactive sputtering method is largely employed for the deposition of resistors, semiconductors, and dielectrics.

For example, NiO_X based HTM layers were successfully deposited on the surface of perovskite film using a reactive DC magnetron sputtering technique. The sputtering was performed with 80 W power and under oxygen pressure of ~3 mTorr, gaining good surface coverage and junction between the perovskite film and the NiO_X layer (Abdollahi Nejand et al. 2015). Another research group (Meng et al. 2015) reported the growth of TiO_2 nanorod films on the ITO substrates by DC reactive magnetron sputtering technique. These films were efficiently employed as DSC photoanodes.

Sacco et al. reported the sputter deposited nanostructured zinc films which were further converted into ZnO by subjecting the zinc films to a simple post heat treatment in ambient atmosphere (Sacco et al. 2012). This heat treatment process is capable of preserving the morphology of the zinc film after the conversion of metallic to oxide phase. Later on, the formation of polycrystalline ZnO nanostructures was proven by the structural analysis as no metallic phase could be spotted (Lamberti et al. 2014). The optimum deposition was achieved at 100 W of RF power under Argon gas pressure of 5 mTorr, which produced a nanostructured zinc film exhibiting sponge-like morphology with larger pores. This method was found suitable for the fabrication of DSC photoanodes. In another study (Chaoumead and Jittham 2014), the deposition of a porous Ti layer was reported which can be prepared by RF power 300 W under substrate temperature of 250ºC and Ar gas pressure of 8 mTorr. These Ti layers exhibited low sheet resistance (~1.7 Ω/sq.) as compared to FTO with the same transparency.

There are numerous advantages of using these sputtering methods for the deposition of nanostructured oxide films. The most important is that the low substrate temperatures can be employed to realize porous structures on flexible substrates like polymeric/plastic types. Recently, ITO electrodes were deposited on PET substrates using a low temperature RF magnetron sputtering of In_2O_3:SnO_2 (95:5 wt%) for the application in PSC electrodes (Qiu et al. 2015). Moreover, sputtering techniques are easy to be implemented in industrial cell fabrication with the potential to cover large-area substrates. An additional aspect of this deposition is that it requires no catalysts during the film growth.

Arc PVD. This is another powerful evaporation PVD technique where the target material is evaporated by applying cathodic arc. A cathode spot is produced avoiding the anode spot formation, while an arc discharge is created under medium/high vacuum, generally ~10^{-5} Torr or better. The cathode spot (with high temperature ~1500ºC) is very active which literally blasts/evaporates the ions from the cathode material (target) resulting in the formation of dense plasma. As the ejected ions are highly energetic, they tend to deposit upon reaching the substrate surface by forming a solid film. Since the ion source (target) of this cathodic arc method is normally solid, then, no crucible is required. Principally, no gas supply is necessary. Nevertheless, for reactive deposition of metal oxide thin films, cathodic arc deposition using the desired cathode material is performed under an appropriate pressure of oxygen gas. Consequently, metal oxide thin films of various stoichiometric compositions can be deposited on a variety of substrates.

Arc-PVD is one of the techniques employed for the fabrication of DSC electrodes such as N_2 doped diamond-like carbon films based DSC counter electrode as reported by Wang and co-workers (Wang et al. 2011). Chen et al. has successfully deposited Ti metallic layer which was further converted into TiO_2-nanotube array for DSC photoanode (Chen et al. 2010) and TiO_2 nanoparticles based DSC photoanode preparation was reported by another research group (Sanjay et al. 2015). Arc PVD has not yet been employed to deposit any layers in the fabrication of perovskite solar cells, as far as the author's knowledge is concerned.

Chemical vapor deposition (CVD)

CVD is a very old technology. In the 1800s, it was used in refining refractory metals; during the early 1900s, it was employed to produce carbon filaments for Edison's incandescent lamps. In the 1950s, it had importance in hard metal coatings, and in the 1960s, CVD based preparation of semiconductor material became significant (Seshan 2002). Today, CVD has emerged as the leading synthesis technique in the fabrication of innovative 1D/2D carbon nanomaterials (carbon nanotubes and graphene) (Pang et al. 2015) and other novel 2D semiconductors like sulfides/selenides of Mo and W (Baek et al. 2015, Wang et al. 2016). CVD is basically the creation of stable solids by decomposition of gaseous/vaporous chemicals using heat (thermal CVD), ultraviolet (photo/laser CVD), plasma (plasma enhanced or plasma assisted CVD), or some other energy source, or a mix of the sources. It fits in the class of atomistic vapor-transfer routes which implies that the depositing species will be the atoms or molecules or both (Pierson 1999).

Thermal CVD. In a typical CVD process (known as thermal CVD), thermal energy is used to heat up the gases in the deposition chamber to initiate the chemical reaction. It has numerous drawbacks; a major one is the required high temperatures (of 600°C and above) to get the most versatile yields, whereas several substrates are thermally unstable at these high temperatures. Fundamentally, laser CVD involves the same deposition mechanism as the typical thermal CVD and hypothetically, the same wide range of materials can be deposited but the choice of substrates is limited practically. In this scenario, the development of 2-zone CVD, metalorganic CVD and plasma enhanced CVD have partially compensated this shortcoming. Some researchers have recently introduced 2-zone CVD as a promising technology to deposit perovskite absorber layer for large-area solar cell fabrication. Tavakoli et al. have employed one-step CVD process for perovskite layer deposition on TiO_2-compact-layer-coated FTO substrates, where CH_3NH_3I and PbX_2 (X = I and Cl) powders were evaporated by heating them in quartz crucibles, placed separately in high temperature zone furnace. The vapors were then simultaneously transported with the Ar carrier gas onto the TiO_2-coated FTO substrates (80°C) in down flow low temperature zone. For the best device efficiency, perovskite absorber layers were optimized by varying different key parameters such as deposition time, temperature, gas flow rate, and the amount of two precursors. Immediately after the deposition, perovskite layers were subjected to *in situ* annealing in the low temperature zone of the furnace at 100°C for 60 minutes. In another work (Leyden et al. 2016), Leyden and co-workers have showed that fabricating FAI based PSC modules using 2-zone CVD furnace appears to be a propitious technique because the growth process is scalable and the resulting modules can uphold a high steady state power at larger areas. They have first prepared lead halide layers onto substrates by thermal evaporation and these films were subsequently converted into perovskite by CVD deposition of MAI or FAI. In another study (Yang et al. 2015), researchers reported the graphene growth up to few-layers on copper foil at 950°C by thermal CVD and it was transferred from copper foil to FTO-coated glass for the application as DSC counter electrode. The incorporation of nitrogen (N) atoms into this graphene counter electrode to enhance its catalytic property was performed by nitrogen plasma ion irradiation. The increasing plasma treatment time has decreased the charge transfer resistance for triiodide reduction by graphene. Such a decrease in resistance can be attributed to the increase in the number of catalytic sites.

Plasma Enhanced CVD. The chemical reaction is activated by the plasma in this CVD process and the deposition temperature is considerably lower. It is also known as plasma assisted CVD (PACVD) or plasma-enhanced CVD (PECVD). This technique involves chemical as well as physical process. Thus, it is analogous to the PVD processes which operate in a chemical environment, for instance, reactive sputtering. It first emerged in the 1960s as a powerful tool for semiconductor applications. Since then the quantity and the diversity of applications have expanded impressively and it has now become a key process at par with the thermal CVD (Seshan 2002). In a PECVD process, when a diatomic gas like hydrogen is heated above a predetermined temperature, the molecules are dissociated into atoms which ultimately become ionized, yielding a plasma which consists of negatively charged electrons, positively charged ions and neutral atoms that have not been ionized. However, a large amount of energy is required to achieve plasma as the ionization temperatures are generally very high (> 5000 K). This energy is attained by glow-discharge, electric arc and electron cyclotron resonance (through the right mix of magnetic and electric fields) (Pierson 1999). Several metal oxide nanostructures deposited by PECVD have been reported. For example, Lee and Kim reported the TiO_2 passivation layer grown between mesoporous TiO_2 photoanode and FTO glass using PECVD. The thickness of this compact layer was optimized to reduce the recombination of electrons and to enhance the electric connection with the surface of FTO electrode, thus improving the efficiency of a DSC (Lee and Kim 2012). Two years later, the same research group investigated the growth of ZnO layers by same technique which were successfully employed as the DSC photoanodes (Lee and Kim 2014). Besides photoanode, fabrication of counter electrode using plasma CVD has also been reported in literature. Mahpeykar et al. have reported for the very first time that the vertically aligned carbon nanofibers, directly grown on ordinary TCO coated glass using low-temperature PECVD can be successfully employed as counter electrode for DSC fabrication (Mahpeykar et al. 2013).

Metalo-Organic CVD (MOCVD). It is relatively a newcomer technology, as the early experiments were performed in late 1960s which demonstrated the possible deposition of critical semiconductor materials at lower temperature as compared to conventional thermal CVD. Since then, the sophistication of the equipment, the purity and variety of the precursor materials and quality of the products have gradually improved and MOCVD has now become a specialized field of CVD. It is being used on a large scale, mainly in optoelectronic and semiconductor applications (Stringfellow and Craford 1997). The suitable temperature range for most of MOCVD reactions is 300–800°C and the pressure should be between atmospheric and 1 Torr. MOCVD technique can be used to grow several metal oxide nanostructures. ZnO nanowires can be deposited by MOCVD without using a catalyst and the crystal growth depends on the substrate position and temperature, pressure of the precursor and hydrogen flow rate. Baxter and Aydil employed ZnO nanowires in fabrication of DSC photoanode and they observed that such morphology can provide a direct path for the electron transport from the injection point to the collecting electrode. Therefore, nanowires may offer improved electronic transport as compared to nanoparticle films where electrons must cross many interfaces to reach the substrate (Baxter and Aydil 2006). ZnO thin film composed of innovative nanostructures such as nanotips and dendritic nanowires, synthesized by MOCVD, which is suitable as DSC photoanode have also been reported in literature (Zhang et al. 2009). To achieve dendritic nanowires, at first, ZnO nanowires with 100 nm diameter are grown with secondary nanowire branches of 20 nm diameter which are nucleated from the primary nanowire backbone. Then, the "second generation" growth is continued on the substrate having both primary nanowire backbone and the secondary branches. During this second generation growth, the stretched out secondary branches of primary nanowires perform as nucleation sites for the further nanowire growth. For dendrite-like nanostructures of ZnO, the growth can also be sustained for the third and fourth generations.

Atomic Layer Deposition (ALD). ALD is another versatile deposition technique for depositing a variety of nanostructured thin films on almost all kinds of substrates. It is comparatively a new growth technique which is capable of operating at low process temperatures. Even though it is a vapor phase deposition technique like CVD, there are two major differences that mark it as a prevailing route for the growth of very thin and conformal films. Firstly, since the film growth is realized by sequential introduction of normally two precursors into the deposition zone, therefore, precursors do not get together in the gas phase being separated by an inert gas purge. Secondly, precursor molecules do not react with themselves under deposition temperatures, but only with the surface of substrate (Kääriäinen et al. 2013). ALD is basically a two-step process where one reactant has to form a self-limiting monolayer on the substrate surface in the first step. The reactant molecules tend to undergo either physisorption or chemisorption while reacting with the surface of substrate. Particularly, the chemisorbed species have a tendency to readily form the self-limiting monolayers, which can be reliably stable at adequate temperatures and are of interest for ALD. After the formation of first reactant monolayer, the introduction of a second reactant into the deposition zone will simply transform the first reactant to the film of a required solid material (Knez 2012).

In a recent work, the composite photoanode based on ZnO nanobelts and the TiO_2 nanoparticles was coated by a continuous TiO_2 film by ALD technique, which increased the carrier lifetime by reducing the recombination rate (Lu et al. 2015). This ALD treatment improved the interface contact within the whole porous layer of photoanode, leading to substantial enhancement in cell performance. Another reason for such coatings on ZnO nanostructures is its chemical instability to acidic dyes, and low electron injection efficiency from the Ru-based sensitizers and presence of surface defects, which hinders its progress as DSC photoanodes. Another research group has discussed the use of ALD for such thin and continuous TiO_2 coating on ZnO-based photoanode to increase the energy efficiency of the device (Liu et al. 2016). Recently, pinhole-free nanostructured TiO_2 films of various thicknesses were deposited on FTO-coated glass at 150°C by ALD technology. The precursor of "Ti" was Titanium (IV) isopropoxide (TTIP) and for "O" the precursor was H_2O, and to transport these precursors, high purity nitrogen was used as the carrier gas as well as for purging purposes. Then, the optimized TiO_2 film was employed as ETL in combination with cheap carbon based counter electrode to develop highly efficient HTM-free PSCs with low cost and outstanding long-term stability in ambient atmosphere (Hu et al. 2016).

Wet chemical deposition

An economical alternative of vapor phase growth procedures is the wet chemical deposition. The most commonly employed wet chemical deposition methods include: sol-gel process, hydrothermal growth, chemical bath deposition, electrochemical anodization method and the electro chemical deposition.

Sol-Gel Process. The sol–gel process is proved to be a powerful and versatile method in order to develop nanostructured semiconductor layers exhibiting a wide range of morphologies with novel properties. It bears a long history as it was introduced about four decades ago with the processing of oxide materials like glass and ceramics (Hench and West 1990). This versatile process can easily mold the properties of metal oxide nanostructures optimizing the process conditions during synthesis (Yahya 2010, Srivastava 2013). Electrode fabrication by sol-gel method comprises two stages. Firstly, the phase conversion is realized from the liquid sol into the solid gel, where sols are dispersions of colloids (solid particles with diameters of 1–100 nm) in a liquid and the gel is a porous integrated network of either agglomerated dense colloidal particles or polymeric chains with average length greater than a micrometer. Secondly, thin films of the produced solid gel can be realized using a range of deposition techniques (discussed in the section "Nanopowder-based deposition tools" below) onto a conducting glass/plastic substrate followed by an annealing treatment, which is essential to enhance the crystallinity and to reduce grain boundaries, thus improving electrical conductivity. There are three usual routes adopted for the sol-gel process: (I) gelation of the colloidal powder based solutions; (II) hydrolysis and polycondensation of nitrate or alkoxide precursors followed by supercritical drying of gels; (III) hydrolysis and polycondensation of alkoxide precursors followed by aging and ambient drying (Yahya 2010, Srivastava 2013, Willander 2013). Several researchers have reported sol-gel as successful technique to grow metal oxide nanoparticles for electrode fabrication. The growth of ZnO nanoparticles using sol-gel process was reported by Kandjani and co-worker (Kandjani et al. 2009) and the growth of TiO_2 was discussed by Yuksel and co-worker (Yuksel et al. 2009). Several researchers have grown nanoparticles of TiO_2/ZnO/SnO_2 by sol-gel process and successfully employed them for the fabrication of DSC photoanodes (Chen et al. 2010, Chaoumead and Jittham 2014, Cheng et al. 2011).

Hydrothermal Growth. Hydrothermal is a competent and cost-effective method to obtain inorganic nanocrystalline materials. The solubility, of almost all inorganic substances, in water is utilized in hydrothermal processing to induce crystallization of the dissolved material at high pressures and temperatures. Water at elevated temperatures (generally less than 250ºC), as implied by the name of this technique, plays an important role in the transformation of precursor material as it can produce high vapor pressure (Srivastava 2013). Another synthesis route analogous to the hydrothermal is also evidenced in literature known as solvothermal where it involves the non-aqueous/organic solvents and relatively high reaction temperatures are possible depending on the boiling point of the solvent used. Hydrothermal/solvothermal growth is a simplistic route to prepare highly pure and crystalline oxides with good dispersion under the modest reaction conditions, such as short reaction time and relatively low temperature as compared to other procedures. Such synthesis method can eliminate the post heat treatment process, which is generally essential to transform the amorphous phase into the crystalline one. In this process, the synthesis reaction usually occurs in precursor solutions, consisting of required materials dissolved in water/organic solvents, placed in a steel-made temperature controlled pressure chamber known as autoclave. The vapor saturation pressure is obtained inside the chamber by raising the reaction temperature beyond the boiling point of the solvent. The internal pressure of the autoclave is largely affected by the temperature and the amount of solution added. Elevated temperatures affect the solubility of the reactants in water/organic solvent and also their specific reactivity, thus providing more versatility to control the quality of the nanostructured materials. An appropriate size distribution and a high nucleation rate can be maintained during the processing of nanocrystals, with diverse morphologies and compositions, by controlling the process parameters such as temperature and pressure in the autoclave, pH of the solution, growth time and the relative concentrations of precursors (Khataee and Mansoori 2012). Several metal oxides in diverse nanostructure morphologies such as TiO_2 nanorods (Liu and Aydil 2009), TiO_2 nanoparticles (Jeon et al. 2015), ZnO nanorods/nanotubes (Schlur et al. 2013, Roza et al. 2015), ZnO nanoflowers (Zhang et al. 2009), ZnO nanoplates (Qiu et al. 2010) have been commonly developed using economical hydrothermal/solvothermal techniques and extensively applied in fabrication of DSCs and PSCs electrodes.

Chemical Bath Deposition (CBD). CBD is the simplest and cheapest route to grow nanostructures of various morphologies and structures without providing vacuum and high process temperatures. Such deposition process is usually performed in one or more stages and the entire reaction takes place in one bath where the cation and the anion of the required compound encounter each other. Normally, cations/metallic ions already exist in the solution at ambient temperatures, and to produce anions it is essential to raise the temperature. Consequently, the growth of the desired final product occurs on the substrate surface, which partly works as catalyst in this process. Nonetheless, this growth reaction can also take place in the solution which leads to the formation of crystallites and clusters in the form of agglomerates and as a result, the solution appears visibly opaque. After the reaction is completed, powder composed of nano/micro crystallites can be separated from solvents and used for the deposition of films through powder based deposition techniques, as discussed in the following section, whereas in the presence of substrate, these agglomerates can attach themselves to the substrate strongly influencing the properties of the deposited nanostructures and layers. In order to grow one dimensional nanostructures such as nanorods/nanopillars, a polycrystalline underlayer is formed by nanocrystal seeds (Kim et al. 2014) which provides a contact layer underneath the nanostructure. The dimensions of nanostructures can be controlled by monitoring the parameters of seeding.

The thickness of the grown coatings or the diameter and length of the nanostructures strongly depend on the concentrations and relative ratio of the precursor materials, solution's pH-value and the process temperature. The instability of these process parameters can not only influence the properties of the grown structures but also provides the opportunity to control the synthesis mechanism. Another advantage of the CBD growth is that nanostructures can be synthesized on different substrates—such as glass, ITO/FTO, metal, Si, etc. One major disadvantage of CBD technique is the low usage (typically below 10%) of the precursor, and due to agglomeration of crystallites after a particular process time, it is essential to reintroduce the fresh solution. For that reason, it is also challenging to produce thicker layers (> 8 μm) which need growth duration of up to 20 h. CBD has been widely employed to grow QDs on the mesoporous film (Tian and Cao 2013) and to fabricate ETL for QDSCs (Zhang et al. 2009, Choi et al. 2014), DSCs (Kim et al. 2014, Ma et al. 2015) and also for the PSCs (Kumar et al. 2013, Yamamoto et al. 2014, Liang et al. 2016).

Electrochemical Anodization Method. The controlled oxidation of metallic electrode under electrochemical anodization offers a facile technique to grow nanomaterials. Generally, a two-electrode electrochemical cell is employed to conduct the anodization process in which a potential is applied between a cathode (generally a Pt foil) and an anode (generally a metallic foil) immersed into an electrolytic solution based on organic solvent with some amount of water (Lamberti et al. 2013, Gong et al. 2001, Berger et al. 2010). Well-organized nanoporous or nanotubular (one-dimensional) nanostructures having high aspect ratios could be effectively achieved by optimizing conditions of anodization. The resulting nanostructure morphologies are mainly dependent on the supplied electrical power and the anodization time (Li et al. 2014). During the last decade, the self-assembled nanostructured oxide films were effectively grown by the optimized anodization procedure on a wide range of metals, such as Zr, Al, Nb, Hf, Ta and Ti (Lamberti et al. 2013, Stergiopoulos et al. 2008, Park et al. 2015). It has been observed that the type of material affects the produced morphologies, when the anodization is performed in fluoride-containing electrolytes: morphologies of anodized Nb, Ta, and Al are found to be porous while anodization of Hf, Zr, and Ti gives rise to tubular structures (Park et al. 2015). On the other hand, it has been shown that the resulting morphology of the produced oxide nanostructure is also affected by certain anodization parameters/conditions (Ghicov and Schmuki 2009). Anodization has been predominantly employed in the production of TiO_2 nanotubes using titanium foil for nanostructured electrodes designed for solar energy conversion/storage applications. In fact, electrochemical anodization has been broadly studied after pioneering work by Gong and co-workers in which they performed the electrochemical anodization of titanium foil in presence of HF aqueous electrolyte leading to the realization of titania nanotubes having lengths up to 0.5 mm (Gong et al. 2001). As a result, a number of nanotubular structures have been developed for nanostructured electrode fabrication to be employed in dye-sensitized as well as perovskite solar cells (Lamberti et al. 2013, Stergiopoulos et al. 2008, Wang et al. 2015). Generally, anodization process can be roughly classified into three steps: at first, the electrochemical oxidization of the metal surface is observed resulting in the formation of a preliminary oxide barrier layer; in the second stage, chemical etching of oxide layer by F^- occurs; and the third step involves the growth of nanopore/nanotube. In brief, the nanostructure synthesis is governed by the equilibrium between chemical dissolution and anodic oxidation. The chemical dissolution rate is largely dependent on the F^- concentration, while the anodic oxidation rate is primarily determined by the anodic potential (Lin and Wang 2014).

Electro Deposition (ED). Usually, the electrodeposition is performed in a temperature controlled chemical batch reactor similar to the one employed in CBD and the process takes place under atmospheric pressure at temperatures range of 70–90°C (Tamiya and Taguchi 2015, Meng et al. 2014, Postels et al. 2008). The precursor materials are mixed in deionized water to prepare a clear solution. In this process, DC current flows across a pair of electrodes, placed in the solution, where the particles get charged and move towards opposite electrode. When particles are charged positively, they move to cathode. In contrast, when particles are charged negatively, they move to anode. Kazuaki Tamiya and Kozo Taguchi have employed electrophoretic deposition (EPD) process to fabricate CNTs based counter electrode for DSCs (Tamiya and Taguchi 2015). Several semiconducting oxide thin films have been synthesized by electrodeposition methods and employed as photoanode in DSCs (Meng et al. 2014) and ETL for PSCs (Su et al. 2015). Huang et al. have deposited a uniform PbO film using electro deposition which was further converted into $MAPbI_3$ perovskite layer *via in situ* solid-state reaction with the adjacent MAI layer (Huang et al. 2015).

Some other solution based techniques have also been evidenced from literature: one of them is successive ionic layer adsorption and reaction (SILAR), which has been employed for the growth of QDs (Tian and Cao 2013) to fabricate QDSCs.

Nanopowder-based deposition tools

This is another economical deposition technique for the fabrication of nanostructures-based electrodes by using nanopowders to fabricate nanoparticulate or nanocrystalline films. Such techniques have attracted immense attention of researchers working in the field of photovoltaics. The most common methodologies to deposit nanoparticulate films employing colloidal solution are the spin-coating (Jeon et al. 2015), doctor blade (Cicero et al. 2013, Keis et al. 2000, Keis et al. 2002, Shahzad et al. 2013), screen printing (Yang et al. 2015, Heo et al. 2013), dip coating (Yuksel et al. 2009) and electrospinning (Mali et al. 2015). To evade the formation of aggregates and to stabilize the colloidal solutions, organic binders/additives are supplemented to the solution. A post deposition treatment at high temperatures is required to eliminate the organic supplements from the grown films. It is observed that for the realization of PSCs in layer by layer configuration, spin coating has been largely employed to deposit not only the electron transport layers, but also perovskite light absorber layer, and the hole transport layers (spiro-MeOTAD, PTAA, CNT/polymer composite) (Habisreutinger et al. 2014, Burschka et al. 2013, Kim et al. 2015, Heo et al. 2013). Another nanopowder based technique known as spray pyrolysis has also been evidenced from literature for the deposition of nanostructured electrodes, which has been successfully employed for TiO_2 blocking layer deposition to be employed in PSCs (Lang et al. 2015, Burschka et al. 2013, Heo et al. 2013, Jeon et al. 2015, Eames et al. 2015).

4. Characterization Techniques: To Explore Distinct Properties of Nanomaterials

The morphological, structural, optical and chemical investigation of the various nanostructured materials are performed using numerous characterization techniques:

- *SPM (Scanning Probe Microscopy)* is a family of microscopies which provide surface images by scanning the sample using a physical probe. AFM (Atomic Force Microscopy) is the most commonly employed type of SPM in the field of nanostructured electrodes (Qiu et al. 2015, Hu et al. 2016, Yamamoto et al. 2014, Stergiopoulos et al. 2008, Chen et al. 2015). AFM is capable of locally exploring the three dimensional shape, namely topography, of the nanostructures by raster scanning the sample with a cantilever tip and recording the reaction forces that sample will exert on the tip depending on their mutual separation. Since an AFM does not employ any lenses, therefore, space resolution is not restricted by the diffraction limit and it can demonstrate a topography with resolution in the order of fraction of a nanometer.

- *TEM (Transmission Electron Microscopy) and HRTEM (High Resolution Transmission Electron Microscopy).* TEM is a powerful microscopic technique which employs electron beam instead of light and provides morphological, crystallographic and compositional information about the nanostructures. In a typical TEM, transmitted electron beam is analyzed to get required information. On the other hand, HRTEM is an imaging mode of the TEM, which creates an interference image by using both the transmitted and scattered electron beams and is, therefore, employed to determine the highly resolved crystallographic structure at the nanoscale. TEM/HRTEM analysis has been extensively employed to get reliable information on crystal structure and defects by several researchers (Han et al. 2016, Wang et al. 2016, Mali et al. 2015, Lu et al. 2015, Xiao et al. 2014). Nevertheless, as the analyzed area is in nano-size, a question arises about the statistical significance of the results in case of non-macrocrystalline materials.
- *FESEM (Field Emission Scanning Electron Microscopy)* is an advanced form of SEM which is useful in determining the surface geometry/morphology of materials and the shape of nanoparticles. In a typical SEM, the source of electron beam is based on thermionic emission whereas in FESEM electromagnetic field emission gun is employed as electron beam source and hence, it provides high magnification/resolution images owing to the powerful electron beam. The investigated areas are generally larger in case of FESEM than that analyzed by HRTEM. Henceforth, the caution about the statistical significance of the results still applies but becomes less stern. Existing literature is abundant with examples of the wide variety of applications of this most prominent imaging technique in exploring the nanostructured electrodes for solar cells (Han et al. 2016, Mali et al. 2015, Lu et al. 2015, Xiao et al. 2014, Stoumpos et al. 2013, Ye et al. 2015).
- *STEM (Scanning Transmission Electron Microscopy)* is a valuable characterization technique for nanostructured electrodes, which can work in different imaging modes providing useful information at atomic scale regarding elemental composition and the electronic structure. STEM can take the form of SEM if a simple transmitted electron detector is added to a typical low voltage SEM and it can resemble HRTEM when it works with high accelerating voltages (200–300 kV), providing the supreme spatial resolution and ultimate analytical sensitivity.

- *Specific Surface Area* analysis is performed using gas adsorption and BET (Brunauer-Emmett-Teller) theory. For a number of motives, nitrogen, N_2, is the most commonly used gas. The experimentally recorded amounts of adsorbed gas (expressed as STP volume) are plotted as a function of the corresponding relative pressures (P/P_0) to obtain the required isotherm. The specific surface area, or the amount of surface, is a key factor to determine the adsorption behavior of a solid, which is really an important factor in sensitized solar cells (Sacco et al. 2012, Mali et al. 2015).
- *XPS (X-ray Photoelectron Spectroscopy)* is used to investigate the chemical properties of the sample surface such as elemental composition and chemical/electronic state of the constituent elements. This characterization technique uses an X-ray beam to irradiate the sample surface and the ejected photoelectrons are counted as a function of their kinetic energies to produce a spectrum. The detected energies and corresponding intensities of the photogenerated electron peaks can be used to identify and quantify the elements present on the surface. As the surface chemistry of nanostructured electrodes plays an important role in performance of emerging solar cells, the use of XPS analysis becomes stringent (Qiu et al. 2015, Mali et al. 2015). Even though the decomposition of the elemental peak is not direct and simple all the times, an exhaustive information can be obtained about the amount of the different bonding and the presence of diverse elements on the surface.
- *SEM-EDS (Energy-dispersive X-ray spectroscopy in the SEM mode)* can offer qualitative, or with suitable criteria, quantitative analysis of the elemental composition of samples with 1–2 microns depth. When the electron beam interacts with the sample atoms, X-rays are emitted as a result of shell transitions. As the energy of the emitted X-ray is characteristic to the parent element, elemental analysis is possible by detection and measurement of this energy. Compositional maps, showing the elemental distribution on a sample surface, are also accessible via EDS.
- *STEM-EDS (Energy-dispersive X-ray spectroscopy in the STEM mode)* gives the elemental compositional maps with high spatial resolution. When the sample is raster-scanned by a highly-focused electron beam, various kinds of scattering are detected as a function of position and the collection of electrons transmitted at high scattering angle which can produce chemically sensitive atomic number (Z) contrast images with high-resolution. This technique helps in exploring the localised elemental compositions of nanostructured materials (Mali et al. 2015).

- *XRD (X-ray Diffraction)* is an invaluable tool to explore the crystalline structure of nanomaterials (films as well as powders) and significant information about the grain size can also be attained using Debye-Scherrer method (Lamberti et al. 2015) even though attention is needed for the nanoparticles that are non-spherical in shape. The investigation of structural properties of crystalline nanomaterials is most often carried out using X-ray Diffractometers as reported in literature (Han et al. 2016, Mali et al. 2015, Xiao et al. 2014, Stoumpos et al. 2013, Bush et al. 2016).
- *Selected Area Electron Diffraction (SAED)* is a diffraction technique. It can be employed to recognize crystal structures and to witness the presence of crystal defects. It is analogous to XRD, but the uniqueness of this technique lies in the fact that the examined areas are as small as some hundred nanometers in size, while typically sampled areas by XRD are several centimeters in size. It is very useful to explore structure of nano-crystalline materials (Han et al. 2016, Wang et al. 2016, Xiao et al. 2014, Stoumpos et al. 2013).
- *Micro-Raman Spectroscopy* represents the most significant tool for the structural analysis of nanostructured electrodes (Meng et al. 2015, Zhang et al. 2009, Stergiopoulos et al. 2008, Mali et al. 2015). Micro-Raman uses a variation of the Raman spectroscopy set-up, in which the excited light is coupled to the sample by means of a microscope. It is used to obtain information about molecular chemical bonds and symmetry. The Raman spectrum of a molecule is unique and it provides a fingerprint by which the molecule can be identified. Nowadays, Raman spectroscopy has a very important role in solid state physics—it is used to characterize nanomaterials, to measure temperature (the ratio between Stokes and antiStokes bands depends on temperature) and to find the crystallographic orientation of a sample using polarized laser light.
- *UV-Vis Spectroscopy*, in the simplest form, is based on the measurement of light absorption by a sample using a spectrophotometers. It was initially developed for dissolved liquid solution material to determine concentration, absorbance and other properties. The most commonly employed wavelength range is from about 200 nm to 800 nm, corresponding to the UV and visible part of electromagnetic spectrum. However, nowadays it is widely used for solid materials characterization as well. In order to extend the measurement beyond 800 nm, UV-Vis-NIR spectrophotometers equipped with integrating sphere are recommended, which are capable of measuring absorption as well as diffused reflectance spectra for nanostructured films (Jeon et al. 2015, Meng et al. 2015, Stoumpos et al. 2013, Shahzad et al. 2016) and powders.

- *Photoluminescence Spectroscopy* is a nondestructive, contactless method to probe both intrinsic and extrinsic electronic structure of semi-insulating and semiconducting materials. When light is incident on a sample, it gets absorbed by the material and imparts excess energy into it, called photo-excitation which causes electrons to move into allowed excited states. Then this excess energy is dissipated through the emission of light or luminescence because of electrons returning to their equilibrium states. In the case of photo-excitation, the resulting luminescence is called photoluminescence (PL). When it is collected at liquid helium temperatures, a PL spectrum provides a superb picture of overall quality and purity of the crystal. It can determine impurity concentrations, identify defect complexes, and measure the band gap of semiconductors. Performing PL mapping with a strongly focused laser beam at room temperature, the micrometer-scale variations in crystal quality are measurable. This technique also finds significant applications in exploring optoelectronic properties of nanostructured electrodes (Postels et al. 2008, Stoumpos et al. 2013).
- *Cyclic Voltammetry (CV)* is a very diverse electrochemical technique which permits the exploration of the transport properties of a system in solution and the mechanics of the redox mediator. The CV setup is composed of a three electrode arrangement whereby the potential is applied at a working electrode relative to some reference electrode while the subsequent current flows through the counter (or auxiliary) electrode and this current is examined in a quiescent solution. Using this technique, rapid detection of redox couples present in a system is possible. It is a promising technique which is widely employed in the study of electrolytes and electro-catalysts for emerging solar cells.

A number of exhaustive reviews and books on the subject have been written and reader can refer to them for an insight on the techniques (Yahya 2010, Gaponenko 1998, Zhang 2009).

5. Conclusion and Perspective

Traditional crystalline silicon solar cells represent mature photovoltaic technologies today and the obtained efficiencies (~25%) at laboratory scale are quite close to the theoretical limit (33%), therefore, a very small window is left to be explored. Besides this fact, it is believed that the cost reduction for traditional PV electricity by getting more and more experience is a risky approach, whereas taking risks for radical improvement in efficiency of emerging PV technologies by innovation would be less risky. These emerging solar cells show tremendous performance in diffuse light and so they retain a competitive edge over silicon based PV in delivering electricity for both indoor and outdoor solutions. Application of these solar cells in building integrated PV is already on track and will grow into a fruitful field of commercial development in future.

The low cost and the simplicity of fabrication techniques usually employed for these emerging solar cells are well fitting for a whole domain of applications, ranging from the low power market to large-scale commercialization. Not only the ease and variety of these techniques are vital but also the ready availability of the precursor materials plays an important role in cheaper cost realization of these nanostructured emerging solar cells. Since there is no theoretical single band gap limit on maximum achievable efficiencies for these types of solar cells, a big window is there to be explored and to improve the device performance. Untill today, a number of characterization techniques have been developed to investigate the light harvesting properties of nanostructured and photoactive materials. These innovative techniques have helped a lot in enhancing device stability and power conversion efficiencies through pin pointing the degradation/loss processes at nanoscale level. On the basis of these characterization tools, on every other day, novel state-of-the-art ideas are floated into research arena where deeper insights of working mechanisms of the nanostructured electrodes and improved device performance are presented. In short, it can be claimed that reaching beyond 22% and showing prospect for further enhancement in efficiencies of these emerging solar cells, they have proven their potential as viable contenders for future large scale solar energy conversion devices, owing to the low-cost fabrication, environmental compatibility as well as availability of the materials.

Reference

Abdollahi Nejand, B., V. Ahmadi and H.R. Shahverdi. 2015. New physical deposition approach for low cost inorganic hole transport layer in normal architecture of durable perovskite solar cells. ACS Appl. Mater. Inter. 7: 21807–21818.

Baek, S.H., Y. Choi and W. Choi. 2015. Large-Area growth of uniform single-layer mos_2 thin films by chemical vapor deposition. Nanosc. Res. Lett. 10: 388.

Bauer, C., G. Boschloo, E. Mukhtar and A. Hagfeldt. 2002. Ultrafast studies of electron injection in Ru Dye sensitized SnO_2 nanocrystalline thin film. Int. J. Photoenergy 4: 17–20.

Baxter, J.B. and E.S. Aydil. 2006. Dye-sensitized solar cells based on semiconductor morphologies with ZnO nanowires. Sol. Energ. Mat. Sol. C. 90: 607–622.

Berger, S., R. Hahn, P. Roy and P. Schmuki. 2010. Self-organized TiO_2 nanotubes: Factors affecting their morphology and properties. Phys. Status. Solidi (b) 247: 2424–2435.

Burschka, J., N. Pellet, S.J. Moon, R. Humphry-Baker, P. Gao, M.K. Nazeeruddin et al. 2013. Sequential deposition as a route to high-performance perovskite-sensitized solar cells. Nature 499: 316–9.

Bush, K.A., C.D. Bailie, Y. Chen, A.R. Bowring, W. Wang, W. Ma et al. 2016. Thermal and environmental stability of semi-transparent perovskite solar cells for tandems enabled by a solution-processed nanoparticle buffer layer and sputtered ITO electrode. Adv. Mater 28: 3937–3943.

Chaoumead, A. and V. Jittham. 2014. Preparation of nanoporous Ti/TiO_2 layers deposition by RF-magnetron sputtering/sol-gel combustion procedure for Dye-sensitized solar cell. Enrgy. Proced. 56: 219–227.

Chen, C., Y. Cheng, Q. Dai and H. Song. 2015. Radio frequency magnetron sputtering deposition of TiO_2 thin films and their perovskite solar cell applications. Sci. Rep. 5: 17684.

Chen, C.-H., K.-C. Chen and J.-L. He. 2010. Transparent conducting oxide glass grown with TiO2-nanotube array for dye-sensitized solar cell. Curr. Appl. Phys. 10: S176–S179.

Chen, H., D. Bryant, J. Troughton, M. Kirkus, M. Neophytou, X. Miao et al. 2016. One-Step facile synthesis of a simple hole transport material for efficient perovskite solar cells. Chem. Mater. 28: 2515–2518.

Chen, J., C. Li, D.W. Zhao, W. Lei, Y. Zhang, M.T. Cole et al. 2010. A quantum dot sensitized solar cell based on vertically aligned carbon nanotube templated ZnO arrays. Electrochem. Commun. 12: 1432–1435.

Chen, Q. and D. Xu. 2009. Large-scale, noncurling, and free-standing crystallized TiO_2 nanotube arrays for dye-sensitized solar cells. The Journal of Physical Chemistry C 113: 6310–6314.

Chen, W., Y. Qiu and S. Yang. 2010. A new ZnO nanotetrapods/SnO_2 nanoparticles composite photoanode for high efficiency flexible dye-sensitized solar cells. Phys. Chem. Chem. Phys. 12: 9494–9501.

Cheng, C., Y. Shi, C. Zhu, W. Li, L. Wang, K.K. Fung et al. 2011. ZnO hierarchical structures for efficient quasi-solid dye-sensitized solar cells. Phys. Chem. Chem. Phys. 13: 10631–10634.

Choi, Y., M. Seol, W. Kim and K. Yong. 2014. Chemical bath deposition of stoichiometric CdSe quantum dots for efficient quantum-dot-sensitized solar cell application. J. Phys. Chem. C 118: 5664–5670.

Chrisey, D.B. and G. Hubler. 1994. Pulsed Laser Deposition of Thin Films. New York: Wiley & Sons.

Cicero, G., G. Musso, A. Lamberti, B. Camino, S. Bianco, D. Pugliese et al. 2013. Combined experimental and theoretical investigation of the hemi-squaraine/TiO_2 interface for Dye sensitized solar cells. Phys. Chem. Chem. Phys. 15: 7198–7203.

Eames, C., J.M. Frost, P.R. Barnes, B.C. O'Regan, A. Walsh and M.S. Islam. 2015. Ionic transport in hybrid lead iodide perovskite solar cells. Nat. Commun. 6: 7497.

Eason, R. 2007. Pulsed Laser Deposition of Thin Films. Hoboken, New Jersey: Wiley & Sons.

Etgar, L., P. Gao, Z. Xue, Q. Peng, A.K. Chandiran, B. Liu et al. 2012. Mesoscopic CH3NH3PbI3/ TiO_2 Heterojunction solar cells. J. Am. Chem. Soc. 134: 17396–17399.

Fujihara, K., A. Kumar, R. Jose, S. Ramakrishna and S. Uchida. 2007. Spray deposition of electrospun TiO_2 nanorods for dye-sensitized solar cell. Nanotechnology 18: 365709.

Gaponenko, S.V. 1998. Optical Properties of Semiconductor Nanocrystals vol. 23. Cambridge, UK.: Cambridge University Press.

Ghicov, A. and P. Schmuki. 2009. Self-ordering electrochemistry: a review on growth and functionality of TiO_2 nanotubes and other self-aligned MOX structures. Chem. Commun. (Camb) 2009: 2791–808.

Ghosh, R., M.K. Brennaman, T. Uher, M.R. Ok, E.T. Samulski, L.E. McNeil et al. 2011. Nanoforest Nb_2O_5 photoanodes for Dye-sensitized solar cells by pulsed laser deposition. ACS Appl. Mater. Inter. 3: 3929–3935.

Gong, D., C.A. Grimes, O.K. Varghese, W. Hu, R.S. Singh, Z. Chen et al. 2001. Titanium oxide nanotube arrays prepared by anodic oxidation. J. Mater. Res. 16: 3331–3334.

Grätzel, M. 2001. Photoelectrochemical cells. Nature 414: 338–344.

Green, M.A. Third Generation Photovoltaics: Advanced Solar Energy Conversion. Berlin, Germany: Springer-Verlag Berlin Heidelberg, 2003.

Habisreutinger, S.N., T. Leijtens, G.E. Eperon, S.D. Stranks, R.J. Nicholas and H.J. Snaith. 2014. Carbon nanotube/polymer composites as a highly stable hole collection layer in perovskite solar cells. Nano. Lett. 14: 5561–8.

Han, B.S., S. Caliskan, W. Sohn, M. Kim, J.K. Lee and H.W. Jang. 2016. Room temperature deposition of crystalline nanoporous ZnO nanostructures for direct use as flexible DSSC photoanode. Nanosc. Res. Lett. 11: 221.

Hench, L.L. and J.K. West. 1990. The Sol-Gel Process. Chem. Rev. 90: 33–72.

Heo, J.H., S.H. Im, J.H. Noh, T.N. Mandal, C.-S. Lim, J.A. Chang et al. 2013. Efficient inorganic–organic hybrid heterojunction solar cells containing perovskite compound and polymeric hole conductors. Nat. Photonics 7: 486–491.

Hu, H., B. Dong, H. Hu, F. Chen, M. Kong, Q. Zhang et al. 2016. Atomic layer deposition of TiO_2 for a high-efficiency hole-blocking layer in hole-conductor-free perovskite solar cells processed in ambient air. ACS Appl. Mater. Inter. 8: 17999–18007.

Huang, C., R. Shi, A. Amini, Z. Wu, S. Xu, L. Zhang et al. 2015. Hierarchical ZnO nanostructures with blooming flowers driven by screw dislocations. Sci. Rep. 5: 8226.

Huang, J.H., K.J. Jiang, X.P. Cui, Q.Q. Zhang, M. Gao, M.J. Su et al. 2015. Direct conversion of $CH_3NH_3PbI_3$ from electrodeposited PbO for highly efficient planar perovskite solar cells. Sci. Rep. 5: 15889.

Hui-Seon Kim, Lee, Chang-Ryul, J.-H. Im, K.-B. Lee, T. Moehl, A. Marchioro et al. 2012. Lead Iodide Perovskite Sensitized All-Solid-State Submicron Thin Film Mesoscopic Solar Cell with Efficiency Exceeding 9%. Sci. Rep. 2: 591.

Hwang, I., I. Jeong, J.H. Lee, M.J. Ko and k. Yong. 2015. Enhancing stability of perovskite solar cells to moisture by the facile hydrophobic passivation. ACS Appl. Mater. Inter. 7: 17330–17336.

Im, J.-H., C.-R. Lee, J.-W. Lee, S.-W. Park and N.-G. Park. 2011. 6.5% Efficient perovskite quantum-dot-sensitized solar cell. Nanoscale 3: 4088–4093.

Jeon, N.J., J.H. Noh, W.S. Yang, Y.C. Kim, S. Ryu, J. Seo et al. 2015. Compositional engineering of perovskite materials for high-performance solar cells. Nature 517: 476–480.

Kääriäinen, T., D. Cameron, M.-L. Kääriäinen and A. Sherman. 2013. Atomic Layer Deposition: Principles, Characteristics, and Nanotechnology Applications, 2nd ed. Salem, Massachusetts, USA: Scrivener Publishing LLC, 2013.

Kalyanasundaram, K. 2010. Dye sensitized solar cells. France: CRC Press, 2010.

Kandjani, A.E., A. Shokufar, M.F. Tabriz, N.A.Arefia and M.R. Vaezi. 2009. Optical properties of Sol-Gel prepared nano ZnO. The effects of aging period and synthesis temperature. J. Optoelectron. Adv. M. 11: 289–295.

Keis, K., J. Lindgren, S.-E. Lindquist and A. Hagfeldt. 2000. Studies of the adsorption process of ru complexes in nanoporous ZnO electrodes. Langmuir 16: 4688–4694.

Keis, K., E. Magnusson, H. Lindstrom, S.-E. Lindquist and A. Hagfeldt. 2002. A 5% efficient photoelectrochemical solar cell based on nanostructured ZnO electrodes. Sol. Energ. Mat. Sol. C. 73: 51–58.

Khan, A. 2010. Raman spectroscopic study of the ZnO nanostructures. J. Pak. Mater. Soc. 4: 5–9.

Khataee, A. and G.A. Mansoori. 2012. Nanostructured Titanium Dioxide Materials: Properties, Preparation and Applications. Toh Tuck Link, Singapore: World Scientific Publishing Co. Pte. Ltd., 2012.

Kim, J.H., P.W. Liang, S.T. Williams, N. Cho, C.C. Chueh, M.S. Glaz et al. 2015. High-performance and environmentally stable planar heterojunction perovskite solar cells based on a solution-processed copper-doped nickel oxide hole-transporting layer. Adv. Mater. 27: 695–701.

Kim, K., K. Utashiro, Y. Abe and M. Kawamura. 2014. Structural properties of zinc oxide nanorods grown on al-doped zinc oxide seed layer and their applications in Dye-sensitized solar cells. Materials 7: 2522–2533.

Knez, N.P.a.M. 2012. Atomic Layer Deposition of Nanostructured Materials, 2nd ed. Weinheim, Germany: Willey-VCH Verlag & Co.
Kojima, A., K. Teshima, Y. Shirai and T. Miyasaka. 2009. Organometal halide perovskites as visible-light sensitizers for photovoltaic cells. J. Am. Chem. Soc. 131: 6050–6051.
Kumar, M.H., N. Yantara, S. Dharani, M. Graetzel, S. Mhaisalkar, P.P. Boix et al. 2013. Flexible, low-temperature, solution processed ZnO-based perovskite solid state solar cells. Chem. Commun. 49: 11089–11091.
L´opez, A. B. e. C. o., A.M.ı. Vega and A.L. L´opez. 2012. Next Generation PV: New Concepts. Berlin, Germany: Springer-Verlag Berlin Heidelberg.
Laban, W.A. and L. Etgar. 2013. Depleted hole conductor-free lead halide iodide heterojunction solar cells. Energy Environ. Sci. 6: 3249–3253.
Labis, J.P., M. Hezam, A. Al-Anazi, H. Al-Brithen, A.A. Ansari, A. Mohamed El-Toni et al. 2015. Pulsed laser deposition growth of 3D ZnO nanowall network in nest-like structures by two-step approach. Sol. Energ. Mat. Sol. C. 143: 539–545.
Lamberti, A., A. Sacco, S. Bianco, D. Manfredi, M. Armandi, M. Quaglio et al. 2013. An easy approach for the fabrication of TiO_2 nanotube-based transparent photoanodes for dye-sensitized solar cells. Sol. Energy 95: 90–98.
Lamberti, A., R. Gazia, A. Sacco, S. Bianco, M. Quaglio, A. Chiodoni et al. 2014. Coral-Shaped ZnO nanostructures for dye-sensitized solar cell photoanodes. Prog. Photovoltaics Res. Appl. 22: 189–197.
Lamberti, A., A. Chiodoni, N. Shahzad, S. Bianco, M. Quaglio and C.F. Pirri. 2015. Ultrafast Room-Temperature Crystallization of TiO_2 Nanotubes Exploiting Water-Vapor Treatment. Sci. Rep. 5: 78081–6.
Lang, F., M.A. Gluba, S. Albrecht, J. Rappich, L. Korte, B. Rech et al. 2015. Perovskite solar cells with large-area CVD-graphene for tandem solar cells. J. Phys. Chem. Lett. 6: 2745–2750.
Lee, J.W., D.J. Seol, A.N. Cho and N.G. Park. 2014. High-efficiency perovskite solar cells based on the black polymorph of $HC(NH_2)_2 PbI_3$. Adv. Mater. 26: 4991–4998.
Lee, K.T., L.J. Guo and H.J. Park. 2016. Neutral- and multi-colored semitransparent perovskite solar cells. Molecules 21: 475 (1-21).
Lee, M.M., J. Teuscher, T. Miyasaka, T.N. Murakami and H.J. Snaith. 2012. Efficient hybrid solar cells based on meso-superstructured organometal halide perovskites. Science 338: 643–647.
Lee, S.Y. and S.H. Kim. 2012. Deposition of TiO $_2$ passivation layer by plasma enhanced chemical vapor deposition between the transparent conducting oxide and mesoporous TiO^2 electrode in dye sensitized solar cells. Jpn. J. Appl. Phys. 51: 10NE19.
Lee, S.Y. and S.H. Kim. 2014. Deposition of Zinc Oxide photoelectrode using plasma enhanced chemical vapor deposition for dye-sensitized solar cells. J. Nanosci. Nanotech. 14: 9272–9278.
Leyden, M.R., Y. Jiang and Y. Qi. 2016. Chemical vapor deposition grown formamidinium perovskite solar modules with high steady state power and thermal stability. J. Mater. Chem. A.
Li, B., X. Gao, H.-C. Zhang and C. Yuan. 2014. Energy modeling of electrochemical anodization process of titanium dioxide nanotubes. ACS Sustain. Chem. Eng. 2: 404–410.
Liang, C., Z. Wu, P. Li, J. Fan, Y. Zhang and G. Shao. 2016. Chemical bath deposited rutile TiO2 compact layer toward efficient planar heterojunction perovskite solar cells. Appl. Surf. Sci. DOI: 10.1016/j.apsusc.2016.06.171.
Liang, Y., Y. Yao, X. Zhang, W.-L. Hsu, Y. Gong, J. Shin et al. 2016. Fabrication of organic-inorganic perovskite thin films for planar solar cells via pulsed laser deposition. AIP Adv. 6: 015001 (1-7).
Lin, Z. and J. Wang. 2014. Low-cost Nanomaterials: Towards Greener and More Efficient Energy Applications. London, UK: Springer-Verlag London.

Liu, B. and E.S. Aydil. 2009. Growth of oriented single-crystalline rutile TiO_2 nanorods on transparent conducting substrates for dye-sensitized solar cells. J. Am. Chem. Soc. 131: 3985–90.

Liu, D. and T.L. Kelly. 2014. Perovskite solar cells with a planar heterojunction structure prepared using room-temperature solution processing techniques. Nat. Photon 8: 133–138.

Liu, M., M.B. Johnston and H.J. Snaith. 2013. Efficient planar heterojunction perovskite solar cells by vapour deposition. Nature 501: 395–398.

Liu, P., Z. Yu, N. Cheng, C. Wang, Y. Gong, S. Bai et al. 2016. Low-cost and efficient hole-transport-material-free perovskite solar cells employing controllable electron-transport layer based on P25 nanoparticles. Electrochem. Acta 213: 83–88.

Liu, X., J. Fang, Y. Liu and T. Lin. 2016. Progress in nanostructured photoanodes for dye-sensitized solar cells. Front. Mater. Sci. 10: 225–237.

Lu, H., W. Tian, J. Guo and L. Li. 2015. Interface engineering through atomic layer deposition towards highly improved performance of dye-sensitized solar cells. Sci. Rep. 5: 12765.

Ma, C.-W., C.-M. Chang, P.-C. Huang and Y.-J. Yang. 2015. Sea-Urchin-Like ZnO nanoparticle film for dye-sensitized solar cells. J. Nanomater. 2015: Article ID 679474.

Mahmood, A., N. Ahmed, Q. Raza, T. Muhammad Khan, M. Mehmood, M.M. Hassan et al. 2010. Effect of thermal annealing on the structural and optical properties of ZnO thin films deposited by the reactive e-beam evaporation technique. Physica. Scripta. 82: 065801.

Mahpeykar, S.M., M.K. Tabatabaei, H. Ghafoori-fard, H. Habibiyan and J. Koohsorkhi. 2013. Low-temperature self-assembled vertically aligned carbon nanofibers as counter-electrode material for dye-sensitized solar cells. Nanotechnology 24: 435402.

Mali, S.S., C.S. Shim and C.K. Hong. 2015. Highly porous Zinc Stannate (Zn_2SnO_4) nanofibers scaffold photoelectrodes for efficient methyl ammonium halide perovskite solar cells. Sci. Rep. 5: 11424.

Mart´i, A. and A. Luque. 2004. Next Generation Photovoltaics: High efficiency through full spectrum utilization. Bristol and Philadelphia: Institute of Physics Publishing.

Meng, L., H. Chen, C. Li and M.P. dos Santos. 2015. Preparation and characterization of dye-sensitized TiO2 nanorod solar cells. Thin Solid Films 577: 103–108.

Meng, Y., Y. Lin and Y. Lin. 2014. Electrodeposition for the synthesis of ZnO nanorods modified by surface attachment with ZnO nanoparticles and their dye-sensitized solar cell applications. Ceram. Int. 40: 1693–1698.

O'Regan, B. and M. Grätzel. 1991. A Low-cost, high-efficiency solar cell based on dye-sensitized colloidal TiO_2 films. Nature 353: 737–740.

Ono, L.K., S.R. Raga, S. Wang, Y. Kato and Y. Qi. 2015. Temperature-dependent hysteresis effects in perovskite-based solar cells. J. Mater. Chem. A 3: 9074–9080.

Pang, J., A. Bachmatiuk, I. Ibrahim, L. Fu, D. Placha, G.S. Martynkova et al. 2015. CVD growth of 1D and 2D sp^2 carbon nanomaterials. J. Mater. Sci. 51: 640–667.

Park, J.H., J. Seo, S. Park, S.S. Shin, Y.C. Kim, N.J. Jeon et al. 2015. Efficient CH3 NH3 PbI3 perovskite solar cells employing nanostructured p-Type NiO electrode formed by a pulsed laser deposition. Adv. Mater. 27: 4013–4019.

Park, N.-G. 2015. Perovskite Solar Cells: an Emerging Photovoltaic Technology. Mater. Today 18: 65–72.

Park, Y.J., J.M. Ha, G. Ali, H.J. Kim, Y. Addad and S.O. Cho. 2015. Controlled fabrication of nanoporous oxide layers on zircaloy by anodization. Nanosc. Res. Lett. 10: 377.

Pierson, H.O. 1999. Handbook of Chemical Vapor Deposition (CVD): Principles, Technology, and Applications, 2nd ed. Norwich, New York, U.S.A.: William Andrew Publishing, LLC.

Ponseca, C.S., Jr., T.J. Savenije, M. Abdellah, K. Zheng, A. Yartsev, T. Pascher et al. 2014. Organometal halide perovskite solar cell materials rationalized: ultrafast charge generation, high and microsecond-long balanced mobilities, and slow recombination. J. Am. Chem. Soc. 136: 5189–5192.

Postels, B., A. Bakin, H.H. Wehmann, M. Suleiman, T. Weimann, P. Hinze et al. 2008. Electrodeposition of ZnO nanorods for device application. Appl. Phys. A 91: 595–599.

Qiu, W., U.W. Paetzold, R. Gehlhaar, V. Smirnov, H.-G. Boyen, J.G. Tait et al. 2015. An electron beam evaporated TiO_2 layer for high efficiency planar perovskite solar cells on flexible polyethylene terephthalate substrates. J. Mater. Chem. A 3: 22824–22829.

Qiu, Y., W. Chen and S. Yang. 2010. Facile hydrothermal preparation of hierarchically assembled, porous single-crystalline ZnO nanoplates and their application in dye-sensitized solar cells. J. Mater. Chem. 20: 1001.

Roza, L., K.A.J. Fairuzy, P. Dewanta, A.A. Umar, M.Y.A. Rahman and M.M. Salleh. 2015. Effect of molar ratio of zinc nitrate: hexamethylenetetramine on the properties of ZnO thin film nanotubes and nanorods and the performance of dye-sensitized solar cell (DSSC). J. Mater. Sci.: Mater. Electr. 26: 7955–7966.

Sacco, A., A. Lamberti, R. Gazia, S. Bianco, D. Manfredi, N. Shahzad et al. 2012. High efficiency dye-sensitized solar cells exploiting sponge-like ZnO nanostructures. Phys. Chem. Chem. Phys. 14: 16203–8.

Saito, M. and S. Fujihara. 2008. Large photocurrent generation in dye-sensitized ZnO solar cells. Energy Environ. Sci. 1: 280–283.

Sanjay, S., V. Naresh, S. Ranbir, A.K. Swarnkar, S. Manauti and B. Tejashree. 2015. Enhancing the efficiency of flexible dye-sensitized solar cells utilizing natural dye extracted from Azadirachta indica. Mater. Res. Express 2: 105903.

Schlur, L., A. Carton, P. Lévêque, D. Guillon and G. Pourroy. 2013. Optimization of a new ZnO nanorods hydrothermal synthesis method for solid state dye sensitized solar cells applications. J. Phys. Chem. C 117: 2993–3001.

Seshan, K. 2002. Handbook of Thin-Film Deposition Processes and Techniques: Principles, Methods, Equipment and Applications, 2nd ed. Norwich, New York, U.S.A.: William Andrew Publishing.

Shahzad, N., F. Risplendi, D. Pugliese, S. Bianco, A. Sacco, A. Lamberti et al. 2013. Comparison of hemi-squaraine sensitized TiO_2 and ZnO photoanodes for DSSC applications. The J. Phys. Chem. C 117: 22778–22783.

Shahzad, N., A. Lamberti, D. Pugliese, M.I. Shahzad and E. Tresso. 2016. Real time monitoring of ultrafast sensitization for dye-sensitized solar cell photoanodes. Sol. Energy 130: 74–80.

Singh, E. and H.S. Nalwa. 2015. Graphene-based dye-sensitized solar cells: A review. Sci. Adv. Mater. 7: 1863–1912.

Srivastava, A.K. 2013. Oxide Nanostructures: Growth, Microstructures, and Properties. Boca Raton, Florida, USA: Pan Stanford Publishing, CRC Press.

Stergiopoulos, T., A. Ghicov, V. Likodimos, D.S. Tsoukleris, J. Kunze, P. Schmuki et al. 2008. Dye-sensitized solar cells based on thick highly ordered TiO_2 nanotubes produced by controlled anodic oxidation in non-aqueous electrolytic media. Nanotechnology 19: 235602.

Stoumpos, C.C., C.D. Malliakas and M.G. Kanatzidis. 2013. Semiconducting tin and lead iodide perovskites with organic cations: phase transitions, high mobilities, and near-infrared photoluminescent properties. Inorg. Chem. 52: 9019–9138.

Stringfellow, G.B. and M.G. Craford. 1997. High Brightness Light Emitting Diodes: Semiconductors and Semimetals vol. 48. Massachusetts, USA: Academic Press.

Su, T.S., T.Y. Hsieh, C.Y. Hong and T.C. Wei. 2015. Electrodeposited Ultrathin TiO_2 Blocking Layers for Efficient Perovskite Solar Cells. Sci. Rep. 5: 16098.

Tamiya, K. and K. Taguchi. 2015. Experimental investigation of carbon nanotubes counter electrodes for dye-sensitized solar cells. International Int. J. Pharm. Biol. Sci. 4: 227–231.

Tian, J. and G. Cao. 2013. Semiconductor quantum dot-sensitized solar cells. Nano Rev. 4: 22578.

Unger, E.L., E.T. Hoke, C.D. Bailie, W.H. Nguyen, A.R. Bowring, T. Heumüller et al. 2014. Hysteresis and transient behavior in current–voltage measurements of hybrid-perovskite absorber solar cells. Energy Environ. Sci. 7: 3690–3698.

van Reenen, S., M. Kemerink and H.J. Snaith. 2015. Modeling anomalous hysteresis in perovskite solar cells. J. Phys. Chem. Lett. 6: 3808–3814.

Wang, S.-F., K.K. Rao, T.C.K. Yang and H.-P. Wang. 2011. Investigation of nitrogen doped diamond like carbon films as counter electrodes in dye sensitized solar cells. J. Alloys Compd. 509: 1969–1974.

Wang, X., L. Zhi and K. Mullen. 2008. Transparent, conductive graphene electrodes for dye-sensitized solar cells. Nano Lett. 8: 323–327.

Wang, X., Z. Li, W. Xu, S.A. Kulkarni, S.K. Batabyal, S. Zhang et al. 2015. TiO_2 nanotube arrays based flexible perovskite solar cells with transparent carbon nanotube electrode. Nano Energy 11: 728–735.

Wang, Z., P. Liu, Y. Ito, S. Ning, Y. Tan, T. Fujita et al. 2016. Chemical Vapor Deposition of Monolayer Mo1-xWxS2 Crystals with Tunable Band Gaps. Sci. Rep. 6: 21536.

Willander, M. 2013. Zinc Oxide Nanostructures: Advances and Applications. Boca Raton, Florida, USA: Pan Stanford Publishing, CRC Press.

Xiao, M., F. Huang, W. Huang, Y. Dkhissi, Y. Zhu, J. Etheridge et al. 2014. A fast deposition-crystallization procedure for highly efficient lead iodide perovskite thin-film solar cells. Angew. Chem. 126: 10056–10061.

Yahya, N. 2010. Carbon and Oxide Nanostructures: Synthesis, Characterisation and Applications vol. 5. New York, USA: Springer.

Yamamoto, K., Y. Zhou, T. Kuwabara, K. Takahashi, M. Endo, A. Wakamiya et al. 2014. "Low temperature TiOx compact layer by chemical bath deposition method for vapor deposited perovskite solar cells," presented at the 2014 IEEE 40th Photovoltaic Specialist Conference (PVSC), Denver, Colorado.

Yang, W.S., J.H. Noh, N.J. Jeon, Y.C. Kim, S. Ryu, J. Seo et al. 2015. High-performance photovoltaic perovskite layers fabricated through intramolecular exchange. Science 348: 1234–1237.

Yang, W., X. Xu, Z. Tu, Z. Li, B. You, Y. Li et al. 2015. Nitrogen plasma modified CVD grown graphene as counter electrodes for bifacial dye-sensitized solar cells. Electrochem. Acta 173: 715–720.

Yang, Z. and W.-H. Zhang. 2014. Organolead halide perovskite: A rising player in high-efficiency solar cells. Chinese J. Catal. 35: 983–988.

Ye, M., X. Wen, M. Wang, J. Iocozzia, N. Zhang, C. Lin et al. 2015. Recent advances in dye-sensitized solar cells: from photoanodes, sensitizers and electrolytes to counter electrodes. Mater. Today 18: 155–162.

Yuksel, B., E.D. Sam, O.C. Aktas, M. Urgen and A.F. Cakir. 2009. Determination of sodium migration in sol- gel deposited titania films on soda-lime glass with r.f. glow discharge optical emission spectroscopy. Appl. Surf. Sci. 255: 4001–4004.

Zhang, D.W., S. Chen, X.D. Li, Z.A. Wang, J.H. Shi, Z. Sun et al. 2009. Cadmium sulfide quantum dots grown by chemical bath deposition for sensitized solar cell applications. P. SPIE—Int. Soc. Opt. Eng. 7518: 751804-1-9.

Zhang, J.Z. 2009. Optical Properties and Spectroscopy of Nanomaterials. New Jersy: World Scientific Publishing Co. Pte. Ltd.

Zhang, Q., C.S. Dandeneau, X. Zhou and G. Cao. 2009. ZnO nanostructures for dye-sensitized solar cells. Adv. Mater. 21: 4087–4108.

Zhou, H., Q. Chen, G. Gang Li, S. Luo, T.-b. Song, H.-S. Duan et al. 2014. Interface engineering of highly efficient perovskite solar cells. Science 345: 542–546.

Zhou, X., C. Bao, F. Li, H. Gao, T. Yu, J. Yang et al. 2015. Hole-transport-material-free perovskite solar cells based on nanoporous gold back electrode. RSC Adv. 5: 58543–58548.

6

Application of Metal-Organic Frameworks (MOFs) for Hydrogen Storage

*Mohammad Jafarzadeh** and *Amir Reza Abbasi*

1. Hydrogen Storage: Opportunities and Limitations

There is a tremendous demand for the development of renewable and clean energy resources to replace fossil fuels. Energy storage is recognized as an important intermediate step towards versatile, clean, and efficient energy applications (Kong et al. 2015). Among the various candidates for energy storage systems, hydrogen (H_2) is a promising alternative energy carrier due to its environmental benefits (zero carbon emissions), high gravimetric energy density (33.3 kWh kg^{-1}, compared to hydrocarbons with 12.4–13.9 kWh kg^{-1}), light-weight, renewability, and abundance. Since its oxidation product is water, hydrogen is a green alternative to non-renewable fuel sources because it can react with oxygen to generate electricity without byproducts (Kong et al. 2015).

The main technical barrier to the commercialization of hydrogen-based fuel-cell vehicles is on-board hydrogen storage. There is concern over the efficiency and safety issues due to the difficulty of storing and using gaseous hydrogen in a container under high pressure (Perles et al. 2005). Hydrogen

Faculty of Chemistry, Razi University, Kermanshah 67149-67346, Iran.
* Corresponding author: mjafarzadeh1027@yahoo.com

has very low volumetric density, both under normal conditions and down to its boiling point at 20 K. Therefore, it is necessary for hydrogen to be either (i) liquefied; (ii) compressed under high pressure; and/or (iii) stored in a solid state (in solid storage and upon surface adsorption, molecular hydrogen dissociates into atomic form and generates a solid hydride solution on the surface of active metals) (Rowsell and Yaghi 2005). Each of the three methods for developing on-board hydrogen storage has their own limitations such as: (i) high energy consumption for liquefaction (cryogenic storage) and the continuous boil-off of hydrogen; (ii) high energy consumption and high pressure for compressed hydrogen, in addition to the safety issues of using a high-pressure storage tank; and (iii) requiring high temperature (e.g., over 373 K) to release the stored hydrogen from its solid phase, and possible susceptibility to impurities (Rowsell and Yaghi 2005, Hu and Zhang 2010). Although, solid storage has difficulties in hydrogen release, it can be applied to long-term storage with good safety (Latroche et al. 2006).

In practical instances of H_2 storage, the volume and mass of the storage tank are considered less important for stationary storage. However, for onboard applications, tank volume and mass, and the necessary heavy and expensive means of heat exchange (e.g., cryogenic cooling for liquid and compressed H_2 and high temperature for solid stored H_2) can prohibitively affect the cost and weight (Dincã and Long 2008, Murray et al. 2009). Moreover, it is reasonable to expect that a real vehicle can be driven for 300 miles with a filled tank, operated at ambient conditions, and refueled quickly and safely (Rowsell and Yaghi 2005). To solve these technical difficulties and to reduce costs, the storage of hydrogen in solid materials with high porosity has been proposed. Porous materials for storage should provide the following features: (i) large storage capacity for high gravimetric and volumetric storage density (the fuel tanks should be light and small); (ii) good thermodynamics (adsorption/desorption occur at a reasonable pressure and temperature); and (iii) fast reaction kinetics (fast tank refilling) (Li and Thonhauser 2012). Therefore, the US Department of Energy (DOE) has set a technical target of 2020 for the development of H_2 storage systems. The vision for near future technology is a 5.5 wt% gravimetric and 40 g L^{-1} volumetric storage capacity of H_2 at an operating temperature of 40–60°C under pressure below 100 atm (Yan et al. 2014).

It has been found that liquid hydrogen has a weight density of 70.8 kg m^{-3} at 20 K at atmospheric pressure and 5 kg of H_2 is known to occupy a volume of 56 m^3 under standard conditions (Suh et al. 2012). Therefore, searching for adsorbent materials with a larger storage capacity, a high structural affinity to hydrogen molecules, and reasonable durability is required.

2. Porous Materials

Highly porous solid materials, such as zeolites, carbon materials, polymers, and metal-organic frameworks (MOFs) provide high surface area, large pore size, and large pore volumes that are appropriate for the adsorption and storage of gases, e.g., hydrogen, methane, carbon dioxide, etc. The storage of gas in solid sorbents is technologically important due to the potential application of hydrogen for energy generation, the use of methane as a hydrocarbon fuel, and carbon dioxide as a main component of greenhouse gases. The following advantages of storing gas inside a porous material rather than physically filling a bottle or tank are proposed: (i) high storage density (due to large porosity and surface area per gram); (ii) better safety (high pressure is common in tank filling); and (iii) easier handling (Morris and Wheatley 2008). There are some critical issues in the usage and design of porous materials that should be considered. For example, achieving sufficient adsorption capacity, quick adsorption/desorption cycles, processability, robustness, and durability with reasonable lifetimes (Morris and Wheatley 2008, Slater and Cooper 2015).

Zeolites and activated carbons exhibit poor gas storage/separation. Zeolites are highly-crystalline porous materials with regular channels but with a low porosity (they are microporous). Activated carbons are also highly porous but have a broad pore size distribution (Kitagawa et al. 2004). Besides inorganic-based porous materials, porous organic polymers (POPs) have been introduced as rigid, highly cross-linked, and amorphous polymers which possess relatively high internal surface areas and micro- and mesoporosity (Kaur et al. 2011). Although POPs are noncrystalline and have nonuniform pores, they can be synthesized readily (*cf.* other solid inorganic materials) "by incorporating di/multitopic building blocks (monomers) into well-known step-growth and chain-growth polymerization processes to provide cross-links between propagating polymer chains, yielding three-dimensional (3D)-network materials" (Kaur et al. 2011). These materials have been found to be promising as a support for catalytic systems. POPs can be used as catalysts in three different ways: "POPs that incorporate rigid well-defined homogeneous catalysts as building blocks, POPs that can be modified post-synthesis, and POPs that encapsulate metal particles" (Kaur et al. 2011). These microporous polymers are also known by different names: polymers of intrinsic microporosity (PIMs), porous organic frameworks (POFs), conjugated microporous polymers (CMPs), and porous aromatic frameworks (PAFs) (Kaur et al. 2011).

Most recently, Slater and Cooper (2015) published an interesting review paper in *Science* which can be useful for researchers in the field of porous materials. They reported a study comparing certain effective parameters such as porosity, crystallinity, stability, modularity, processing, and designability in different porous solid materials (e.g., zeolites, metal-organic frameworks, covalent organic frameworks, porous organic polymers, and porous molecular solids).

3. Porous Coordination Polymers and Reticular Chemistry

Porous coordination polymers (PCPs) are built from metallic/non-metallic nodes connected by organic linkers to extend well-ordered periodic/crystalline materials (Morris and Wheatley 2008). The main characteristic of these materials is porosity, with internal surface areas exceeding 5000 $m^2 g^{-1}$ compared to only several hundred $m^2 g^{-1}$ for zeolites. PCPs can be easily synthesized from molecular units (connectors and linkers) under mild conditions *via* a reticular chemistry approach. Transition-metal ions are generally used as connectors and their coordination numbers can range from 2 to 7 to give various geometries, e.g., linear, T- or Y-shaped, tetrahedral, square-planar, square-pyramidal, trigonal-bipyramidal, octahedral, trigonal-prismatic, and pentagonal-bipyramidal (Kitagawa et al. 2004).

Reticular chemistry has been defined and developed by Yaghi and his group: "with the linking of molecular building blocks (organic molecules, inorganic clusters, dendrimers, peptides, and proteins) into predetermined structures in which such units are repeated and are held together by strong bonds" (Yaghi and Li 2009). Through this approach, interesting classes of crystalline porous materials have been designed, including covalent organic frameworks, zeolitic imidazolate frameworks, and metal-organic frameworks (MOFs).

4. Covalent Organic Frameworks

Covalent organic frameworks (COFs) are porous crystalline materials made by the linkage of organic secondary building units (SBUs) through strong covalent bonds to extended periodic structures (Wan et al. 2011). COFs are built from light elements (thus far C, N, O, B, Si) held together by strong covalent bonds (B–O, C–N, B–N, and B–O–Si) (Waller et al. 2015). The first COF (COF-1, $(C_3H_2BO)_6 \cdot (C_9H_{12})_1$) was synthesized by condensation reactions of phenyl diboronic acid $\{C_6H_4[B(OH)_2]_2\}$ and hexahydroxytriphenylene $[C_{18}H_6(OH)_6]$ *via* a simple one-pot procedure under mild reaction conditions (Côté et al. 2005). COF-1 (Fig. 1) exhibited rigid structures, permanent porosity (surface areas: 711 $m^2 g^{-1}$, pore size: 7–20 Å), and high thermal stability (up to 600°C). Different topologies, two-dimensional (2D) (Côté et al. 2007) and 3D (El-Kaderi et al. 2007) materials are constructed by controlling the degree of connectivity, geometry, and functionality. The 2D COFs have been designed by the stacking of organic layers (Wan et al. 2011),

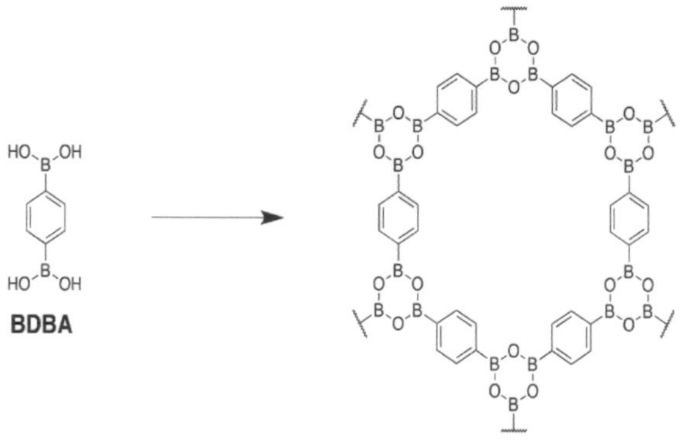

Figure 1. A synthetic method for COF-1. Reprinted with permission from (Waller et al. 2015). Copyright (2015) American Chemical Society.

while 3D COFs were made by targeting two nets based on triangular and tetrahedral nodes (El-Kaderi et al. 2007). The 3D COFs (COF-102 and COF-103) exhibited high surface areas of 3472–4210 $m^2 g^{-1}$ and extremely low densities of 0.17 g cm^{-3} (El-Kaderi et al. 2007) and are, therefore, attractive candidates for gas storage of H_2 (Han et al. 2008), CH_4 and CO_2 (Furukawa and Yaghi 2009), and NH_3 (Doonan et al. 2010). Since COFs have a high density of Lewis acid boron sites in their structure, they are able to strongly interact with Lewis base ammonia (Doonan et al. 2010).

5. Zeolite Imidazole Frameworks

Zeolitic imidazolate frameworks (ZIFs) are a class of porous and crystalline materials that have the advantages of inorganic zeolites (high stability) and MOFs (high porosity and organic functionality) (Hayashi et al. 2007). In ZIFs, extended three-dimensional structures are constructed from tetrahedral metal ion (e.g., Zn, Co) nodes connected through ditopic imidazolate ($C_3N_2H_3^-$ = Im) or functionalized Im bridges (Phan et al. 2010, Nguyen et al. 2014). The resulting structure exhibits topologies resembling microporous zeolites, as the M-Im-M angle is similar to the Si-O-Si angle (145°) in zeolites. A large variety of ZIFs with different topologies (structure symbols: sod, cag, mer, crb, dft, gis, rho, gme and lta) (Wang et al. 2008) can be made (refer Fig. 2) by using metal salts with imidazole (ImH) using, for example, mechanochemistry (Beldon et al. 2010) and solvent-assisted linker exchange (Karagiaridi et al. 2012) methods. ZIFs exhibit permanent porosity, and high thermal and chemical stability (Park et al. 2006), therefore having potential applications in catalysis (Stephenson et al. 2015), gas separation, and the storage (Thornton et al. 2012) of CO_2 (Banerjee et al. 2009) and

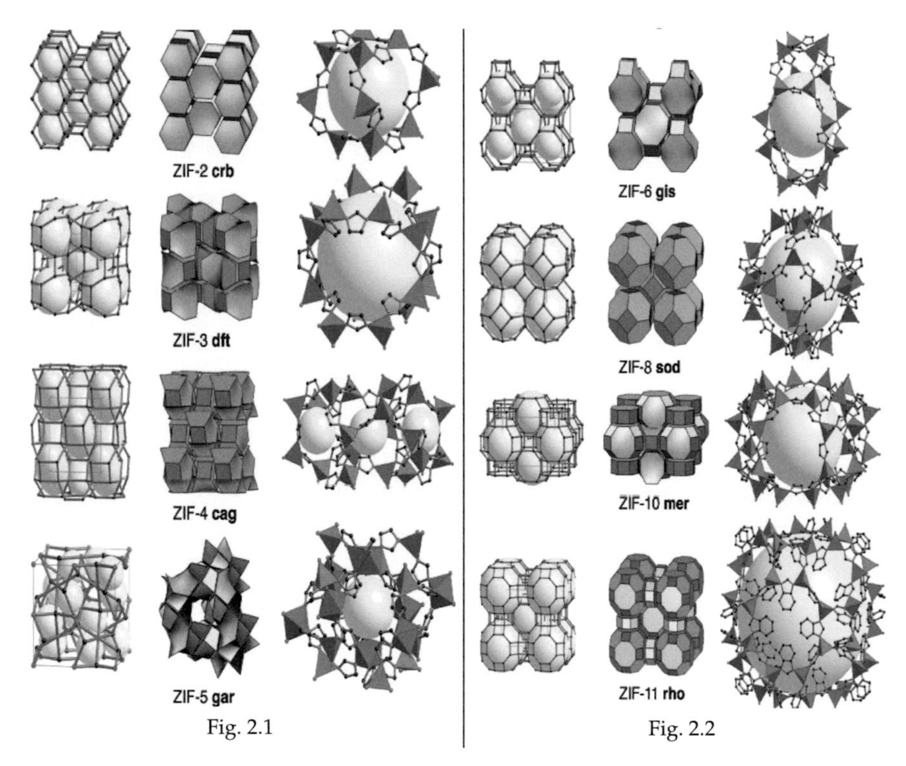

Figure 2. ZIFs with different topologies (Park et al. 2006). Copyright (2006) National Academy of Sciences, U.S.A.

CH_4 (Houndonougbo et al. 2013). By precise control of the pore size and structural functionality, gas storage could be improved (Banerjee et al. 2009).

6. Metal-Organic Frameworks

Metal-organic frameworks (MOFs) are hybrid periodic materials with infinite crystalline lattices that are generally made of transition metal ions or cluster nodes bridged with multitopic organic struts to form ordered pores and channels (Fig. 3) with relatively rigid structures (Zhang and Shreeve 2014, Farha et al. 2008). MOFs possess large internal surface areas (exceeding ca. 7000 $m^2 g^{-1}$ experimentally and 14000 $m^2 g^{-1}$ computationally) (Mondloch et al. 2013), ultralow densities, high available free volume (up to 90% free volume), the potential for multi-functionality, phase purity, and simplicity in synthesis.

Numerous effective parameters of metal ions, ligand geometries, temperature, solvent, pH, and stoichiometry can play the key roles in modulating MOF network topology and dimensionality (Zhang and

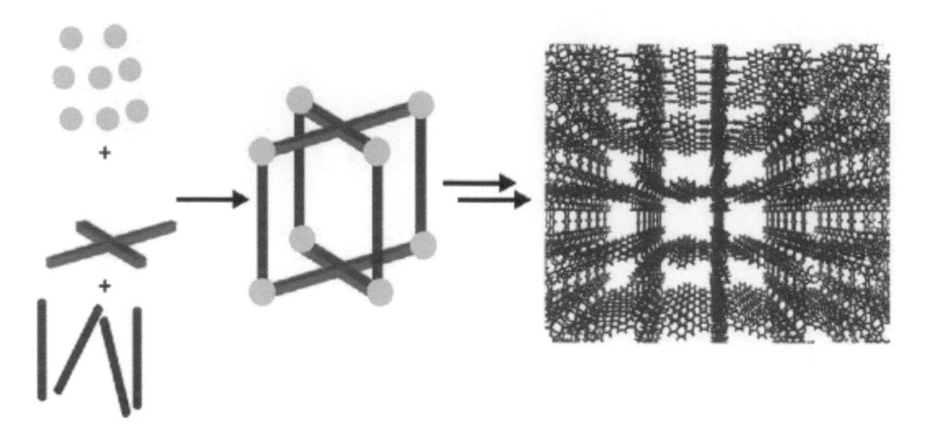

Figure 3. General scheme for MOF construction; sphere, thick stick and thin stick represent metallic node, ligand, and organic strut, respectively. Reprinted with permission from (Farha and Hupp 2010). Copyright (2010) American Chemical Society.

Shreeve 2014). In addition, it has been found that solubility of the organic link and metal salt, solvent polarity, and ionic strength of the medium are among other factors that affect the character of the resulting MOFs (Rosi et al. 2002). Isolation of pure crystalline MOFs from some impurities in synthesis media (organic-strut and/or metal-ion building blocks, catenated networks) can affect the performance of MOF structures (Farha et al. 2008). Removal of the solvent (trapped during the synthesis) without damaging the framework and collapsing pores can activate the structure (Farha and Hupp 2010) to allow potential application of gas storage, since the accessible metal sites can act as Lewis acids (Jiang and Yaghi 2015). MOFs can be activated through different approaches such as (i) conventional heating and vacuum; (ii) solvent-exchange; (iii) supercritical CO_2 (scCO_2) exchange; (iv) freeze-drying; and (v) chemical treatment (Mondloch et al. 2013).

A challenge in the preparation of MOFs is the generation of micropores within the structures. Micropores do not generally allow fast molecular diffusion and mass transfer and, consequently, access to the internal pores by catalyst and drug molecules for their applications (catalytic reaction and drug delivery) is significantly restricted (Song et al. 2012). On the other hand, large pores are susceptible to easy collapse during applications under pressure and high temperature (Song et al. 2012). Therefore, a precise adjustment of the pore formation is essential for the aforementioned applications.

Structural modification has been proposed as a useful method to create functionalized MOF materials with tunable mesoporosity (Song et al. 2012) and the capability of selective adsorption and separation of small molecules in MOF structures (McGuire and Forgan 2015). There are different

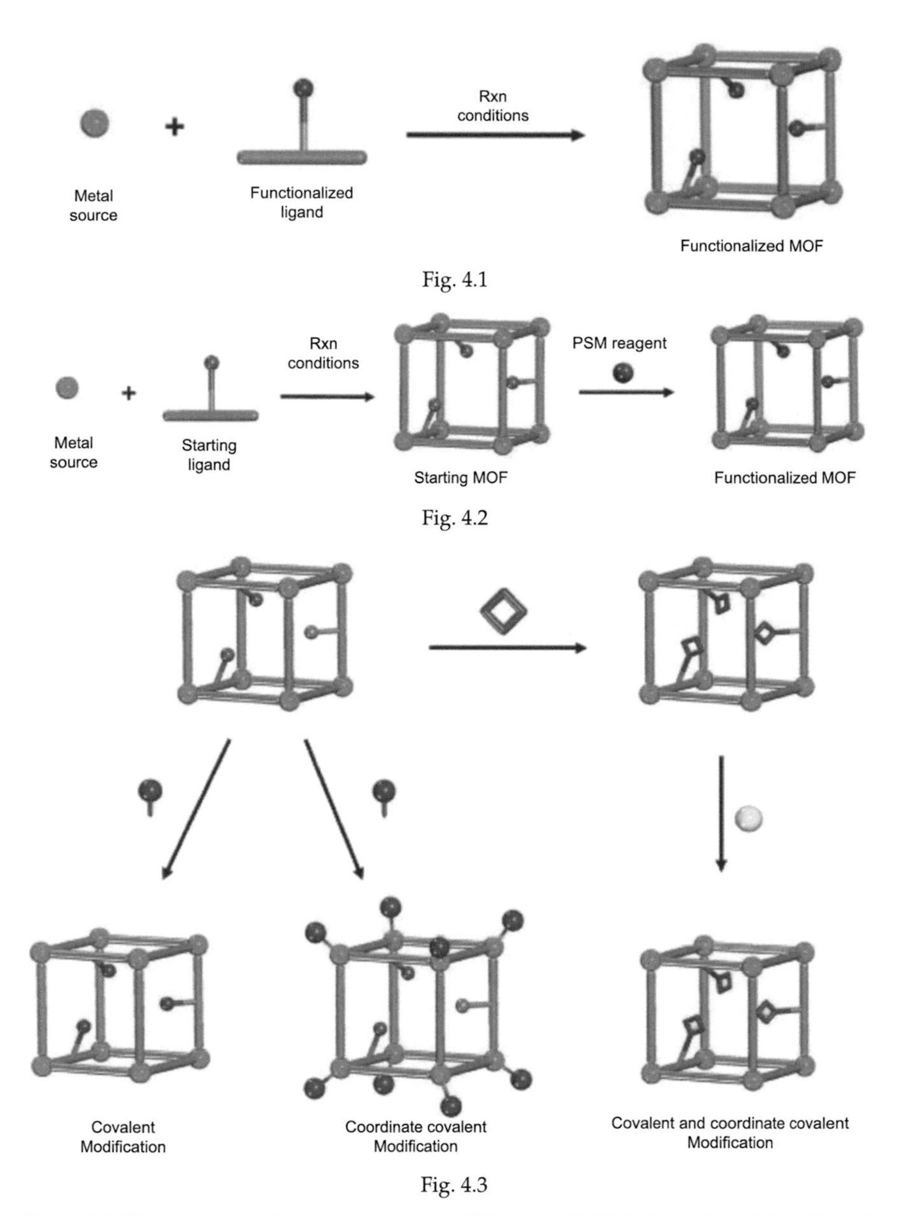

Figure 4. Different approaches for structural modification of MOFs. Reproduced from (Tanabe and Cohen 2011) with permission from The Royal Society of Chemistry, Copyright (2011).

approaches to modification (Fig. 4) such as "(i) surface modification during MOF synthesis through coordination modulation, (ii) post-synthetic surface modification, and (iii) MOFs grown directly on the surfaces of others in the form of epitaxial and surface growth" (McGuire and Forgan 2015). Post-synthetic modification (PSM) has attracted great attention due to its ease and versatility in modifying the external and internal available surface of the MOF without significant undesirable pore blockage. This technique also allows precise control over the types and number of incorporated functional groups into the framework without affecting its structural stability, as compared to pre-functionalization approaches (Tanabe and Cohen 2011).

PSM can be performed in two ways: "(i) modification of linkers (covalent modification, deprotection of linker functionality, and electron addition) and (ii) modification of metal-containing nodes (incorporation of pendant ligands to unsaturated metal sites, alkyl or silyl grafting to oxygen atoms in metal-oxide nodes, and attachment of metal ions or complexes at node oxygen sites *via* atomic layer deposition and/or with organometallic species)" (Deria et al. 2014). Post-synthesis based on building block replacement (BBR) was recently developed through (i) solvent-assisted linker exchange (SALE); (ii) non-bridging ligand replacement; and (iii) transmetalation at nodes or within linkers (Deria et al. 2014).

Multivariate (MTV) functionalization is another approach for structural heterogeneity since multiple metals and/or organic linkers are incorporated into the MOF while the original structure is retained (Jiang and Yaghi 2015). There are different approaches for achieving structural heterogeneity such as "mixing of organic linkers within the MOF backbone, mixing of the metal-containing secondary building units (SBUs), mixing of both the SBUs and organic linkers within the same MOF backbone, mixing of functional groups along the backbone, MOFs with random and ordered defects, attaching MOFs to functional surfaces, combining inorganic nanocrystals and MOFs, MOFs with heterogeneous pores" (Furukawa et al. 2015). The types of metal ion and organic ligand can affect the topology (pore size and shape) of the framework and its chemical functionality for applications in the area of gas storage, catalysis, and separation (Jiang and Yaghi 2015). Multivariate metal-organic frameworks (MTV-MOFs) can enhance gas storage capacity and their selectivity relative to the mixtures of gases (Kong et al. 2013). It has been found that MTV-MOF-5-EHI had an enhanced selectivity of up to 400% for carbon dioxide over carbon monoxide (Deng et al. 2010).

A subset of MOFs, metal–organic polyhedra (MOPs) have also been developed and synthesized as SBUs topological polygons linked to form closed structures. "The polyhedral molecule may be either composed of one kind of SBU linked by a ditopic linker, or of two SBUs with more than two points of extension" (Tranchemontagne et al. 2008a). Their internal volume is tailorable in size, functionality, and active metal sites, making these structures a strong potential candidate for gas storage.

6.1 Synthetic approaches for MOFs

There are various available methods for the preparation of MOFs described in the literature: solvothermal (Li et al. 2014b) and room temperature synthesis (Tranchemontagne et al. 2008b), hydrothermal (Yaghi and Li 1995), ultrasound-assisted synthesis (Safarifard and Morsali 2015), microwave-assisted hydrothermal (Jhung et al. 2007, Klinowski et al. 2011), solid-state mechanochemical methods: ball milling (Yuan et al. 2010) and solvent-free grinding (Fujii et al. 2010), liquid-assisted grinding (LAG) and ion- and liquid-assisted grinding (ILAG) (Friscic 2012), and electrochemical synthesis (Khazalpour et al. 2015). Although conventional synthetic methods (*de novo* synthesis) offer great capability to control the surface area and porosity in MOFs (Farha et al. 2010a), certain drawbacks such as the formation of undesirable topologies, low solubility of precursors, loss of functionality of linkers (by coordinating to the metal centers), formation of amorphous by-products, and the formation of undesirable phases (e.g., catenated) (Karagiaridi et al. 2014a) could limit the development of these techniques. As an alternative, liquid-assisted grinding has been introduced as a rapid and environmental-friendly technique for the preparation of different MOFs using simple and inexpensive metal oxide precursors (Friscic and Fábián 2009, Friscic et al. 2010). Solvent-assisted linker exchange (SALE) has also been developed, involving a heterogeneous reaction of parent MOF crystals with a concentrated solution of linkers (So et al. 2015a). SALE is also known as "bridging-linker replacement" or "post-synthetic exchange" (Karagiaridi et al. 2014a). By introducing SALE, problems associated with conventional (*de novo*) synthesis methods can be solved and effective parameters of pore volume, functionality, and aperture size can be significantly controlled (Karagiaridi et al. 2014a). Recently, a solid-phase polypeptide synthesis technique was used for the preparation of heterometallic metal–organic complex arrays (MOCAs). In this technique, amino acids functionalize, metalate, and then link to create MOCAs (Sajna et al. 2015).

6.2 Applications of MOFs

MOFs have found emerging applications in a wide variety of areas due to their unique properties of large surface area, large pore volume, and tunable functionality. They have been used for catalysis (Valvekens et al. 2013) and photocatalysis (Wang and Wang 2015), supercapacitors (Choi et al. 2014), electronic and opto-electronic devices (Stavila et al. 2014), luminescent materials (Hu et al. 2015), light-harvesting (So et al. 2015b) and dye-sensitized solar cells (Maza et al. 2016), adsorption of toxic gas

(Glover et al. 2011) and chemical warfare agents (Mondloch et al. 2015), removal of pollutants (Howarth et al. 2015) and heavy metals (Tahmasebi et al. 2015), gas detection (Hwang et al. 2013) and chemical sensing (Kreno et al. 2012), electrochemical applications (Morozan and Jaouen 2012), and drug delivery (Orellana-Tavra et al. 2015). MOFs can also be used as precursors for the preparation of nanomaterials (Masoomi and Morsali 2012). Recently, a nanocomposite material of MOF with magnetic particles (so-called magnetic framework composites—MFCs) was developed and found to exhibit a fast response to an external magnetic field (Ricco et al. 2013). It is expected that these composites will find applications in targeted-drug delivery and magnetic-assisted separation techniques. Recently, Furukawa and co-workers (Furukawa et al. 2015) reported that the number of synthesized MOFs and the number of patents (based on MOFs) used in industry were rapidly increased from 2000 to 2012. Therefore, MOFs have attracted considerable attention from both academic research and industrial application since 2000. Moreover, adsorption of gases has been a primary area of application (one-third of all application) focused on by academia and industry.

6.3 MOFs for gas adsorption and separation

The storage of gas in MOFs was first studied by Japanese researchers in 1997 (Kondo et al. 1997). Since then, many efforts have been made to capture other gases such as N_2, Ar, H_2 (Sudik et al. 2005), CH_4 (Peng et al. 2013), CO_2 (Sumida et al. 2012, Zhang et al. 2014b), H_2O (Furukawa et al. 2014), and NH_3 (Morris et al. 2011) in MOF structures. Studies have revealed that post-synthetic modification (Morris et al. 2011) and the introduction of proper functionalities (Zhang et al. 2015) can efficiently improve gas sorption. Aliphatic and cyclic amines, e.g., diamine (Lee et al. 2014) and azolium (Lee et al. 2009) demonstrate good affinity towards gas molecules, particularly CO_2. The creation of defects *via* the linker fragmentation approach provides vacant sites inside the frameworks that can enhance the number of active sites for interaction with guest molecules (Barin et al. 2014). In addition, transmetalation on the structure of pillared-paddle wheel MOFs is another approach used to enhance surface area *via* treatment with water to displace 50% of MOF pillars (Karagiaridi et al. 2014b). Cho et al. (2015) reported that capillary condensation in the pores of MOFs can lead to more gas adsorption *via* the formation of "extra adsorption domains".

Post-modification of MOFs can be efficiently employed for selective gas separation (Sapchenko et al. 2015). Many successful separations of particular gas molecules in the presence of other gases can be found in literature such as the selective separation of alkanes (Chen et al. 2006), carbon dioxide versus methane (Farha et al. 2010b) and light hydrocarbons (Wang et al.

2015a), CO_2 in the presence of water (Fracaroli et al. 2014), harmful gases (e.g., sulfur dioxide, ammonia, chlorine, tetrahydrothiophene, benzene, dichloromethane, ethylene oxide, and carbon monoxide) (Britt et al. 2008), and CO_2 and CH_4 in the presence of N_2 (He et al. 2015). The efficiency of CO_2 separation in the presence of N_2 could be enhanced by introducing a polar functional group ($-CF_3$) to the ligand (Bae et al. 2009). Furthermore, a cation exchange by lithium in MOFs facilitates easy and selective separation of CO_2 in the presence of CH_4 in metal-organic frameworks containing cations (Bae et al. 2011).

7. MOFs for Hydrogen Storage

Advanced porous materials are highly desirable for safe and effective physisorption of large amounts of hydrogen under mild conditions (common temperature and relatively low pressure), also providing fast kinetics and reversible gas sorption (Kong et al. 2015). In addition, these materials should be low-cost, light-weight, and with a large hydrogen capacity and simple recyclability (Song et al. 2011). Various porous materials have been used so far such as activated carbons, carbon nanostructures, zeolites, porous polymers, and MOFs.

MOFs are a class of crystalline coordination polymers consisting of metallic polyhedrons linked with organic ligands by strong coordination bonds. MOFs are promising porous organic-inorganic hybrid materials for H_2 storage owing to their broad structural diversity, high internal surface area, large free volumes, tunable properties (topologies, pore size, geometry, and functionality), high thermal stability (up to 400°C in some cases), high purity, light weight, low cost, ease of modular preparation, and availability for mass production (Yan et al. 2014, Sun et al. 2013, Langmi et al. 2014). MOFs are generally available in loose powders with low packing densities and low thermal conductivities. For their practical applications in gas storage, some properties of MOFs such as structural stability, thermal conductivity, and hydrogen storage capability must be improved (Ren et al. 2015a). Enhanced thermal conductivity can promote mass transfer of hydrogen and heat transfer of H_2 adsorption. In addition, for long term usage, particularly in industry, MOFs must exhibit high moisture tolerance (Ren et al. 2015a).

Yaghi and co-workers were the first to use MOF structures (IRMOF-8) for H_2 storage (2.0 wt%, at 293 K and 10 bar) (Rosi et al. 2003). Since then, vast advances in the area have been made by the scientific community. Many efforts have been made to understand the adsorption/desorption phenomena of guest-host systems, to synthesize versatile and efficient porous materials, and to develop the relevant technological processes. Many relevant books (Walker 2008, Ma et al. 2009, Collins et al. 2010, Tomas 2011,

Dailly 2011) and review papers (Rowsell and Yaghi 2005, Collins and Zhou 2007, Zhao et al. 2008, Dincă and Long 2008, Tomas 2009, Han et al. 2009, Murray et al. 2009, Hu and Zhang 2010, Sculley et al. 2011, Suh et al. 2012, Getman et al. 2012, Li and Xu 2013, Kumar et al. 2015, Liu et al. 2016a) can be found in the literature. These resources are centered on the fundamental aspects and recent scientific and technological progresses made in the field.

There are two approaches to H_2 storage: chemisorption (using metal hydrides, amine boranes, etc.) and physisorption (adsorption on surfaces). In chemisorption, due to the formation of strong covalent bonds and the existence of high interaction energy, high heat is involved during H_2 absorption, which limits fast refueling. Moreover, hydrides are too heavy and thus require high temperatures for hydrogen desorption. Chemisorption may be irreversible due to the possible high activation within the adsorption/desorption process (Suh et al. 2012). The physisorption of H_2 is a fast process due to the existence of weak interactions (van der Waals forces) between adsorbent and adsorbate (less than 10 kJ mol^{-1}) (Han et al. 2009) and thus, high storage capacities can be achieved at low temperatures (around 77 K) if a suitable adsorbent with high surface area and porosity is used and possesses H_2 binding energy of 0.2–0.4 eV (Hirscher 2011).

Despite the advantages of MOFs in gas storage, some critical problems still remain that must be resolved. The interactions between molecular hydrogen and MOFs are relatively weak and this can negatively affect the long-time storage of hydrogen near ambient temperatures. MOFs are generally able to uptake hydrogen of 7 wt% at 77 K but the hydrogen capacities drop to less than 2 wt% at ambient conditions (Hu and Zhang 2010). Moreover, the preservation of MOFs against decomposition and degradation (during handling processes) gives rise to the concern in their application (Sun et al. 2013).

There are numerous effective parameters in the H_2 storage capacities of MOFs which include: (i) sample preparation and activation (removal of trapped solvent in MOF structures); (ii) morphology and size; (iii) surface area and pore volume; (iv) interpenetration (catenation); (v) ligand structure and functionalization (post-synthetic modification); (vi) unsaturated metal sites; (vii) chemical doping; (viii) spillover; and (ix) adsorption mechanism (Sculley et al. 2011, Hu et al. 2012). To enhance the H_2 storage capacities at ambient temperatures, there are various approaches described in literature such as tuning the ligand (elongation of ligands and modification with more polarizable constituents), increasing surface area with narrow pore size, increasing the number of available and open metal sites, and embedding metal nanoparticles and cations with strong electrostatic fields in the cavity (Nouar et al. 2009, Hirscher 2011, Fairen-Jimenez et al. 2012).

Hydrogen uptake is proportional to the heat of adsorption at low pressures and H_2 storage can be appreciable in smaller pore diameters at low pressures (Sculley et al. 2011). Furthermore, by catenation (self-assembly of two separate frameworks within each other) (Ryan et al. 2008) the pore size becomes smaller and the number of metal sites per unit volume increases, leading to stronger MOF–H_2 interactions, higher heats of adsorption, and more hydrogen uptake at low temperatures and pressures. Therefore, catenation is not advantageous for storage under ambient conditions. On the other hand, non-catenated structures are more suitable for higher uptake capacities, particularly at high pressure, due to the greater availability of free volume (Sculley et al. 2011). It is noted that the interaction between hydrogen and MOFs contributes to three attractive potential energies: Van der Waals, charge–quadrupole, and induction (Hu and Zhang 2010). Solvents trapped in the structure of MOFs having a Lewis acid/base interaction with metallic nodes occupy some active and binding sites of MOFs. The removal of such small molecules by heating without any deformation in the structure generates open metal sites acting as host for guest (H_2) molecules (Rowsell and Yaghi 2005).

For gas storage, there are two active sites in MOFs: metal–oxygen clusters with preferential adsorption sites for hydrogen at ambient conditions and organic linkers for high pressure storage (due to weaker binding energy under normal conditions). A neutron powder diffraction and density functional theory (DFT) study showed that the former site is primarily responsible for hydrogen adsorption in the aspect of binding energy (Hu and Zhang 2010), while organic linker sites offer greater capacity for H_2 loading (Rowsell et al. 2005). Inelastic neutron scattering is an effective and powerful technique to determine the type and nature of H_2 binding sites (rotational and translational modes of adsorbed H_2) at a molecular level in porous structures (Pham et al. 2016). The affinity of ligands towards hydrogen can be enhanced through functionalization with electron-donor groups such as amine (Si et al. 2011), tetrazole, and azulene (Barman et al. 2010), and groups that offer zwitterionic environments such as *N*-heterocyclic azolium (Lalonde et al. 2013). Tranchemontagne et al. (2012) reported that smaller pores and polarized linkers in MOFs (e.g., MOF-324 and IRMOF-62) could provide stronger interactions with hydrogen compared to potential interactions between the nodes of the framework and hydrogen molecules. Furthermore, the number of aromatic rings in the linker enhanced the hydrogen uptake in IRMOF-11 compared to IRMOFs-1 and IRMOF-8 (Rowsell et al. 2004). Post-synthetic modification can also tune the surface area, pore volume, and pore size of the MOFs (Suh et al. 2012). Simulation studies showed that for H_2 storage at low, intermediate and high pressures, the H_2 uptake correlates with the isosteric heat (Q_{st})

of adsorption, the surface area, and the free volume, respectively (Getman et al. 2012). In other words, H_2 storage by physisorption depends on the surface area and the porosity of the adsorbent. For materials with multiple pores, the smaller pores fill first, then the larger pores are occupied under higher pressure (Suh et al. 2012).

In the following section, we discuss the progress made over the last four years in the area of hydrogen capture/storage using MOFs. Different effective parameters such as surface area, pore volume, mass density, ligand functionality, and metal/ion doping on the gravimetric/volumetric H_2 uptake and heat of H_2 adsorption are also discussed.

8. Recent Advances in Hydrogen Storage by MOFs

Linker type was found to affect the architecture of MOFs with regards to structural stability and porosity for their potential application in gas storage. A {$Cu(II)_2$} paddle wheels-based MOF, constructed using a series of tetra-, hexa-, and octacarboxylate aromatic linkers (isophthalate units), with high structural stability (aromatic ligands cause rigidity in frameworks) and robust porosity was used for H_2 storage (Yan et al. 2014). By variation of organic linkers, different MOFs with various pore sizes, geometry, and functionality were obtained, demonstrating different gas storage capacities. The best result in H_2 storage was found to be 77.8 mg g^{-1} at 77 K, 60 bar. A correlation was found between the pressure and structural properties of absorbent in gas storage. The results revealed a significant effect on the H_2 uptake, as H_2 adsorption at low, medium, and high pressures was correlated with the isosteric heat of adsorption, surface area, and pore volume, respectively. The frameworks with hexacarboxylate linkers exhibited higher surface areas and pore volumes, and higher H_2 storage capacities than those using tetracarboxylates or octacarboxylates (Yan et al. 2014).

Lalonde et al. (2013) reported that the interaction between H_2 and MOF structures can be efficiently improved with the introduction of a ligand that provides a zwitterionic environment inside the framework. The MOFs of NU-301 and NU-302 exhibited negatively charged nodes of $Zn_2(CO_2)_5$ and positively charged linkers of imidazole tetra acid (imidazolium ring). Enhanced isosteric heats of adsorption of 7.0 and 6.9 kJ mol^{-1} were found for NU-301 and NU-302, respectively, under low H_2 loading. Hydrogen uptakes of 4.50 mmol g^{-1} and 2.99 mmol g^{-1} were observed for a zwitterionic NU-301 MOF at temperatures of 77 K and 87 K, respectively under 1 bar pressure. The charge transfers between the metal cations and the linkers generate short-range interactions between H_2 molecules and the metal cation sites of the MOF, causing the enhanced H_2 adsorption. The authors used a simulation study to determine the H_2 binding sites in the zwitterionic

MOFs. Their findings were that "H_2 prefers to adsorb in the zwitterionic region between two negatively charged $Zn_2(CO_2)_5$ nodes, in front of an imidazole ring; in between the Zn nodes and the free carboxylic acid groups; and near the free carboxylic acid groups" (Lalonde et al. 2013). Binding energies of –8.33 and –8.20 kJ mol^{-1} were calculated for NU-301 and NU-302, respectively. Barman et al. (2014) found the polarization effect of the azulene linker (2,6-azulenedicarboxylate) in MOF-650 with high isosteric heat of adsorption of 6.8 kJ mol^{-1} and with H_2 uptake of 14.8 mg g^{-1} at 77 K and 1 bar.

The effects of synthesis temperature (80, 110 and 140°C) and solvent amount (50, 75 and 100 mL DMF) on the structure and also H_2 storage were studied for Cu–BTC (Copper (II) benzene-1,3,5-tricarboxylate) MOF (Khoshhal et al. 2015). Increasing the synthesis temperature primarily reduced the degree of crystallinity of the MOFs and subsequently decreased their total pore volume, micropore volume, and BET surface area. Moreover, by increasing the amount of solvent within the MOF synthesis, the amount of hydrogen uptake was increased. The highest H_2 uptake was found to be 0.7 wt% for the MOF prepared under conditions of 80°C with 75 mL DMF. The isosteric heat of adsorption of hydrogen was found to decrease with increasing the surface loading. The maximum adsorption heat was calculated (using the Freundlich equation) to be 20.5 kJ mol^{-1} at 0.05 wt% uptake from the corresponding isotherms at 288, 298, and 308 K. The order of kinetic model (1.59) revealed that the adsorption process could be performed by using a combination of physisorption and chemisorption (Khoshhal et al. 2015).

The type of cations incorporated with corresponding ligands is another effective parameter in H_2 sorption. SNU-200 MOFs with 18-crown-6 ether moiety as a specific binding site for cations such as K^+, NH_4^+, and methyl viologen-(MV^{2+}) were used for hydrogen storage. K^+ was found to be the best candidate among cations for enhancing the isosteric heat of the H_2 adsorption (9.92 kJ mol^{-1}, H_2 uptake: 0.78% at 87 K), resulting in accessible open metal sites on the K^+ ion. NH_4^+ and MV^{2+} showed isosteric heat of 6.41 (H_2 uptake: 0.71 wt% at 87 K) and 7.62 kJ mol^{-1} (H_2 uptake: 0.54 wt% at 87 K), respectively (Lim et al. 2014).

Boranes have gained much attention in H_2 capture due to their intrinsic affinity to hydrogen molecules. A Cu–carborane-based MOF (NU-135) with a quasi-spherical polyhedral *para*-carborane cluster in the structure, possessing a pore volume of 1.02 cm^3 g^{-1}, and a gravimetric and volumetric BET surface area of 2600 m^2 g^{-1} and 1900 m^2 cm^{-3}, respectively, was used for gas storage of H_2. A value of 49 g L^{-1} was obtained for volumetric capacity of H_2 at a temperature of 77 K and under pressure of 55 bar. In addition, this structure was able to capture other gases such as CH_4 and CO_2 (Kennedy et al. 2013). Wang et al. (2015a) introduced the MOF of UiO-66-NH_2 for hydrogen storage *via* the formation of amino-borane complex by reacting borane with amino groups of UiO-66. Hydrogen could be released by thermolysis of UiO-66-NH_2/BH_3 at a temperature of 78°C without any by-products, and it could be easily reused. The released hydrogen enabled the reduction of aldehyde derivatives to their corresponding alcohols with excellent selectivity for aliphatic aldehydes compared to aromatic ones.

$$H_3NBH_3 \xrightarrow{\text{Heat}} [NB] + 3H_2$$

Ni-based MOFs using a benzene-1,3,5-tricarboxylic acid (H_3BTC) ligand were also used as a support for ammonia borane, and then for hydrogen desorption (*via* a dehydrogenation process) (Kong et al. 2015). Ammonia has a high hydrogen content (17.6 wt%) which can produce hydrogen *via* a decomposition reaction by non-noble metallic (e.g., Fe, Ni, Co) and bimetallic catalysts (Bell and Torrente-Murciano 2016). Hydrolytic dehydrogenation of ammonia borane (NH_3BH_3) to a stoichiometric amount of hydrogen is an efficient method for the chemical storage of hydrogen (Lu et al. 2012). Ammonia borane (NH_3BH_3) as a solid-state hydrogen storage medium has the interesting properties of satisfactory stability, relatively low molecular mass, and high energy density. Its practical application is still limited due to the poor kinetics of hydrogen generation below 85°C and the release of impurities that are detrimental to the fuel cells (Kong et al. 2015). It was found that the dehydrogenation reaction can be promoted using a synergistic effect of bimetallic Au-Co nanoparticles (NPs) supported on silica nanospheres (Lu et al. 2012). Ammonia borane-confined boron nitride nanopolyhedra were used for chemical hydrogen storage. The B–N–H composite possessed a BET specific surface area of 200 m^2 g^{-1}, a total pore volume of 0.287 cm^3 g^{-1}, and a low density of 2.27 g cm^{-3}. The organic scaffold provided a gravimetric hydrogen storage capacity of 3.4 wt% at 80°C (Moussa et al. 2014). Yoo et al. (2014) used hydrous hydrazine as a candidate for efficient hydrogen storage. The hydrous hydrazine can be decomposed to hydrogen and nitrogen in the presence of the bimetallic catalyst of Rh and Ni on hollow silica microspheres with 99% H_2 selectivity at 25°C. Formic acid (HCO_2H) has also been introduced as a candidate for reversible liquid hydrogen storage (H_2 content of 4.4 wt%), as the storage/release is performed *via* catalytic hydrogenation (of CO_2 to formic acid) and dehydrogenation (of formic acid) processes (Grasemann and Laurenczy 2012). Formic acid is a kinetically stable liquid at room temperature. Its dehydrogenation is thermodynamically favorable at room temperature and the addition of a base (e.g., amine) accelerates the reaction through the formation of the corresponding formates (Mellmann et al. 2016). Koh et al. (2014) used Pd NPs supported on amine-functionalized silica for the dehydrogenation of formic acid.

Zhang et al. (2014a) reported that hydrogen storage using aluminohydrides could be facilitated by introducing Ti-based MOFs. MIL-125(Ti) was added into $NaAlH_4$ to improve the dehydrogenation–hydrogenation kinetics. By adding an optimum amount of 5 wt% MIL-125(Ti) to $NaAlH_4$, the operating temperatures reduced significantly and the kinetics of hydrogen storage were improved. The starting temperature for hydrogen desorption was found to be lowered by 50°C in the presence of MIL-125(Ti). The apparent activation energies of the first two dehydrogenation steps in MIL-125(Ti)/$NaAlH_4$ were found to be 98.9 and 96.3 kJ mol^{-1}, respectively, which were lower than the case without MOF. At the optimum amount of the MOF, approximately 4.66 wt% of hydrogen was released from the $NaAlH_4$ at 140°C, while less than 0.6 wt% of the released hydrogen was obtained in the absence of the MOF. Mechanistic studies showed that the catalytic activity of the Ti (in Ti–Al species) and the nanopore effects of the dehydrogenation reduced the operating temperatures and enhanced the rate of the dehydrogenation–hydrogenation process. Furthermore, the authors speculated that by reacting between MIL-125(Ti) and $NaAlH_4$, zero-valent Ti and carboxyl ligands might be formed during the ball milling process and the carbon and Ti elements as active centers could be dispersed on the surface of $NaAlH_4$ to facilitate the dissociation and recombination of molecular hydrogen.

$$3NaAIH_4 \longrightarrow Na_3AIH_6 + 2AI + 3H_2 \ (3.7 \text{ wt\%})$$

$$Na_3AIH_6 \longrightarrow 3NaH + Al + 3/2H_2 \ (1.9 \text{ wt\%})$$

$$NaH \longrightarrow Na + 1/2H_2 \ (1.9 \text{ wt\%})$$

Fu et al. (2016) reported that the dehydrogenation of $NaBH_4$ and $LiBH_4$ can be improved using ionic liquid (IL) of 1-butyl-3-methylimidazolium bis(trifluoromethylsulfonyl)imide (bmimNTf$_2$). The dehydrogenation was exothermic in the IL with theoretical hydrogen release of 70% at temperatures below 180°C, while these values could be obtained at temperatures of 370°C and 500°C for $LiBH_4$ and $NaBH_4$, respectively, in a solid state under an endothermic process. The role of IL was to destabilize the borohydrides through charge transfer from BH_4^- to the imidazolium cation of the IL.

It was found that the incorporation of metallic NPs (e.g., Pt, Pd) into the pores of MOFs may enhance hydrogen storage *via* atomic hydrogen adsorption rather than molecular H_2 (Langmi et al. 2014). Li et al. (2014a) reported a considerable enhancement in the capacity and kinetics of hydrogen storage in Pd-incorporated HKUST-1 (copper(II) 1,3,5-benzenetricarboxylate) MOF. The incorporated MOF showed roughly twice (~74%) the storage capacity compared to that of bare Pd nanocrystals.

The surface reactivity of the nanocrystals was a result of enhancement in the adsorption of molecular hydrogen on the surface of Pd (Takata et al. 2014). The content of Pd in MOFs was another effective parameter in hydrogen storage. Pd-doped MIL-101 MOFs with different Pd content (1, 3, 5 wt%) have showed reversible hydrogen storage capacities at room temperature (under pressure of 7.3 MPa) by a factor of 1.5–2.3 enhancement (0.42–0.64 wt%) rather than that in un-doped MIL-101 (0.28 w%). By the addition of different amounts of Pd, different sizes of Pd NPs were produced (2.4, 2.9, and 5.8 nm using 1, 3, 5 wt% of $PdCl_2$, respectively). The size and the doping content of Pd were found to be two effective parameters in hydrogen storage. H_2 uptake values of 0.42, 0.64, and 0.48 wt% were obtained for 1, 3, 5 wt% of Pd/MIL-101, respectively. Pd in the form of $PdCl_2$-doped MIL-101 was also employed for efficient toluene uptake (1285 mg g^{-1}) (Qin et al. 2014). Kim et al. (2015) introduced platinum (Pt) NPs onto the outer surface of MOF-5 to improve the hydrogen storage capacity of the MOF at room temperature. In addition, the Pt NPs were coated with hydrophobic microporous carbon black (CB) to improve the hydro-stability of the MOF under ambient conditions. The composite MOF (CB/Pt/MOF-5) exhibited 40% enhancement (at 298 K and 100 bar) in hydrogen storage (0.62 wt%) over that of pristine MOF-5 (0.44 wt%). The Pt NPs of different sizes and surface areas were formed and dispersed onto the outer surface of MOF-5 by varying Pt contents (1, 4, 7, and 10 wt%). 4Pt/MOF-5 (with 4 wt% Pt) showed high surface area (1862 m^2 g^{-1}) and a high dispersion of Pt NPs that consequently led to improved hydrogen storage capacity *via* hydrogen spillover effects.

Zr-based MOFs (e.g., UiO-66) exhibit higher chemical and thermal stabilities than other varieties of metal-organic frameworks. Zhao et al. (2013) utilized UiO-66 in octahedral crystalline form with sizes of 150–200 nm and a specific surface area of 1358 m^2 g^{-1} for hydrogen storage. The highly pure MOFs were prepared in DMF at 50°C. Various solvents such as ethanol, acetone, and *n*-heptane were also used for the MOF synthesis but sensible impurities were observed within the synthesis that could influence the surface area and performance of the MOFs for gas storages. A hydrogen uptake of 3.35 wt% was achieved by using UiO-66 at 77 K and 1.8 MPa. A theoretical study using grand canonical Monte Carlo simulation was in good agreement with the experimental results (Zhao et al. 2013). In parallel work, Ren et al. (2014) introduced a modification procedure for the synthesis of crystalline Zr-MOFs with a shorter reaction time (2 h) by using formic acid as a modulator. It was found that formic acid efficiently accelerated the formation of nuclei and subsequently, the formation of crystalline Zr-MOFs with an octahedral shape. By using 10 eq formic acid, the MOFs of size 100–200 nm, surface area of 918 m^2 g^{-1}, and pore volume of 0.42 cm^3 g^{-1} were obtained and then used for H_2 uptake (1.1 wt%). Larger MOFs (using

100 eq formic acid) with size of 1–3 μm (surface area of 1367 $m^2 g^{-1}$, and pore volume of 0.56 $cm^3 g^{-1}$) were used for higher uptake of H_2 (1.5 wt%). For technological usage in gas storage, the absorbent in powder must form an appropriate shape. For this purpose, Zr-MOF powder was shaped into spherical pellets of size of 0.5–15 mm using 10 wt% sucrose as a binder. Zr-MOF pellets can be efficiently produced at scales of a kilogram within 30 min *via* a granulation method instead of conventional mechanical pressing. The pellets were then packed into a small tank for hydrogen storage. The pellets showed reasonable physical stability, with zero breakage observed after 70 consecutive drops at a height of 0.5 m (drop test), and 5% breakage after 60 min of tumbling time at a speed of 25 rpm (simulated tumbler drum test). Zr-MOF pellets with a surface area of 674 $m^2 g^{-1}$ and a pore volume of 0.34 $cm^3 g^{-1}$ exhibited 0.85 wt% H_2 uptake at 77 K under 1 bar pressure (Ren et al. 2015b). Theoretical studies (using the Dubinine-Astakhov adsorption model) on the hydrogen adsorption/desorption process with MOF-5 powder (0.13 g cm^{-3}) and its compacted tablet (0.30 g cm^{-3}) were carried out by Xiao and coworkers (Xiao et al. 2013). The compacted MOF-5 showed higher hydrogen storage at lower pressure and with less temperature rise. Therefore, storage capacity in technological applications depends on the packing density of the MOFs powder (Hirscher 2011).

To understand the relation between guest molecules (e.g., H_2) and host structure (e.g., MOF) in gas storage, extensive efforts have been carried out through simulation and modeling approaches. Sumida et al. (2013) used a technique of variable-temperature infrared spectroscopy to find the physical/chemical interaction of H_2 with metal cations in the MOF structure. The effect of metal cation types and their corresponding anions in the MOF of $M_3[(M_4Cl)_3(BTT)_8]_2$ (M-BTT; M^{2+} = Mn, Cu, Fe, Zn; BTT^{3-} = 1,3,5-benzenetristetrazolate) for storage of H_2 were investigated. M-BTT MOFs provide a high density of exposed metal cation sites on the pore surface that is accessible to the guest molecules. It was found that Zn-BTT has stronger binding energy toward H_2 and therefore, the adsorption of H_2 was enhanced. The corresponding anions were found to be another factor in H_2 binding energy, as more diffuse counter-anions (Br^- compared to Cl^- and F^-) project the metal ions away from the center of the cluster and eventually lead to stronger M–H_2 interactions (by enhancement in the enthalpy of adsorption).

It is believed that "a large portion of voids in MOFs remain unutilized for hydrogen storage owing to weak interactions between the walls of MOFs and H_2 molecules" (Rallapalli et al. 2013). Thus, microporous activated carbon (AC) was used to incorporate into chromium-based MOF MIL-101(Cr) to reduce its unutilized voids and to enhance its volumetric capacity in hydrogen storage. With the addition of 0.63 wt% of AC into MIL-101, the highest H_2 uptake of 10.1 wt% was obtained for the composite MOF at

77.4 K and 6 MPa pressure (Rallapalli et al. 2013). In another study, hybrid MIL-101/Pt/C composite materials were used for high pressure hydrogen storage at ambient temperature. The results showed that the presence of platinum could efficiently promote the hydrogen storage capacity under high pressure. The maximum H_2 uptake was achieved with 6.1 wt% at 298 K for the composite by increasing the pressure from 25 to 250 bar, whereas in the pristine MIL-101 the H_2 storage reached a maximum of 4.5 wt% at 40 bar and then decreased gradually as pressure increased (Klyamkin et al. 2014). Adhikari et al. (2015) modified the pristine MIL-100 with 5 wt% Pt-loaded activated carbon to achieve the hydrogen spillover effect. The H_2 uptake of modified MIL-100 was enhanced to 0.41 wt% from 0.25 wt% for pristine MIL-100 at 298 K and 31 bar. Xie et al. (2014) employed an external electric field to induce the adsorption of H_2 in MOF structures. They synthesized percolative composites by direct deposition of IRMOF8 (*iso*-reticular metal-organic frameworks) onto activated carbon *via* heterogeneous nucleation. The dielectric constant of the composites MOF was found to be enhanced by increasing carbon content due to insulator-conductor transitions. Therefore, H_2 storage was significantly improved (~31.5%) due to a stronger electrostatic interaction between IRMOF8 and hydrogen molecules caused by field polarization (using PMN-PT; single crystal lead magnesium niobate-lead titanate). The rapid adsorption/desorption kinetics could be controlled by the presence/absence of an external electric field.

^2H solid-state nuclear magnetic resonance (NMR) was employed to study the adsorption behavior and dynamics of D_2 in several MOFs of UiO-66, M-MOF-74 (M = Zn, Mg, Ni), and a-$M_3(COOH)_6$ (M = Mg, Zn) (Lucier et al. 2016). The type of metal center was found to influence the binding of hydrogen to the open metal sites. The strong D_2 binding in Mg-MOF-74, Ni-MOF-74, a-$Mg_3(COOH)_6$, and a-$Zn_3(COOH)_6$ caused broadening of the ^2H NMR patterns, while weak D_2 adsorption (narrow ^2H resonances) was found in UiO-66 and Zn-MOF-74 due to the rapid reorientation of the D_2 molecules. A new series of biocompatible MOF, M-MOF-74 (M = Mg, Fe, Co, Ni, and Zn), have been prepared using the drug olsalazine (H_4olz) as a ligand (Levine et al. 2016). The M_2(olz) frameworks with large hexagonal pores (27 Å in diameter) were efficiently used for hydrogen storage. *In situ* infrared spectroscopy has been applied to determine the nature and strength of the interaction between H_2 and open metal centers in M_2(olz) frameworks, since the binding of H_2 to metal sites may cause red-shift in the stretching vibration of H–H. The magnitude of red-shift was found to be a function of H_2 binding strength. At a loading of one H_2 per metal, a doublet in the spectrum appeared at 4097 cm^{-1} and 4091 cm^{-1} relating to the vibrational modes of ortho-H_2 and para-H_2, respectively. A greater red-shift was observed for Ni_2(olz) (Q_{st} = −12.1 kJ mol^{-1}) compared to Mg_2(olz)

(Q_{st} = −10.8 kJ mol^{-1}) due to the stronger binding of H_2 to open metal sites of Ni_2(olz).

Naeem et al. (2016) introduced UBMOF-31 for hydrogen absorption. UBMOF-31 has been prepared by mixing two linkers: 4,4′-stilbenedicarboxylic acid and 2,2′-diamino-4,4′-stilbenedicarboxylic acid. This MOF possessed high thermal stability, moderate water stability, and a surface area of 2500 m^2 g^{-1}. The strong attractive interactions between the H_2 molecules and the amino groups in the linkers led to 4.9 wt% hydrogen uptake at 4.6 MPa. Based on the adsorption isotherm, chemisorption of H_2 at the amine sites was the primary phenomenon at the early stage of adsorption, whereas physisorption on the internal surface of the MOF occurred latter. A chromium(II)-based MOF $Cr_3[(Cr_4Cl)_3(BTT)_8]_2$ (Cr-BTT; BTT^{3-} = 1,3,5-benzenetristetrazolate) with BET surface areas of 2300 m^2 g^{-1} was also used for H_2 storage (Bloch et al. 2016). Hydrogen uptake was found to be 12 mmol g^{-1} (2.4 wt%) at 1 bar. Neutron powder diffraction showed moderately strong Cr–H_2 interactions (Q_{st} of H_2= −10.0 kJ mol^{-1}) in the coordinatively unsaturated metal cation sites of the MOF. Liu et al. (2016b) introduced a (3,3,5)-c polyhedral MOF with high thermal stability for hydrogen storage. This MOF has been synthesized *via* solvothermal reaction of cupric nitrate and tri-isophthalate ligand of 3,3′,3″,5,5′,5″-pyridyl-1,3,5-triylhexabenzoic acid (H_6PHB). The gdm-MOF showed high gravimetric H_2 uptakes of 240 cm^3 g^{-1} (2.16 wt%, 10.8 mmol g^{-1}) at 77 K, and 164 cm^3 g^{-1} (1.46 wt%, 7.3 mmol g^{-1}) at 87 K under 1 bar pressure.

9. Computational and Modelling Studies

Molecular simulation and modelling have been used extensively for a deeper understanding of hydrogen storage in MOFs. Quantum mechanics calculations are able to predict the behavior of trapped hydrogen (hydrogen–hydrogen interactions) and its binding energies to MOFs (hydrogen–MOF interaction) (Basdogan and Keskin 2015). Goldsmith et al. (2013) analyzed 20,000 MOFs from the Cambridge Structural Database to find the relationship between hydrogen storage properties and structural characteristics (surface area and porosity) of as-reported MOFs. They found that H_2 uptake is not only correlated with surface area, but other parameters such as mass density of MOFs and the structural stability of frameworks (upon solvent removal) should also be considered. Density started to decrease with increasing MOF surface area, indicating a trade-off between gravimetric and volumetric H_2 storage that is concave downward. The theoretical H_2 uptakes were found to be higher than the corresponding empirical values in gravimetry and volumetry due to the incomplete removal of solvents from the structure under experimental usages.

Meng et al. (2013) investigated the performance of catenated and noncatenated MOFs for H_2 separation and H_2 adsorption using the grand canonical Monte Carlo simulation method. Catenated MOFs, IRMOF-11 and IRMOF-13 exhibited efficient separation of H_2 from the mixture of CH_4, CO and CO_2, compared to noncatenated MOFs, IRMOF-12, and IRMOF-14. Furthermore, catenated IRMOF-9 doped with Li achieved an impressive hydrogen storage performance of 4.91 wt% and 36.6 g L^{-1} at 243 K and 100 bar. In other words, the metal doping can be a promising alternative strategy to improve hydrogen storage in MOFs. It has been reported that doping of the structure of absorbents with metal ions such as Li^+ and Mg^{2+} can generate strong electrostatic fields within the pores and provide effective interactions between H_2 and the host structure to enhance H_2 uptake *via* spillover (Yan et al. 2014). A grand canonical Monte Carlo simulations conducted by Volkova et al. (2014) showed that an enhancement of binding energy between hydrogen and lithium atoms in MOF-C30 could lead to a hydrogen uptake of 5.5 wt% at room temperature and 100 bar, compared to a hydrogen storage of 3.1 wt% in the unmodified MOF. Storage of H_2 in IRMOF-12 and IRMOF-14 doped by Li atoms and impregnated with fullerenes was simulated also using a grand canonical Monte Carlo method. Based on the simulation, both Li-C_{60}@Li-IRMOF-12 and Li-C_{60}@Li-IRMOF-14 theoretically achieved the US Department of Energy targets for 2017 in both gravimetric density and volumetric density at 243 K and 100 bar. H_2 uptakes of 5.6 wt% (40.0 g L^{-1}) and 5.77 wt% (40.6 g L^{-1}) were calculated for Li-C_{60}@Li-IRMOF-12 and Li-C_{60}@Li-IRMOF-14, respectively. The simulation results also revealed that Li enhanced adsorption capacity and C_{60} impregnation improved the H_2 uptake. The effects of adsorption enthalpy, crystal density, surface area, and free volume on the H_2 uptake were also investigated. The adsorption enthalpy was found to be the dominant factor in H_2 storage (Rao et al. 2013). A combination of atomistic simulation and continuum modeling revealed that beryllium has a strong interaction with hydrogen (adsorption enthalpy 5.5 kJ mol^{-1}) in MOF beryllium benzene tribenzoate (Be-BTB). The interaction of H_2 with Be was found to be dependent on the geometry of the cavity in the Be-BTB MOF, as H_2 had stronger interaction with the Be-ring cavity than the spherical cavity. Moreover, the pore size and the framework mass were other effective factors in reaching a superior hydrogen storage of 2.3 wt% at 298 K. This simulation was used to compare the performance of Be-BTB with a compressed H_2 tank and MOF-5 and MOF-177-based fuel cells. The Be-BTB fuel cell achieved greater energy release per volume and mass than the other material-based storage tanks due to the pore effect in Be-BTB. The uptake of H_2 was correlated with pore volume for weak adsorption (MOFs without metal sites) and surface area and pore volume for stronger adsorption (MOFs with metal sites) at room temperature (Lim et al. 2013).

To extend these studies, 3d transition metal atoms such as Sc, Ti, V, and Cr were used to make a metal complex with pyridine linker in the structure of MOFs. Among the aforementioned metals, the Sc-pyridine complex showed higher H_2 binding energies *via* strong Kubas interaction (Kubas et al. 1984) to adsorb H_2 on the surface of scandium, reaching a maximum storage of 7.55 wt% at ambient conditions (Bora and Singh 2014). Brand et al. (2013) employed grand canonical Monte Carlo simulations to study hydrogen adsorption in Li, Be, Mg, Mn, Fe, Ni, Cu, and Zn alkoxide functionalized MOFs. IRMOF-16, NU-100, and UiO-68 functionalized with Mg or Fe catecholates on the linkers demonstrated the best H_2 storage properties. It was found that H_2 uptake correlated with MOF free volume, pore size, and heat of adsorption at low loading. In MOFs with large free volume, gravimetric storage capacity was increased, while a trade-off was seen in small pore sizes with enhanced volumetric storage capacity. To optimize the heat of adsorption (Q_{st}), functionalization of the MOFs could be carried out with multiple alkoxide groups per linker. IRMOF-16 functionalized with three Fe alkoxide groups per linker revealed a Q_{st} of 23 kJ mol^{-1} at low loading, with gravimetric and volumetric H_2 storage capacities of 6 wt% and 20 g L^{-1}, respectively at 243 K and 100 bar. In Mn-MOF (MOF-73), the computational calculation revealed that d_{xy} orbitals in manganese are bonded to the hydrogen molecule *via* H–H sigma-bonds (Rydén et al. 2013).

Density functional theory (DFT) was used to study the energetic and thermodynamic phenomena in the initial stage of hydrogen storage on IRMOF-1. It was found that the addition of the first hydrogen in the pure IRMOF-1 performed poorly due to weaker binding energy between them. This energy barrier could be reduced by hole doping (for example, with Zn vacancies). Hydrogen storage by spillover was not observed in other periodic frameworks such as COF even with hole doping (Ganz and Dornfeld 2014). It was reported that DFT calculation is faster than conventional simulation methods, therefore, it is a promising technique for the large-scale screening of MOFs for H_2 storage (Fu et al. 2015). Ding and Yazaydin (2013) calculated the geometric surface areas and pore volumes of MOFs with ultrahigh surface areas (NU-100, NU-108, NU-109, NU-110, MOF-180, MOF-200, MOF-210 and MOF-399) for hydrogen storage capacity using grand canonical Monte Carlo simulations. They found that MOF-399 exhibited the highest gravimetric surface area and pore volume (exceeding 7100 m^2 g^{-1} and 7.55 cm^3 g^{-1}), with considerable hydrogen storage predictable in the structure. This structure was also efficiently used for gas storage of CH_4.

Molecular simulation studies by Gómez-Gualdrón et al. (2016) exhibited that hydrogen deliverable capacity can be significantly enhanced when cryogenic conditions operated for H_2 adsorption. Highest volumetric deliverable capacity was found to be 57 g L^{-1} of MOF at 77 K and 100 bar.

This value is promising compared to the incumbent technology in hydrogen compressing (deliverable capacity of 37 g L^{-1} of tank under 700 bar pressure at room temperature). The authors also found that the hydrogen storage capacity in cryogenic conditions depends on the topology of MOFs.

Recently, Y. Liu et al. (2016) studied the entropy of H_2 adsorption in MOFs (e.g., HKUST-1, MOF-5, MOF-505 and ZIF-8), and its correlations with other thermodynamic properties such as uptake, isosteric heat, and adsorption degree. The DFT results showed that the dependence of entropy on isosteric heat was stronger than other properties (e.g., uptake and adsorption degree). It was found that H_2 molecules were localized in the 8 independent cages (for one unit cell) of HKUST-1 with 3-dimensional confinement and a 0-dimensional degree of freedom while for MOF- 505, H_2 was localized in the 1-dimensional pore with 2-dimensional confinement. In other words, H_2 was more localized and more ordered in HKUST-1 (lower entropy) compared to MOF-505 at a low pressure, but under a high pressure, H_2 could be more delocalized to form a crystal-like structure (higher ΔS) (Liu et al. 2016).

10. Outlook

Metal–organic frameworks with an exceptional internal surface area and excellent thermal and moisture stability are promising candidates for efficient and safe hydrogen storage with potential application in real on-board usage. Easy and modular synthesis are attractive features of MOFs in the production of a large variety of frameworks with tunable size, morphology, surface area, porosity, density, functionality, and structural stability. However, much potential still exists to develop and modify the structure and properties of MOFs for the storage and release of H_2 at ambient temperature and pressure. Long-term H_2 storage in MOFs is an important issue that needs further consideration. Although the strength of the interaction between H_2 molecules and porous materials can efficiently improve the storage of hydrogen (exothermic), the release process may require higher temperatures (endothermic), which can affect the kinetics and thermodynamics of the adsorption/desorption phenomena.

Alternatively, the technology of tank design must be further developed for simple and quick charge/discharge cycles of the fuel. Moreover, the design of cooling and pressure-controlling systems should meet the criteria of safety and efficiency. The fashion of packing (size, shape, thickness and density) of porous materials is another important parameter in the technology. It is expected that the scientific contributions and technological efforts in this area will only increase in order to reach the DOE's ultimate target of hydrogen storage and usage in the near future.

11. List of Abbreviation

BBR	:	Building Block Replacement
BET	:	Brunauer–Emmett–Teller
CMPs	:	Conjugated Microporous Polymers
COFs	:	Covalent Organic Frameworks
Cu–BTC	:	Copper Benzenetricarboxylate
DFT	:	Density Functional Theory
DMF	:	Dimethylformamide
HKUST-1	:	Hong Kong University of Science and Technology
IL	:	Ionic Liquid
ILAG	:	Ion– and Liquid–Assisted Grinding
IRMOF	:	Isoreticular Metal Organic Framework
LAG	:	Liquid–Assisted Grinding
MFCs	:	Magnetic Framework Composites
MIL-101	:	Matérial Institut Lavoisier
MOCAs	:	Metal–Organic Complex Arrays
MOFs	:	Metal–Organic Frameworks
MOPs	:	Metal–Organic Polyhedra
MTV	:	Multivariate
NMR	:	Nuclear Magnetic Resonance
NPs	:	Nanoparticles
NU-31	:	Northwestern University
PAFs	:	Porous Aromatic Frameworks
PCPs	:	Porous Coordination Polymers
PIMs	:	Polymers of Intrinsic Microporosity
POFs	:	Porous Organic Frameworks
POPs	:	Porous Organic Polymers
PSM	:	Post–Synthetic Modification
SALE	:	Solvent–Assisted Linker Exchange
SBUs	:	Secondary Building Units
SNU-200	:	Seoul National University
UBMOF	:	University of Bradford MOF
UiO-66	:	Universitetet i Oslo
ZIFs	:	Zeolitic Imidazolate Frameworks

12. Acknowledgment

The authors gratefully acknowledge Razi University and Iran Nanotechnology Initiative Council for financial supports.

References

Adhikari, A.K., K.-S. Lin and C.-S. Chang. 2015. Improved hydrogen storage capacity by hydrogen spillover and fine structural characterization of MIL-100 metal organic frameworks. Res. Chem. Intermed. 41: 7655–7667.

Bae, Y.-S., O.K. Farha, J.T. Hupp and R.Q. Snurr. 2009. Enhancement of CO_2/N_2 selectivity in a metal-organic framework by cavity modification. J. Mater. Chem. 19: 2131–2134.

Bae, Y.-S., B.G. Hauser, O.K. Farha, J.T. Hupp and R.Q. Snurr. 2011. Enhancement of CO_2/CH_4 selectivity in metal-organic frameworks containing lithium cations. Microporous Mesoporous Mater. 141: 231–235.

Banerjee, R., H. Furukawa, D. Britt, C. Knobler, M. O'Keeffe and O.M. Yaghi. 2009. Control of pore size and functionality in isoreticular zeolitic imidazolate frameworks and their carbon dioxide selective capture properties. J. Am. Chem. Soc. 131: 3875–3877.

Barin, G., V. Krungleviciute, O. Gutov, J.T. Hupp, T. Yildirim and O.K. Farha. 2014. Defect creation by linker fragmentation in metal–organic frameworks and its effects on gas uptake properties. Inorg. Chem. 53: 6914–6919.

Barman, S., H. Furukawa, O. Blacque, K. Venkatesan, O.M. Yaghi and H. Berke. 2010. Azulene based metal–organic frameworks for strong adsorption of H_2. Chem. Commun. 46: 7981–7983.

Barman, S., A. Khutia, R. Koitz, O. Blacque, H. Furukawa, M. Iannuzzi et al. 2014. Synthesis and hydrogen adsorption properties of internally polarized 2,6-azulenedicarboxylate based metal–organic frameworks. J. Mater. Chem. A 2: 18823–18830.

Basdogan, Y. and S. Keskin. 2015. Simulation and modelling of MOFs for hydrogen storage. CrystEngComm. 17: 261–275.

Beldon, P.J., L. Fábián, R.S. Stein, A. Thirumurugan, A.K. Cheetham and T. Friscic. 2010. Rapid room-temperature synthesis of zeolitic imidazolate frameworks by using mechanochemistry. Angew. Chem. Int. Ed. 49: 9640–9643.

Bell, T.E. and L. Torrente-Murciano. 2016. H_2 Production via ammonia decomposition using non-noble metal catalysts: A review. Top Catal. 59: 1438–1457.

Bloch, E.D., W.L. Queen, M.R. Hudson, J.A. Mason, D.J. Xiao, L.J. Murray et al. 2016. Hydrogen storage and selective, reversible O_2 adsorption in a metal–organic framework with open chromium(II) sites. Angew. Chem. Int. Ed. 55: 8605–8609.

Bora, P.L. and A.K. Singh. 2014. Enhancing hydrogen storage capacity of pyridine-based metal organic framework. Int. J. Hydrogen Energy 39: 9293–9299.

Brand, S.K., Y.J. Colón, R.B. Getman and R.Q. Snurr. 2013. Design strategies for metal alkoxide functionalized metal–organic frameworks for ambient temperature hydrogen storage. Microporous Mesoporous Mater. 171: 103–109.

Britt, D., D. Tranchemontagne and O.M. Yaghi. 2008. Metal-organic frameworks with high capacity and selectivity for harmful gases. PNAS 105: 11623–11627.

Chen, B., C. Liang, J. Yang, D.S. Contreras, Y.L. Clancy, E.B. Lobkovsky et al. 2006. A microporous metal–organic framework for gas-chromatographic separation of alkanes. Angew. Chem. Int. Ed. 45: 1390–1393.

Cho, H.S., H. Deng, K. Miyasaka, Z. Dong, M. Cho, A.V. Neimark et al. 2015. Extra adsorption and adsorbate superlattice formation in metal-organic frameworks. Nature 527: 503–507.

Choi, K.M., H.M. Jeong, J.H. Park, Y.-B. Zhang, J.K. Kang and O.M. Yaghi. 2014. Supercapacitors of nanocrystalline metal organic frameworks. ACSNano 8: 7451–7457.

Collins, D.J. and H.-C. Zhou. 2007. Hydrogen storage in metal–organic frameworks. J. Mater. Chem. 17: 3154–3160.

Collins, D.J., S. Ma and H.-C. Zhou. 2010. Hydrogen and methane storage in metal-organic frameworks. *In*: MacGillivray, L.R. (ed.). Metal-Organic Frameworks: Design and Application. John Wiley & Sons, Inc., Hoboken, NJ, USA.

Côté, A.P., A.I. Benin, N.W. Ockwig, M. O'Keeffe, A.J. Matzger and O.M. Yaghi. 2005. Porous, crystalline, covalent organic frameworks. Science 310: 1166–1170.

Côté, A.P., H.M. El-Kaderi, H. Furukawa, J.R. Hunt and O.M. Yaghi. 2007. Reticular synthesis of microporous and mesoporous 2D covalent organic frameworks. J. Am. Chem. Soc. 129: 12914–12915.

Dailly, A. 2011. Research status of metal–organic frameworks for on-board cryo-adsorptive hydrogen storage applications. *In*: Farrusseng, D. (ed.). Metal-Organic Frameworks: Applications from Catalysis to Gas Storage. Wiley-VCH Verlag GmbH & Co. KGaA, Weinheim, Germany.

Deng, H., C.J. Doonan, H. Furukawa, R.B. Ferreira, J. Towne, C.B. Knobler et al. 2010. Multiple functional groups of varying ratios in metal-organic frameworks. Science 327: 846–850.

Deria, P., J.E. Mondloch, O. Karagiaridi, W. Bury, J.T. Hupp and O.K. Farha. 2014. Beyond post-synthesis modification: evolution of metal–organic frameworks via building block replacement. Chem. Soc. Rev. 43: 5896–5912.

Dincă, M. and J.R. Long. 2008. Hydrogen storage in microporous metal–organic frameworks with exposed metal sites. Angew. Chem. Int. Ed. 47: 6766–6779.

Ding, L. and A.O. Yazaydin. 2013. Hydrogen and methane storage in ultrahigh surface area metal–organic frameworks. Microporous Mesoporous Mater. 182: 185–190.

Doonan, C.J., D.J. Tranchemontagne, T.G. Glover, J.R. Hunt and O.M. Yaghi. 2010. Exceptional ammonia uptake by a covalent organic framework. Nature Chem. 2: 235–238.

El-Kaderi, H.M., J.R. Hunt, J.L. Mendoza-Cortés, A.P. Côté, R.E. Taylor, M. O'Keeffe et al. 2007. Designed synthesis of 3D covalent organic frameworks. Science 316: 268–272.

Fairen-Jimenez, D., Y.J. Colón, O.K. Farha, Y.-S. Bae, J.T. Hupp and R.Q. Snurr. 2012. Understanding excess uptake maxima for hydrogen adsorption isotherms in frameworks with rht topology. Chem. Commun. 48: 10496–10498.

Farha, O.K., K.L. Mulfort, A.M. Thorsness and J.T. Hupp. 2008. Separating solids: Purification of metal-organic framework materials. J. Am. Chem. Soc. 130: 8598–8599.

Farha, O.K. and J.T. Hupp. 2010. Rational design, synthesis, purification, and activation of metal-organic framework materials. Acc. Chem. Res. 43: 1175–1166.

Farha, O.K., A.Ö. Yazaydın, I. Eryazici, C.D. Malliakas, B.G. Hauser, M.G. Kanatzidis et al. 2010a. *De novo* synthesis of a metal–organic framework material featuring ultrahigh surface area and gas storage capacities. Nature Chem. 2: 944–948.

Farha, O.K., Y.-S. Bae, B.G. Hauser, A.M. Spokoyny, R.Q. Snurr, C.A. Mirkin et al. 2010b. Chemical reduction of a diimide based porous polymer for selective uptake of carbon dioxide versus methane. Chem. Commun. 46: 1056–1058.

Fracaroli, A.M., H. Furukawa, M. Suzuki, M. Dodd, S. Okajima, F. Gańdara et al. 2014. Metal–organic frameworks with precisely designed interior for carbon dioxide capture in the presence of water. J. Am. Chem. Soc. 136: 8863–8866.

Friscic, T. and L. Fábián. 2009. Mechanochemical conversion of a metal oxide into coordination polymers and porous frameworks using liquid-assisted grinding (LAG). CrystEngComm. 11: 743–745.

Friscic, T., D.G. Reid, I. Halasz, R.S. Stein, R.E. Dinnebier and M.J. Duer. 2010. Ion- and liquid-assisted grinding: Improved mechanochemical synthesis of metal–organic frameworks reveals salt inclusion and anion templating. Angew. Chem. Int. Ed. 49: 712–715.

Friscic, T. 2012. Supramolecular concepts and new techniques in mechanochemistry: cocrystals, cages, rotaxanes, open metal–organic frameworks. Chem. Soc. Rev. 41: 3493–3510.

Fu, H., Y. Wu, J. Chen, X. Wang, J. Zheng and X. Li. 2016. Promoted hydrogen release from alkali metal borohydrides in ionic liquids. Inorg. Chem. Front. 3: 1137–1145.

Fu, J., Y. Liu, Y. Tian and J. Wu. 2015. Density functional methods for fast screening of metal–organic frameworks for hydrogen storage. J. Phys. Chem. C 119: 5374–5385.

Fujii, K., A.L. Garay, J. Hill, E. Sbircea, Z. Pan, M. Xu et al. 2010. Direct structure elucidation by powder X-ray diffraction of a metal–organic framework material prepared by solvent-free grinding. Chem. Commun. 46: 7572–7574.

Furukawa, H. and O.M. Yaghi. 2009. Storage of hydrogen, methane, and carbon dioxide in highly porous covalent organic frameworks for clean energy applications. J. Am. Chem. Soc. 131: 8875–8883.

Furukawa, H., F. Gańdara, Y.-B. Zhang, J. Jiang, W.L. Queen, M.R. Hudson et al. 2014. Water adsorption in porous metal–organic frameworks and related materials. J. Am. Chem. Soc. 136: 4369–4381.

Furukawa, H., U. Müller and O.M. Yaghi. 2015. Heterogeneity within order in metal–organic frameworks. Angew. Chem. Int. Ed. 54: 3417–3430.

Ganz, E. and M. Dornfeld. 2014. Energetics and thermodynamics of the initial stages of hydrogen storage by spillover on prototypical metal–organic framework and covalent–organic framework materials. J. Phys. Chem. C 118: 5657–5663.

Getman, R.B., Y.-S. Bae, C.E. Wilmer and R.Q. Snurr. 2012. Review and analysis of molecular simulations of methane, hydrogen, and acetylene storage in metal organic frameworks. Chem. Rev. 112: 703–723.

Glover, T.G., G.W. Peterson, B.J. Schindler, D. Britt and O. Yaghi. 2011. MOF-74 building unit has a direct impact on toxic gas adsorption. Chem. Eng. Sci. 66: 163–170.

Goldsmith, J., A.G. Wong-Foy, M.J. Cafarella and D.J. Siegel. 2013. Theoretical limits of hydrogen storage in metal–organic frameworks: Opportunities and trade-offs. Chem. Mater. 25: 3373–3382.

Gómez-Gualdrón, D.A., Y.J. Colón, X. Zhang, T.C. Wang, Y.-S. Chen, J.T. Hupp et al. 2016. Evaluating topologically diverse metal–organic frameworks for cryo-adsorbed hydrogen storage. Energy Environ. Sci. 9: 3279–3289.

Grasemann, M. and G. Laurenczy. 2012. Formic acid as a hydrogen source—recent developments and future trends. Energy Environ. Sci. 5: 8171–8181.

Han, S.S., H. Furukawa, O.M. Yaghi and W.A. Goddard III. 2008. Covalent organic frameworks as exceptional hydrogen storage materials. J. Am. Chem. Soc. 130: 11580–11581.

Han, S.S., J.L. Mendoza-Cortés and W.A. Goddard III. 2009. Recent advances on simulation and theory of hydrogen storage in metal–organic frameworks and covalent organic frameworks. Chem. Soc. Rev. 38: 1460–1476.

Hayashi, H., A.P. Côté, H. Furukawa, M. O'Keeffe and O.M. Yaghi. 2007. Zeolite A imidazolate frameworks. Nature Mater. 6: 501–506.

He, Y., J. Shang, Q. Gu, G. Li, J. Li, R. Singh et al. 2015. Converting 3D rigid metal–organic frameworks (MOFs) to 2D flexible networks via ligand exchange for enhanced CO_2/N_2 and CH_4/N_2 separation. Chem. Commun. 51: 14716–14719.

Hirscher, M. 2011. Hydrogen storage by cryoadsorption in ultrahigh-porosity metal–organic frameworks. Angew. Chem. Int. Ed. 50: 581–582.

Houndonougbo, Y., C. Signer, N. He, W. Morris, H. Furukawa, K.G. Ray et al. 2013. A combined experimental–computational investigation of methane adsorption and selectivity in a series of isoreticular zeolitic imidazolate frameworks. J. Phys. Chem. C 117: 10326–10335.

Howarth, A.J., Y. Liu, J.T. Hupp and O.K. Farha. 2015. Metal–organic frameworks for applications in remediation of oxyanion/cation-contaminated water. CrystEngComm. 17: 7245–7253.

Hu, D., Y. Song and L. Wang. 2015. Nanoscale luminescent lanthanide-based metal–organic frameworks: properties, synthesis, and applications. J. Nanopart. Res. 17: 310.

Hu, L., P. Zhang, Q. Chen, H. Zhong, X. Hu, X. Zheng et al. 2012. Morphology-controllable synthesis of metal organic framework $Cd_3[Co(CN)_6]_2 \cdot nH_2O$ nanostructures for hydrogen storage applications. Cryst. Growth Des. 12: 2257–2264.

Hu, Y.H. and L. Zhang. 2010. Hydrogen storage in metal–organic frameworks. Adv. Mater. 22: E117–E130.

Hwang, Y., H. Sohn, A. Phan, O.M. Yaghi and R.N. Candler. 2013. Dielectrophoresis-assembled zeolitic imidazolate framework nanoparticle-coupled resonators for highly sensitive and selective gas detection. Nano Lett. 13: 5271–5276.

Jhung, S.H., J.-H. Lee, J.W. Yoon, C. Serre, G. Férey and J.-S. Chang. 2007. Microwave synthesis of chromium terephthalate MIL-101 and its benzene sorption ability. Adv. Mater. 19: 121–124.

Jia, Y., C. Sun, Y. Peng, W. Fang, X. Yan, D. Yang et al. 2015. Metallic Ni nanocatalyst *in situ* formed from a metal–organic-framework by mechanochemical reaction for hydrogen storage in magnesium. J. Mater. Chem. A 3: 8294–8299.

Jiang, J. and O.M. Yaghi. 2015. Brønsted acidity in metal–organic frameworks. Chem. Rev. 115: 6966–6997.

Karagiaridi, O., W. Bury, A.A. Sarjeant, C.L. Stern, O.K. Farha and J.T. Hupp. 2012. Synthesis and characterization of isostructural cadmium zeolitic imidazolate frameworks via solvent-assisted linker exchange. Chem. Sci. 3: 3256–3260.

Karagiaridi, O., W. Bury, J.E. Mondloch, J.T. Hupp and O.K. Farha. 2014a. Solvent-assisted linker exchange: An alternative to the *De novo* synthesis of unattainable metal–organic frameworks. Angew. Chem. Int. Ed. 53: 4530–4540.

Karagiaridi, O., W. Bury, D. Fairen-Jimenez, C.E. Wilmer, A.A. Sarjeant, J.T. Hupp et al. 2014b. Enhanced gas sorption properties and unique behavior toward liquid water in a pillared-paddlewheel metal–organic framework transmetalated with Ni(II). Inorg. Chem. 53: 10432–10436.

Kaur, P., J.T. Hupp and S.T. Nguyen. 2011. Porous organic polymers in catalysis: Opportunities and challenges. ACS Catal. 1: 819–835.

Kennedy, R.D., V. Krungleviciute, D.J. Clingerman, J.E. Mondloch, Y. Peng, C.E. Wilmer et al. 2013. Carborane-based metal–organic framework with high methane and hydrogen storage capacities. Chem. Mater. 25: 3539–3543.

Khazalpour, S., V. Safarifard, A. Morsali and D. Nematollahi. 2015. Electrochemical synthesis of pillared layer mixed ligand metal–organic framework: DMOF-1–Zn. RSC Adv. 5: 36547–36551.

Khoshhal, S., A.A. Ghoreyshi, M. Jahanshahi and M. Mohammadi. 2015. Study of the temperature and solvent content effects on the structure of Cu–BTC metal organic framework for hydrogen storage. RSC Adv. 5: 24758–24768.

Kim, J., S. Yeo, J.-D. Jeon and S.-Y. Kwak. 2015. Enhancement of hydrogen storage capacity and hydrostability of metal–organic frameworks (MOFs) with surface-loaded platinum nanoparticles and carbon black. Microporous Mesoporous Mater. 202: 8–15.

Kitagawa, S., R. Kitaura and S.-I. Noro. 2004. Functional porous coordination polymers. Angew. Chem. Int. Ed. 43: 2334–2375.

Klinowski, J., F.A. Almeida Paz, P. Silva and J. Rocha. 2011. Microwave-assisted synthesis of metal–organic frameworks. Dalton Trans. 40: 321–330.

Klyamkin, S.N., S.V. Chuvikov, N.V. Maletskaya, E.V. Kogan, V.P. Fedin, K.A. Kovalenko et al. 2014. High-pressure hydrogen storage on modified MIL-101 metal–organic framework. Int. J. Energy Res. 38: 1562–1570.

Koh, K., J.-E. Seo, J.H. Lee, A. Goswami, C.W. Yoon and T. Asefa. 2014. Ultrasmall palladium nanoparticles supported on amine-functionalized SBA-15 efficiently catalyze hydrogen evolution from formic acid. J. Mater. Chem. A 2: 20444–20449.

Kondo, M., T. Yoshitomi, H. Matsuzaka, S. Kitagawa and K. Seki. 1997. Three-dimensional framework with channeling cavities for small molecules: $\{[M_2(4,4'\text{-bpy})_3(NO_3)_4]\cdot xH_2O\}n$ (M = Co, Ni, Zn). Angew. Chem. Int. Ed. Engl. 36: 1725–1727.

Kong, S., R. Dai, H. Li, W. Sun and Y. Wang. 2015. Microwave hydrothermal synthesis of Ni-based metal–organic frameworks and their derived yolk–shell NiO for Li-ion storage and supported ammonia borane for hydrogen desorption. ACS Sustainable Chem. Eng. 3: 1830–1838.

Kong, X., H. Deng, F. Yan, J. Kim, J.A. Swisher, B. Smit et al. 2013. Mapping of functional groups in metal-organic frameworks. Science 341: 882–885.

Kreno, L.E., K. Leong, O.K. Farha, M. Allendorf, R.P. Van Duyne and J.T. Hupp. 2012. Metal organic framework materials as chemical sensors. Chem. Rev. 112: 1105–1125.

Kubas, G.J., R.R. Ryan, B.I. Swanson, P.J. Vergamini and H.J. Wasserman. 1984. Characterization of the first examples of isolable molecular hydrogen complexes, $M(CO)_3(PR_3)_2(H_2)$ (M = molybdenum or tungsten; R = Cy or isopropyl). Evidence for a side-on bonded dihydrogen ligand. J. Am. Chem. Soc. 106: 451–452.

Kumar, P., V. Bansal, A. Deep and K.-H. Kim. 2015. Synthesis and energy applications of metal organic frameworks. J. Porous Mater. 22: 413–424.

Lalonde, M.B., R.B. Getman, J.Y. Lee, J.M. Roberts, A.A. Sarjeant, KA. Scheidt et al. 2013. A zwitterionic metal–organic framework with free carboxylic acid sites that exhibits enhanced hydrogen adsorption energies. CrystEngComm. 15: 9408–9414.

Langmi, H.W., J. Ren, B. North, M. Mathe and D. Bessarabov. 2014. Hydrogen storage in metal-organic frameworks: A review. Electrochim. Acta 128: 368–392.

Latroche, M., S. Surblé, C. Serre, C. Mellot-Draznieks, P.L. Llewellyn, J.-H. Lee et al. 2006. Hydrogen storage in the giant-pore metal–organic frameworks MIL-100 and MIL-101. Angew. Chem. Int. Ed. 45: 8227–8231.

Lee, J.Y., J.M. Roberts, O.K. Farha, A.A. Sarjeant, K.A. Scheidt and J.T. Hupp. 2009. Synthesis and gas sorption properties of a metal-azolium framework (MAF) material. Inorg. Chem. 48: 9971–9973.

Lee, W.R., S.Y. Hwang, D.W. Ryu, K.S. Lim, S.S. Han and D. Moon et al. 2014. Diamine-functionalized metal–organic framework: exceptionally high CO_2 capacities from ambient air and flue gas, ultrafast CO_2 uptake rate, and adsorption mechanism. Energy Environ. Sci. 7: 744–751.

Levine, D.J., T. Runcěvski, M.T. Kapelewski, B.K. Keitz, J. Oktawiec, D.A. Reed et al. 2016. Olsalazine-based metal–organic frameworks as biocompatible platforms for H_2 adsorption and drug delivery. J. Am. Chem. Soc. 138: 10143–10150.

Li, G., H. Kobayashi, J.M. Taylor, R. Ikeda, Y. Kubota, K. Kato et al. 2014a. Hydrogen storage in Pd nanocrystals covered with a metal–organic framework. Nature Mater. 13: 802–806.

Li, Q. and T. Thonhauser. 2012. A theoretical study of the hydrogen-storage potential of $(H_2)_4CH_4$ in metal organic framework materials and carbon nanotubes. J. Phys.: Condens. Matter 24: 424204.

Li, S.-L. and Q. Xu. 2013. Metal–organic frameworks as platforms for clean energy. Energy Environ. Sci. 6: 1656–1683.

Li, Z.J., S.K. Khani, K. Akhbari, A. Morsali and P. Retailleau. 2014b. Achieve to easier opening of channels in anionic nanoporous metal–organic framework by cation exchange process. Microporous Mesoporous Mater. 199: 93–98.

Lim, D.-W., S.A. Chyun and M.P. Suh. 2014. Hydrogen storage in a potassium-ion-bound metal–organic framework incorporating crown ether struts as specific cation binding sites. Angew. Chem. Int. Ed. 53: 7819–7822.

Lim, W.-X., A.W. Thornton, A.J. Hill, B.J. Cox, J.M. Hill and M.R. Hill. 2013. High performance hydrogen storage from Be-BTB metal–organic framework at room temperature. Langmuir 29: 8524–8533.

Liu, J., R. Zou and Y. Zhao. 2016a. Recent developments in porous materials for H_2 and CH_4 storage. Tetrahedron Lett. 57: 4873–4881.

Liu, J., G. Liu, C. Gu, W. Liu, J. Xu, B. Li et al. 2016b. Rational synthesis of a novel 3,3,5-c polyhedral metal–organic framework with high thermal stability and hydrogen storage capability. J. Mater. Chem. A 4: 11630–11634.

Liu, Y., F. Guo, J. Hu, S. Zhao, H. Liu and Y. Hu. 2016. Entropy prediction for H_2 adsorption in metal–organic frameworks. Phys. Chem. Chem. Phys. 18: 23998–24005.

Lu, Z.-H., H.-L. Jiang, M. Yadav, K. Aranishi and Q. Xu. 2012. Synergistic catalysis of Au-Co@SiO_2 nanospheres in hydrolytic dehydrogenation of ammonia borane for chemical hydrogen storage. J. Mater. Chem. 22: 5065–5071.

Lucier, B.E.G., Y. Zhang, K.J. Lee, Y. Lu and Y. Huang. 2016. Grasping hydrogen adsorption and dynamics in metal–organic frameworks using ^2H solid-state NMR. Chem. Commun. 52: 7541–7544.

Ma, S.-Q., C.D. Collier and H.-C. Zhou. 2009. Design and construction of metal–organic frameworks for hydrogen storage and selective gas adsorption. *In*: Hong, M.-C. and L. Chen (eds.). Design and Construction of Coordination Polymers. John Wiley & Sons, Inc., Hoboken, NJ, USA.

Masoomi, M.Y. and A. Morsali. 2012. Applications of metal–organic coordination polymers as precursors for preparation of nano-materials. Coord. Chem. Rev. 256: 2921–294.

Maza, W.A., A.J. Haring, S.R. Ahrenholtz, C.C. Epley, S.Y. Lin and A.J. Morris. 2016. Ruthenium(II)-polypyridyl zirconium(IV) metal–organic frameworks as a new class of sensitized solar cells. Chem. Sci. 7: 719–727.

McGuire, C.V. and R.S. Forgan. 2015. The surface chemistry of metal–organic frameworks. Chem. Commun. 51: 5199–5217.

Mellmann, D., P. Sponholz, H. Junge and M. Beller. 2016. Formic acid as a hydrogen storage material–development of homogeneous catalysts for selective hydrogen release. Chem. Soc. Rev. 45: 3954–3988.

Meng, Z., R. Lu, D. Rao, E. Kan, C. Xiao and K. Deng. 2013. Catenated metal-organic frameworks: Promising hydrogen purification materials and high hydrogen storage medium with further lithium doping. Int. J. Hydrogen Energy 38: 9811–9818.

Mondloch, J.E., O. Karagiaridi, O.K. Farha and J.T. Hupp. 2013. Activation of metal–organic framework materials. CrystEngComm. 15: 9258–9264.

Mondloch, J.E., M.J. Katz, W.C. Isley III, P. Ghosh, P. Liao, W. Bury et al. 2015. Destruction of chemical warfare agents using metal–organic frameworks. Nature Mater. 14: 512–516.

Morozan, A. and F. Jaouen. 2012. Metal organic frameworks for electrochemical applications. Energy Environ. Sci. 5: 9269–9290.

Morris, R.E. and P.S. Wheatley. 2008. Gas storage in nanoporous materials. Angew. Chem. Int. Ed. 47: 4966–4981.

Morris, W., C.J. Doonan and O.M. Yaghi. 2011. Postsynthetic modification of a metal organic framework for stabilization of a hemiaminal and ammonia uptake. Inorg. Chem. 50: 6853–6855.

Moussa, G., U.B. Demirci, S. Malo, S. Bernard and P. Miele. 2014. Hollow core@mesoporous shell boron nitride nanopolyhedron-confined ammonia borane: a pure B–N–H composite for chemical hydrogen storage. J. Mater. Chem. A 2: 7717–7722.

Murray, L.J., M. Dincă and J.R. Long. 2009. Hydrogen storage in metal–organic frameworks. Chem. Soc. Rev. 38: 1294–1314.

Naeem, A., V.P. Ting, U. Hintermair, M. Tian, R. Telford, S. Halim et al. 2016. Mixed-linker approach in designing porous zirconium-based metal–organic frameworks with high hydrogen storage capacity. Chem. Commun. 52: 7826–7829.

Nguyen, N.T.T., H. Furukawa, F. Gándara, H.T. Nguyen, K.E. Cordova and O.M. Yaghi. 2014. Selective capture of carbon dioxide under humid conditions by hydrophobic chabazite-type zeolitic imidazolate frameworks. Angew. Chem. Int. Ed. 53: 10645–10648.

Nouar, F., J. Eckert, J.F. Eubank, P. Forster and M. Eddaoudi. 2009. Zeolite-*like* metal-organic frameworks (ZMOFs) as hydrogen storage platform: Lithium and magnesium ion-exchange and H_2-(*rho*-ZMOF) interaction studies. J. Am. Chem. Soc. 131: 2864–2870.

Orellana-Tavra, C., E.F. Baxter, T. Tian, T.D. Bennett, N.K.H. Slater, A.K. Cheetham et al. 2015. Amorphous metal–organic frameworks for drug delivery. Chem. Commun. 51: 13878–13881.

Park, K.S., Z. Ni, A.P. Côté, J.Y. Choi, R. Huang, F.J. Uribe-Romo et al. 2006. Exceptional chemical and thermal stability of zeolitic imidazolate frameworks. PNAS 103: 10186–10191.

Peng, Y., V. Krungleviciute, I. Eryazici, J.T. Hupp, O.K. Farha and T. Yildirim. 2013. Methane storage in metal–organic frameworks: Current records, surprise findings, and challenges. J. Am. Chem. Soc. 135: 11887–11894.

Perles, J., M. Iglesias, M.-Á. Martín-Luengo, M.Á. Monge, C. Ruiz-Valero and N. Snejko. 2005. Metal-organic scandium framework: Useful material for hydrogen storage and catalysis. Chem. Mater. 17: 5837–5842.

Pham, T., K.A. Forrest, B. Space and J. Eckert. 2016. Dynamics of H_2 adsorbed in porous materials as revealed by computational analysis of inelastic neutron scattering spectra. Phys. Chem. Chem. Phys. 18: 17141–17158.

Phan, A., C.J. Doonan, F.J. Uribe-Romo, C.B. Knobler, M. O'Keeffe and O.M. Yaghi. 2010. Synthesis, structure, and carbon dioxide capture properties of zeolitic imidazolate frameworks. Acc. Chem. Res. 43: 58–67.

Qin, W., W. Cao, H. Liu, Z. Li and Y. Li. 2014. Metal–organic framework MIL-101 doped with palladium for toluene adsorption and hydrogen storage. RSC Adv. 4: 2414–2420.

Rallapalli, P.B.S., M.C. Raj, D.V. Patil, K.P. Prasanth, R.S. Somani and H.C. Bajaj. 2013. Activated carbon@MIL-101(Cr): A potential metal-organic framework composite material for hydrogen storage. Int. J. Energy Res. 37: 746–753.

Rao, D., R. Lu, Z. Meng, G. Xu, E. Kan, Y. Liu et al. 2013. Influences of lithium doping and fullerene impregnation on hydrogen storage in metal organic frameworks. Molecular Simulation 39: 968–974.

Ren, J., H.W. Langmi, B.C. North, M. Mathe and D. Bessarabov. 2014. Modulated synthesis of zirconium-metal organic framework (Zr-MOF) for hydrogen storage applications. Int. J. Hydrogen Energy 39: 890–895.

Ren, J., H.W. Langmi, B.C. North and M. Mathe. 2015a. Review on processing of metal–organic framework (MOF) materials towards system integration for hydrogen storage. Int. J. Energy Res. 39: 607–620.

Ren, J., N.M. Musyoka, H.W. Langmi, A. Swartbooi, B.C. North and M. Mathe. 2015b. A more efficient way to shape metal-organic framework (MOF) powder materials for hydrogen storage applications. Int. J. Hydrogen Energy 40: 4617–4622.

Ricco, R., L. Malfatti, M. Takahashi, A.J. Hill and P. Falcaro. 2013. Applications of magnetic metal–organic framework composites. J. Mater. Chem. A 1: 13033–13045.

Rosi, N.L., M. Eddaoudi, J. Kim, M. O'Keeffe and O.M. Yaghi. 2002. Advances in the chemistry of metal–organic frameworks. CrystEngComm. 4: 401–404.

Rosi, N.L., J. Eckert, M. Eddaoudi, D.T. Vodak, J. Kim, M. O'Keeffe et al. 2003. Hydrogen storage in microporous metal-organic frameworks. Science 300: 1127–1129.

Rowsell, J.L.C., A.R. Millward, K.S. Park and O.M. Yaghi. 2004. Hydrogen sorption in functionalized metal-organic frameworks. J. Am. Chem. Soc. 126: 5666–5667.

Rowsell, J.L.C. and O.M. Yaghi. 2005. Strategies for hydrogen storage in metal–organic frameworks. Angew. Chem. Int. Ed. 44: 4670–4679.

Rowsell, J.L.C., J. Eckert and O.M. Yaghi. 2005. Characterization of H_2 binding sites in prototypical metal-organic frameworks by inelastic neutron scattering. J. Am. Chem. Soc. 127: 14904–14910.

Ryan, P., L.J. Broadbelt and R.Q. Snurr. 2008. Is catenation beneficial for hydrogen storage in metal–organic frameworks? Chem. Commun. 4132–4134.

Rydén, J., S. Öberg, M. Heggie, M. Rayson and P. Briddon. 2013. Hydrogen storage in the manganese containing metal–organic framework MOF-73. Microporous Mesoporous Mater. 165: 205–209.

Safarifard, V. and A. Morsali. 2015. Applications of ultrasound to the synthesis of nanoscale metal–organic coordination polymers. Coord. Chem. Rev. 292: 1–14.

Sajna, K., A.M. Fracaroli, O.M. Yaghi and K. Tashiro. 2015. Modular synthesis of metal–organic complex arrays containing precisely designed metal sequences. Inorg. Chem. 54: 1197–1199.

Sapchenko, S.A., D.N. Dybtsev, D.G. Samsonenko, R.V. Belosludov, V.R. Belosludov, Y. Kawazoe et al. 2015. Selective gas adsorption in microporous metal–organic frameworks incorporating urotropine basic sites: an experimental and theoretical study. Chem. Commun. 51: 13918–13921.

Sculley, J., D. Yuan and H.-C. Zhou. 2011. The current status of hydrogen storage in metal–organic frameworks—updated. Energy Environ. Sci. 4: 2721–2735.

Si, X., C. Jiao, F. Li, J. Zhang, S. Wang, S. Liu et al. 2011. High and selective CO_2 uptake, H_2 storage and methanol sensing on the amine-decorated 12-connected MOF CAU-1. Energy Environ. Sci. 4: 4522–4527.

Slater, A.G. and A.I. Cooper. 2015. Function-led design of new porous materials. Science 348: 988.

So, M.C., M.H. Beyzavi, R. Sawhney, O. Shekhah, M. Eddaoudi, S.S. Al-Juaid et al. 2015a. Post-assembly transformations of porphyrin-containing metal–organic framework (MOF) films fabricated via automated layer-by-layer coordination. Chem. Commun. 51: 85–88.

So, M.C., G.P. Wiederrecht, J.E. Mondloch, J.T. Hupp and O.K. Farha. 2015b. Metal–organic framework materials for light-harvesting and energy transfer. Chem. Commun. 51: 3501–3510.

Song, L., J. Zhang, L. Sun, F. Xu, F. Li, H. Zhang et al. 2012. Mesoporous metal–organic frameworks: design and applications. Energy Environ. Sci. 5: 7508–7520.

Song, P., Y. Li, B. He, J. Yang, J. Zheng and X. Li. 2011. Hydrogen storage properties of two pillared-layer Ni(II) metal-organic frameworks. Microporous Mesoporous Mater. 142: 208–213.

Stavila, V., A.A. Talin and M.D. Allendorf. 2014. MOF-based electronic and opto-electronic devices. Chem. Soc. Rev. 43: 5994–6010.

Stephenson, C.J., J.T. Hupp and O.K. Farha. 2015. Pt@ZIF-8 composite for the regioselective hydrogenation of terminal unsaturations in 1,3-dienes and alkynes. Inorg. Chem. Front. 2: 448–452.

Sudik, A.C., A.R. Millward, N.W. Ockwig, A.P. Côté, J. Kim and O.M. Yaghi. 2005. Design, synthesis, structure, and gas (N_2, Ar, CO_2, CH_4, and H_2) sorption properties of porous metal-organic tetrahedral and heterocuboidal polyhedra. J. Am. Chem. Soc. 127: 7110–7118.

Suh, M.P., H.J. Park, T.K. Prasad and D.-W. Lim. 2012. Hydrogen storage in metal organic frameworks. Chem. Rev. 112: 782–835.

Sumida, K., D.L. Rogow, J.A. Mason, T.M. McDonald, E.D. Bloch, Z.R. Herm et al. 2012. Carbon dioxide capture in metal organic frameworks. Chem. Rev. 112: 724–781.

Sumida, K., D. Stück, L. Mino, J.-D. Chai, E.D. Bloch, O. Zavorotynska et al. 2013. Impact of metal and anion substitutions on the hydrogen storage properties of M-BTT metal–organic frameworks. J. Am. Chem. Soc. 135: 1083–1091.

Sun, Y., L. Wang, W.A. Amer, H. Yu, J. Ji, L. Huang et al. 2013. Hydrogen storage in metal-organic frameworks. J. Inorg. Organomet. Polym. 23: 270–285.

Tahmasebi, E., M.Y. Masoomi, Y. Yamini and A. Morsali. 2015. Application of mechanosynthesized azine-decorated zinc(II) metal–organic frameworks for highly efficient removal and extraction of some heavy-metal ions from aqueous samples: A comparative study. Inorg. Chem. 54: 425–433.

Tanabe, K.K. and S.M. Cohen. 2011. Postsynthetic modification of metal–organic frameworks—a progress report. Chem. Soc. Rev. 40: 498–519.

Thomas, K.M. 2009. Adsorption and desorption of hydrogen on metal–organic framework materials for storage applications: comparison with other nanoporous materials. Dalton Trans. 1487–1505.

Thomas, K.M. 2011. Hydrogen adsorption on metal organic framework materials for storage applications. *In:* Bruce, D.W., D. O'Hare and R.I. Walton (eds.). Energy Materials. John Wiley & Sons, Ltd., Chichester, UK.

Thornton, A.W., D. Dubbeldam, M.S. Liu, B.P. Ladewig, A.J. Hill and M.R. Hill. 2012. Feasibility of zeolitic imidazolate framework membranes for clean energy applications. Energy Environ. Sci. 5: 7637–7646.

Tranchemontagne, D.J., Z. Ni, M. O'Keeffe and O.M. Yaghi. 2008a. Reticular chemistry of metal–organic polyhedra. Angew. Chem. Int. Ed. 47: 5136–5147.

Tranchemontagne, D.J., J.R. Hunt and O.M. Yaghi. 2008b. Room temperature synthesis of metal-organic frameworks: MOF-5, MOF-74, MOF-177, MOF-199, and IRMOF-0. Tetrahedron 64: 8553–8557.

Tranchemontagne, D.J., K.S. Park, H. Furukawa, J. Eckert, C.B. Knobler and O.M. Yaghi. 2012. Hydrogen storage in new metal–organic frameworks. J. Phys. Chem. C 116: 13143–13151.

Valvekens, P., F. Vermoortele and D. De Vos. 2013. Metal–organic frameworks as catalysts: the role of metal active sites. Catal. Sci. Technol. 3: 1435–1445.

Volkova, E.I., A.V. Vakhrushev and M. Suyetin. 2014. Improved design of metal-organic frameworks for efficient hydrogen storage at ambient temperature: A multiscale theoretical investigation. Int. J. Hydrogen Energy 39: 8347–8350.

Walker, G. 2008. Solid-State Hydrogen Storage. Woodhead Publishing Ltd., Cambridge, England.

Waller, P.J., F. Gándara and O.M. Yaghi. 2015. Chemistry of covalent organic frameworks. Acc. Chem. Res. 48: 3053–3063.

Wan, S., F. Gándara, A. Asano, H. Furukawa, A. Saeki, S.K. Dey et al. 2011. Covalent organic frameworks with high charge carrier mobility. Chem. Mater. 23: 4094–4097.

Wang, B., A.P. Côté, H. Furukawa, M. O'Keeffe and O.M. Yaghi. 2008. Colossal cages in zeolitic imidazolate frameworks as selective carbon dioxide reservoirs. Nature 453: 207–211.

Wang, D., B. Liu, S. Yao, T. Wang, G. Li, Q. Huo et al. 2015a. A polyhedral metal–organic framework based on the supermolecular building block strategy exhibiting high performance for carbon dioxide capture and separation of light hydrocarbons. Chem. Commun. 51: 15287–15289.

Wang, S. and X. Wang. 2015. Multifunctional metal–organic frameworks for photocatalysis. Small 11: 3097–3112.

Wang, X., L. Xie, K.-W. Huang and Z. Lai. 2015b. A rationally designed amino-borane complex in a metal organic framework: a novel reusable hydrogen storage and size-selective reduction material. Chem. Commun. 51: 7610–7613.

Xiao, J., M. Hu, P. Bénard and R. Chahine. 2013. Simulation of hydrogen storage tank packed with metal-organic framework. Int. J. Hydrogen Energy 38: 13000–13010.

Xie, S., J.-Y. Hwang, X. Sun, S. Shi, Z. Zhang, Z. Peng et al. 2014. Percolative metal-organic framework/carbon composites for hydrogen storage. J. Power Sources 253: 132–137.

Yaghi, O.M. and H. Li. 1995. Hydrothermal synthesis of a metal-organic framework containing large rectangular channels. J. Am. Chem. Soc. 117: 10401–10402.

Yaghi, O.M. and Q. Li. 2009. Reticular chemistry and metal-organic frameworks for clean energy. MRS Bull 34: 682–690.

Yan, Y., S. Yang, A.J. Blake and M. Schröder. 2014. Studies on metal organic frameworks of Cu(II) with isophthalate linkers for hydrogen storage. Acc. Chem. Res. 47: 296–307.

Yoo, J.B., H.S. Kim, S.H. Kang, B. Lee and N.H. Hur. 2014. Hollow nickel-coated silica microspheres containing rhodium nanoparticles for highly selective production of hydrogen from hydrous hydrazine. J. Mater. Chem. A 2: 18929–18937.

Yuan, W., J. O'Connorb and S.L. James. 2010. Mechanochemical synthesis of homo- and hetero-rare-earth(III) metal–organic frameworks by ball milling. CrystEngComm. 12: 3515–3517.

Zhang, Q. and J.M. Shreeve. 2014. Metal–organic frameworks as high explosives: A new concept for energetic materials. Angew. Chem. Int. Ed. 53: 2540–2542.

Zhang, X., Y. Liu, Y. Pang, M. Gao and H. Pan. 2014a. Significantly improved kinetics, reversibility and cycling stability for hydrogen storage in $NaAlH_4$ with the Ti-incorporated metal organic framework MIL-125(Ti). J. Mater. Chem. A 2: 1847–1854.

Zhang, Y.-B., H. Furukawa, N. Ko, W. Nie, H.J. Park, S. Okajima et al. 2015. Introduction of functionality, selection of topology, and enhancement of gas adsorption in multivariate metal–organic framework-177. J. Am. Chem. Soc. 137: 2641–2650.

Zhang, Z., Z.-Z. Yao, S. Xiang and B. Chen. 2014b. Perspective of microporous metal–organic frameworks for CO_2 capture and separation. Energy Environ. Sci. 7: 2868–2899.

Zhao, D., D. Yuan and H.-C. Zhou. 2008. The current status of hydrogen storage in metal–organic frameworks. Energy Environ. Sci. 1: 222–235.

Zhao, Q., W. Yuan, J. Liang and J. Li. 2013. Synthesis and hydrogen storage studies of metal-organic framework UiO-66. Int. J. Hydrogen Energy 38: 13104–13109.

7

Nanomaterials for Electrochemical Capacitor

Pei Yi, Chan

1. Introduction

Ever wondered how the lightning occurred when you were young? And, why the plastic sticks to your hand when you remove it from a package? The curiosity always leads to a discovery. It is all about electrostatic phenomena. Since a long time ago, some materials have been found to attract light objects after rubbing. This observation was not fully understood until the concept of electric charge was developed after numerous efforts had been made. The famous 'kite experiment' performed by Benjamin Franklin led to the naming of electric charge as either 'positive' or 'negative'. The positive and negative charge can interact through electrostatic force. The electrostatic interaction can be noticed in our daily life such as dust collected on television screen, a small electric shock when our hand touches the door knob after rubbing feet on carpet, and even the photocopy makes use of the concept of electrostatics. One of the contributions of the electrostatic interaction is the formation of double layer when an object is exposed to liquid under an electric field. The electrical double layer capacitor (EDLC) employs this idea for charge storage purpose. Based on the adsorption/de-adsorption processes, an EDLC is able to charge/discharge and supply current. It is worth mentioning that EDLC does not involve charge transfer reaction. The charge stored in EDLC is called double layer capacitance while the processes involved in an EDLC are summarized as shown below (Zheng et al. 1997).

Centre for Ionics University of Malaya, Department of Physics, Faculty of Science, University of Malaya, Lembah Pantai, 50603 Kuala Lumpur, Malaysia.
Email: stac_chan@live.com

Positive electrode : $$E_s + A^- \underset{\text{Discharge}}{\overset{\text{Charge}}{\Leftrightarrow}} E_s^+ // A^- + e^- \quad (1.1)$$

Negative electrode : $$E_s + C^+ + e^- \underset{\text{Discharge}}{\overset{\text{Charge}}{\Leftrightarrow}} E_s^- // C^+ \quad (1.2)$$

Overall reaction : $$E_s + E_s + C^+A^- \underset{\text{Discharge}}{\overset{\text{Charge}}{\Leftrightarrow}} E_s^+ // A^- + E_s^- // C^+ \quad (1.3)$$

where E_s depicts the surface of carbon electrode, // illustrates the double layer, A^- represents the electrolyte anion, and C^+ acts as the electrolyte cation. During the charging process, which is when the circuit is connected to a power source, the electrolyte anions and cations are attracted to the opposite charged electrode. The electrons are accumulated at the negative electrodes. When the circuit is connected to a load, the discharging process takes place. Electrons are transferred from the negative electrode to the positive electrode through the load. At the same time, the electrolyte ions are released back to the bulk. A simplified version of EDLC is displayed in Fig. 1. Each interface of electrode/electrolyte acts like a capacitor. Hence, an EDLC can be considered as a combination of two capacitors in series. Hence, the total capacitance for a symmetric capacitor is the sum of two capacitances, Eqs. 1 and 2.

$$C_{cell} = \frac{1}{C_1} + \frac{1}{C_2} \quad \text{(Equation 1)}$$

$$C_{dl} = \frac{\varepsilon A}{4\Pi t} \quad \text{(Equation 2)}$$

C_{cell} and C_{dl} represent cell capacitance and double layer capacitance respectively, ε stands for dielectric constant, A is the surface area of electrode while t acts as the thickness of double layer.

The development and application of EDLC are the ultimate and continuous efforts from Becker, Rightmire, Boos, von Helmholtz, Gouy, Chapman, Stern, Grahame, and others (Becker 1957, Boos 1970, Chapman 1913, Gouy 1910, Grahame 1947, Helmholtz 1853, Rightmire 1966, Stern 1924). The emergence of pseudocapacitance has brought out further development of electrochemical capacitor. Electrochemical capacitor is a big family that comprises of EDLC and pseudocapacitor. As an energy device with intermediate performance between conventional capacitor and battery, an electrochemical capacitor owns higher power density compared to battery, as well as higher energy density compared to conventional capacitor.

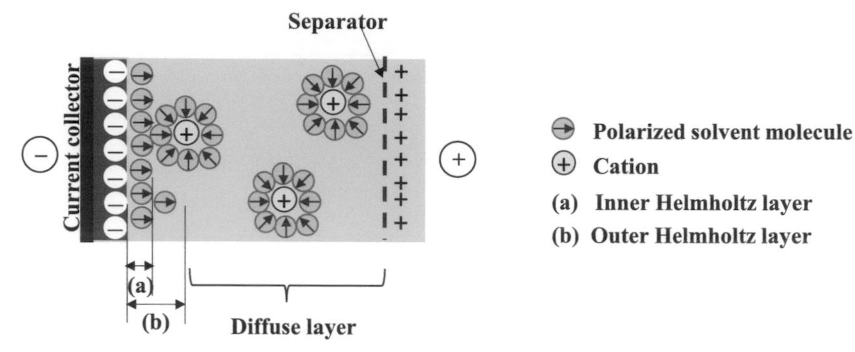

Figure 1. Schematic representation of EDLC.

It is believed that the combination of battery and electrochemical capacitor can meet the start-up power which usually requires high power density. Unlike EDLC, pseudocapacitor stores charge through Faradaic reactions. In other words, the charge transfer reaction takes place at the interface between the electrode film and electrolyte. As double layer capacitance is limited by the surface area, pseudocapacitor can obtain higher value of pseudocapacitance by modifying the morphology and structure of the electrode film.

1.1 Origin of pseudocapacitance

The term 'pseudocapacitive' is used to describe the electrodes that possess similar electrochemical behavior as a capacitor. Both show a linear dependence of the charge stored with varying potential within the potential window tested (Conway 1999). The corresponding cyclic voltammetry curve is rectangular in shape and symmetric. Although the electrochemical signature is similar, their charge storage principles are not the same. The electrochemical behavior of a capacitor is contributed by the ions accumulation on the electrode while pseudocapacitance is conducted by the charge transfer reactions which can be explained *via* band theory (Bard and Faulkner 2001).

Figure 2 describes the relationship between the electronic energy and conductivity. A conduction band is a region where the delocalized electron is present and is responsible for conductivity. For semiconductor, it is also the orbital band that accepts electrons jumping from valence band under thermal excitation. Transition metal oxide possesses semiconductive behavior. Due to the small energy gap, its conductivity can be greatly improved through doping or incorporation with other materials. Based on the band theory, the potential energy of the redox active sites can be

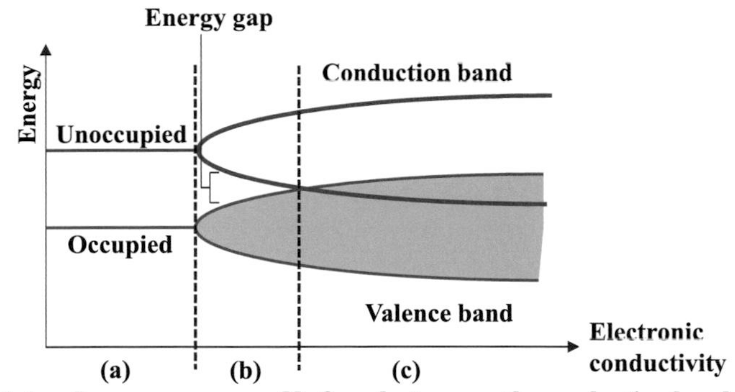

(a) Insulator: Large energy gap. No free electrons as the conduction band is empty.
(b) Semiconductor: Small energy gap.
(c) Conductor: No energy gap. The conduction band is overlapped with valence band.

Figure 2. Band theory for solids.

explained as follows (redox is the site where reaction process occurs on the substance). For well separated redox sites on the film, they will have similar energy states. Therefore, the electron transfer reactions occur at similar potential, leading to the formation of peak-shaped cyclic voltammetry curve. On the contrary, if the redox sites can interact with each other, their energy states will be slightly affected. The dissimilar but continuous energy states can turn into a wide range of potentials when the electron transfer reactions take place. This corresponds to the cyclic voltammetry curve with rectangular envelope.

The capacitance type exhibited in a capacitor is determined by the types of electrode material. EDLC requires electrodes with high surface area. Carbon material, which is high in surface area and conductivity, turns out to be a suitable candidate for EDLC. For pseudocapacitor, a pseudocapacitive material such as transition metal oxide and conducting polymer is needed. Hydrous ruthenium oxide ($RuO_2{\cdot}xH_2O$) and manganese oxide (MnO_x) have been recognized as pseudocapacitive materials (Ardizzone et al. 1990, Conway 1999, Trasatti and Buzzanca 1971). As the first material that disclosed the pseudocapacitance, hydrous ruthenium oxide showed that the proton diffusion can participate in the charge storage process (Arikado et al. 1977, Galizzioli et al. 1974, Zheng and Jow 1995). On the other hand, the pseudocapacitive behavior of manganese oxide was revealed by Lee and co-workers in 1999 (Lee and Goodenough 1999, Lee et al. 1999). Their works showed that capacitance is not only related to the protons, but also the electrolyte cations. The result of the improved capacity with higher amorphousness uncovered the potential of the amorphous materials to be employed in electrochemical capacitor (Lee et al. 1999). The water content

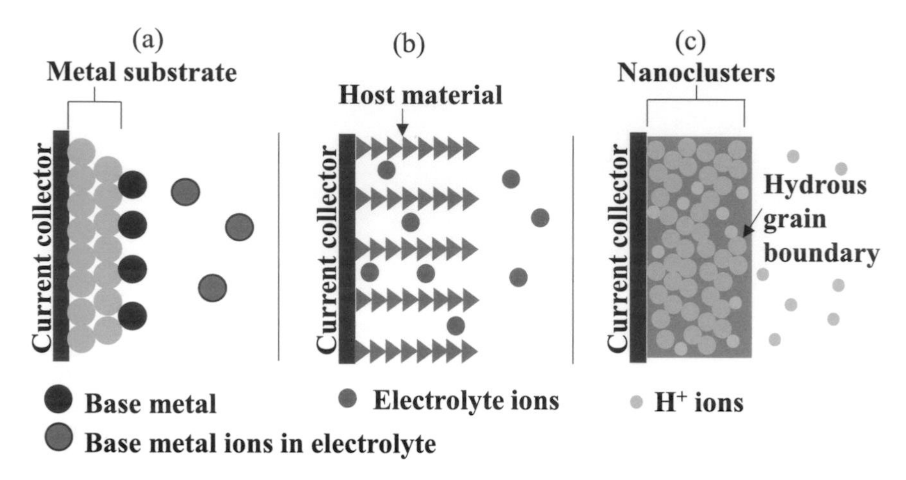

Figure 3. Types of systems that give rise to pseudocapacitance: (a) Underpotential deposition, (b) intercalation pseudocapacitance, and (c) redox pseudocapacitance.

in an amorphous material offers the capacitive storage purpose while the crystalline part allows the insertion processes to occur. There are three main sources of pseudocapacitance: (a) underpotential deposition, (b) intercalation processes, and (c) redox processes as shown in Fig. 3 (Conway et al. 1997). It should be noted that the charge transfer reactions take place in all of the processes mentioned. The underpotential deposition is the result of the adsorption of metal ions onto a different metal electrode surface. The intercalation process, as depicted by its name, involves the intercalation of ions into the lattice (for example, layer- or tunnel-type structure) of material. On the other hand, the redox reactions occur on the surface or sub-surface of the electrode when the ions, which are electrochemically adsorbed onto the electrode surface, contribute to the redox pseudocapacitance.

Based on the meaning of pseudocapacitance defined by Conway, the pseudocapacitance of a material can be described as either intrinsic or extrinsic. The electrochemical signature of an intrinsic pseudocapacitive material does not vary with particle size. The most common intrinsic pseudocapacitive materials are hydrous ruthenium oxide ($RuO_2 \cdot nH_2O$) and manganese oxide (MnO_2) (Ardizzone et al. 1990, Brousse et al. 2015, Brousse et al. 2006, Zheng et al. 1995). On the contrary, an extrinsic pseudocapacitive material is particle size dependent. It shows the capacitive electrochemical signature only when it is downscaled into nanometric form (Augustyn et al. 2014). The materials that possess this behavior are molybdenum disulfide (MoS_2) and molybdenum dioxide (MoO_2) (Cook et al. 2016, Kim et al. 2015).

Pseudocapacitive is more appropriate to be used for describing a single material's behavior. When a pseudocapacitive material is incorporated with

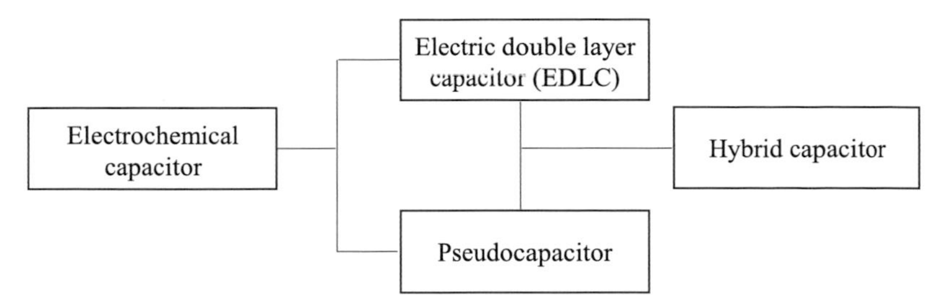

Figure 4: Family of electrochemical capacitor.

capacitive material, the electrochemical behavior shown by this composite electrode is the combination of two charge storage mechanisms. The cyclic voltammetry curve may display a rectangular shape indicating capacitive-like behavior or exhibit broad redox peak designating the participation of charge transfer reactions (Hou et al. 2010, Hwang et al. 2015). A two-terminal capacitor which is made up of capacitive and pseudocapacitive electrodes is called as hybrid capacitor (Fig. 4) (Brousse et al. 2015). As the electrochemical activities cannot be easily identified from the macroscopic view shown by the cyclic voltammetry curve and galvanostatic charge discharge curve, it is not proper to apply the term 'pseudocapacitive'. Thus, an appropriate term usage is crucial and should be chosen wisely in order to not bring about any confusion.

1.2 Nanomaterials for electrochemical capacitor

A generic electrochemical capacitor comprises of electrolyte, electrode, and separator. The electrolyte is responsible for the operating cell voltage of the electrochemical capacitor. As shown in Eqs. 3 and 4, the operating cell voltage plays a bigger role in determining the energy and power density.

Energy density $$E = \frac{1}{2} C(\Delta V)^2 \quad \text{(Equation 3)}$$

Power density $$P = \frac{(\Delta V)^2}{4R} \quad \text{(Equation 4)}$$

In both of the above equations, C represents the total specific capacitance, ΔV depicts the potential window range (operating cell voltage), and R acts as the equivalent series resistance (ESR) of the electrochemical capacitor. In most cases, ESR is the total resistance inclusive of electrolyte resistance, electrode resistance, and contact resistance between the electrode and electrode film. As R is inversely proportional to the power density, a lower ESR can lead to higher power density. One can notice that both the energy and power density respond proportionally to the exponential of two of the operating cell voltage. Hence, they greatly rely on the operating cell voltage.

Typically, the electrolyte performance is mainly dependent on several factors: (1) ion type, (2) ion size, and (3) ion concentration. Different electrolyte has different working potential window range. Due to the water decomposition which occurs at about 1.23 V, an aqueous electrolyte has a potential window of 1.0–1.3 V, while the organic and ionic liquid based electrolytes has potential window ranging from 2.5 to 2.7 V and 3.5 to 4.0 V, respectively. Besides liquid electrolyte, solid electrolyte and quasi-solid-state electrolyte are also being studied for the application in electrochemical capacitor. It is important to ensure that the electrolyte is suited to the electrode used as the electrode is accountable for the total specific capacitance. This parameter will contribute to the energy density of an electrochemical capacitor (Eq. 3). In order to optimize the electrochemical capacitor performance, a suitable electrolyte should be coupled with electrodes with good design. Nowadays, studies on nanomaterials are leading in the material sciences field. Nanostructure is beneficial for providing high surface area, high mechanical strength, and high catalytic activity in various energy devices. It is especially important for pseudocapacitive material as the ionic transportation and electronic transfer becomes more efficient when the material is in nanoscale. From the industry point of view, a nanostructure material marks the eligible production of miniature and light device, hence minimizing the material cost. Raw material consumption will also be reduced. To further understand the potential of nanomaterials in pseudocapacitive devices, the subsequent section is divided into two parts: synthesis and electrochemical performance of nanostructure material.

2. Synthesis of Nanomaterial

Many methods have been developed to prepare nanostructure material. Some of the methods are hydrothermal, solvothermal, chemical reduction, sol-gel, electrodeposition, and chemical vapor deposition (CVD). Hydrothermal synthesis is a technique employed to crystallize material through physical and chemical processes in a closed system at high temperature (> 100°C) and high vapor pressure (> 1 atm). The formation of crystal can be simplified into three steps: crystal dissolution, nucleation, and re-crystallization. Thus, the final product is usually in nanopowder form. The powder then can be coated on any substrates using drop coating method. The shape and particle size can be modified by varying the operating temperature and pressure along with other experimental conditions. For example, RuO_2 nanoparticles with uniform size can be formed under mild hydrothermal condition with temperature of 180°C for 24 h (Chang and Hu 2004, Chang et al. 2007). The nanoparticles have a mean diameter of 2.6 nm and it decreases with shorter hydrothermal time. RuO_2 nanoparticles can also be deposited on carbon nanofiber using hydrothermal method at 150°C for 2 hours (Chuang et al. 2012). The particle size of RuO_2 would decrease after it is incorporated with carbon nanofiber with an average diameter of 2 nm. In order to strengthen the anchoring of nanoparticles on carbon nanofiber, carbon nanofiber could be treated with acetone in order to remove the hydrophobic hydrocarbon layer on the carbon surface. The removal of hydrocarbon layer allows the direct attachment of nanoparticles on nanosize pores at carbon surface which are produced during hydrothermal process. Other than RuO_2, Mn_3O_4 nanorods also can be formed using hydrothermal method. Instead of forming arrays on the graphene sheets, the nanorods are dispersed on the graphene sheet, as illustrated in Fig. 5. This composite shows remarkable cycle stability over 10,000 cycles without significant degradation (Lee et al. 2012). The specific capacitance increased slightly during the first 1,000 cycles and maintained over the next 9,000 cycles.

Apart from hydrothermal, solvothermal method can be employed to produce graphene–Mn_3O_4 nanocomposite. Solvothermal process is similar to hydrothermal process. They are different in terms of the solvent used. The hydrothermal operates with water while solvothermal employs organic as solvent. The formation of nanomaterials *via* hydrothermal and solvothermal is shown in Fig. 6. The morphology and particle size are

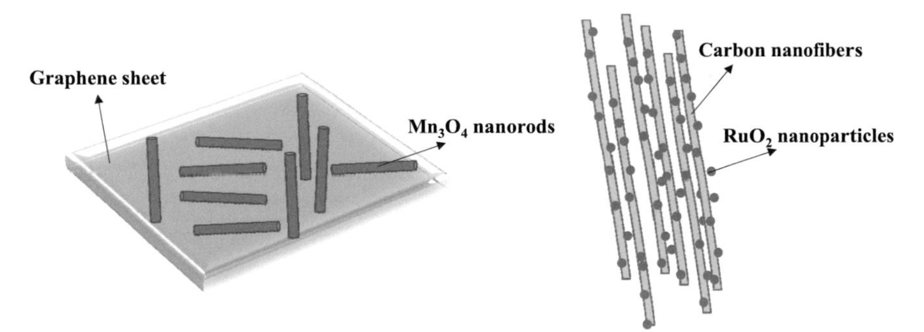

Figure 5. Hydrothermal–synthesized nanocomposite.

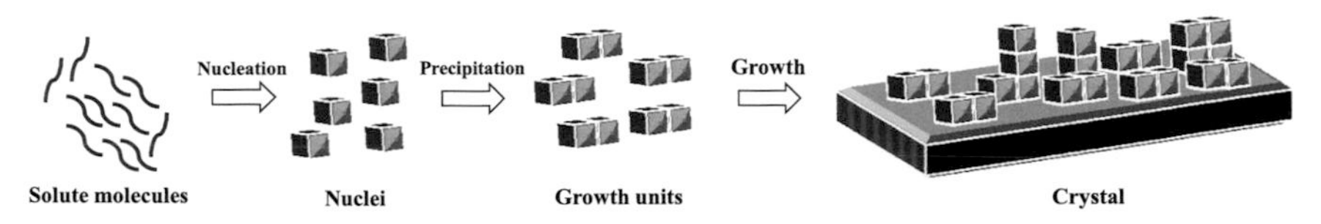

Figure 6. Schematic illustration of crystal growth during hydrothermal or solvothermal process.

determined by the nucleation and growth rate. When the supersaturation state is reached, nucleation is initiated and crystal starts to grow. As the structure and morphology are dependent on the preparation technique, the graphene–Mn_3O_4 nanocomposite fabricated using solvothermal method exhibits a distinct structure. The Mn_3O_4 nanoparticles are formed and it is found to agglomerate on the graphene sheets (Liu et al. 2014). Thus, hydrothermal-synthesized graphene–Mn_3O_4 nanocomposite presents better electrochemical performance.

Electrodeposition allows a direct formation of film on a substrate. Therefore, it has been widely used for electrode fabrication. It originated from the oxidation/reduction of reactant species in the solution under appropriate polarization. Polarization, or overpotential, is the minimum energy required to initiate any chemical, electrochemical, and charge transfer reactions. The ions near the electrode surface undergo reduction and then adsorbed on the electrode surface. At the same time, a concentration gradient is generated which further directs the diffusion of ions from solution to electrode surface. The continuous reduction process favors the formation of film on the electrode surface (refer Fig. 7). After the deposition is completed, a proper post-treatment that is carried out on the electrode film can improve the physical properties of the as-deposited film. The electrodeposition regarding to multiple ions involves complicated mechanisms based on fundamental principles (Paunovic et al. 2010, Schwartz 1994). Hence, one should perform analysis in detail for different combination of materials in order to comprehend the hidden mechanisms.

A desired structure can be guided by template employed during the deposition process. A nanotubular arrays structure can be directed by anodic aluminium oxide (AAO) template. The formation of this structure is conducted by the insertion of ions into the AAO pores. The particles must be strongly stacked on each other in order to maintain the structure after the removal of template. This suggests that the outer diameter of the nanotubes should be comparable to the diameter of AAO. If this nanostructure is accompanied by a suitable post-treatment, this material can show an

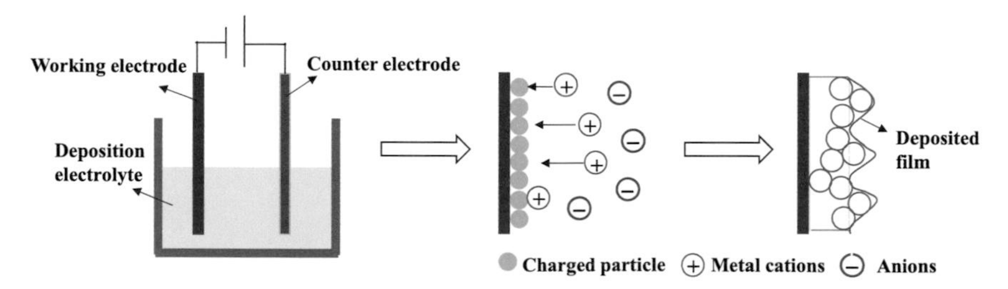

Figure 7. Schematic illustration of electrodeposition process.

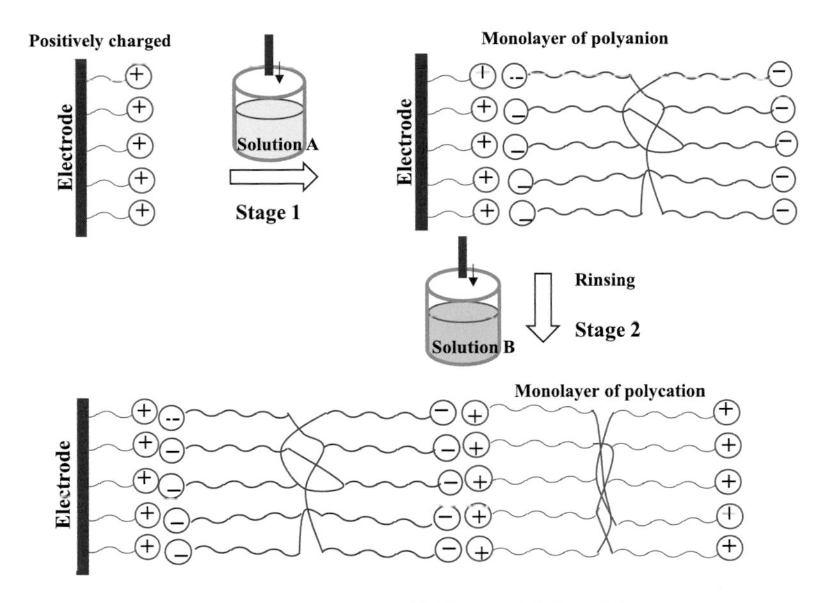

Figure 8. Schematic representation of LBL assembled multilayer structure.

astonishing electrochemical performance. For example, RuO_2 nanotubular arrays achieved 1300 F g^{-1} of specific capacitance after being annealed at 200°C for 2 h (Hu et al. 2006). This is about 76% of increment compared to the electrode before being annealed. Other than metal oxide, conducting polymer such as polyaniline (PANI) nanowires also can be formed using AAO. Nanowire or nanotube structure eases the ion penetration and facilitates the electronic transportation. The advantage of nanostructure is ascertained by the specific capacitance obtained. PANI nanowire and its composite have been reported to obtain specific capacitance in the range of 400–1200 F g^{-1} (Cao and Mallouk 2008, Devarayan et al. 2015, Pan et al. 2016, Wang et al. 2010a, Zhao and Li 2008). In the absence of template, a rod-like structure of nanocomposite can be formed by various deposition techniques as well (Lu et al. 2011c, Santhanagopalan et al. 2013, Wu and Chiang 2004).

Lamellar nanostructure can be constructed *via* layer-by-layer (LBL) self-assembly technique. LBL involves cyclical steps in building a lamellar nanostructure. This repeated procedure can be simplified into two stages, as depicted in Fig. 8 (Decher et al. 1992). Two solutions should be prepared: each contains high concentration of anionic (Solution A) or cationic polyelectrolyte (Solution B). The first step describes the immersion of positively charged electrode into the Solution A, forming a monolayer of polyanion. At this stage, the surface charge is reversed as the high concentration of polyelectrolyte causes the remaining ionic groups being exposed to the solution. Before the second stage is performed, the electrode is rinsed with pure water. The immersion of electrode into Solution B forms a second monolayer of polycation which restores the surface charge of the electrode. By repeating these steps, an alternating multilayer assembly of two polymers is produced. The formation of multilayer can be attributed to the driven force participating during the deposition process such as electrostatic interaction, hydrogen bond, charge transfer reaction, and step-by-step reaction. However, the lamellar nanostructure that is constructed by single charged, electrically neutral, and water-insoluble species cannot be formed based on driven force solely (Zhang et al. 2007). Instead, supramolecular approach is preferable for this kind of multilayer. Overall, LBL technique is time-consuming. Hence, spin coating can be employed to assist the fabrication process and improve the efficiency.

Generally, electrode material with lamellar nanostructure shows good cycle stability as it facilitates the ion insertion into the lattice. Electrode made up of manganese oxide (MnO_2), poly(diallyldimethylammonium) (PDDA), and poly(sodium 4-styrenesulfonate) (PSS) mediated graphene sheets with multilayer structure shows specific capacitance retention of 90% compared to the initial value after 1000 cycles (Li et al. 2011). A poly(ethylenimine) (PEI)–manganese oxide nanosheets electrode film also exhibits similar cycle stability performance (Zhang et al. 2008b). The fall of specific capacitance value upon cycling is considered to be caused by inevitable oxide dissolution. PEI-modified graphene sheets with multiwalled carbon nanotubes (MWCNTs) film also can be formed using LBL technique (Yu and Dai 2010). The nanoscale pores are found to be present on the multilayer structure. This network is shown to provide better ionic and electronic transportation. The growth of RuO_2 nanosheets in lamellar structure can be achieved with self-assembly interaction between anionic RuO_2 nanosheets and cationic poly(vinyl amine)–poly(vinyl alcohol) (PVAm–PVA) copolymer (Sato et al. 2010).

For the preparation of nanostructure material, sol-gel technique offers good homogeneity, high purity, controlled texture, and controlled stoichiometry for the product (Brinker and Scherer 2013). The nanostructure formed *via* sol-gel method has an integrated network (*gel*) resulting from the polycondensation reactions of monomer in a colloidal solution (*sol*) at low temperature (Fig. 9). A *sol* is a liquid that contains colloidal particles that do not agglomerate and sediment. When the water content is lost to environment, the viscosity of the solution increased. A continuous network is formed during the process. *Gel* is said to be completely formed when there is no fluid leaking out while the container is turned upside down. During the drying process, the construction of network is affected by any event that happens, which will result in two kinds of gel. If the network structure is

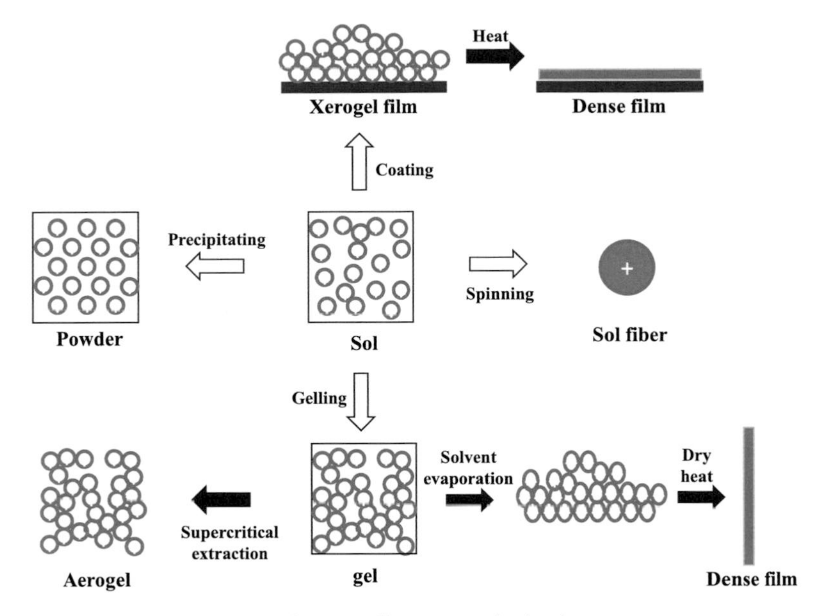

Figure 9. Schematic illustration of sol-gel process.

not disrupted, an aerogel is formed. In contrast, the gel with the collapsed network is named as xerogel.

Since the process does not involve high temperature, the degradation that can be induced by high temperature process is avoided. It turns out to be a great advantage for the preparation of graphene-based nanocomposite. The conditions of graphene formed in a graphene-based nanocomposite influence the overall performance of the composite. For the graphene-like material synthesized using chemical reduction route, which is termed as reduced graphene oxide (RGO), defect is unavoidable. A sol-gel prepared graphene-silica nanocomposite film that composed of graphene sheets and silica film is clear of aggregation (Innocenzi et al. 2014). The incorporation of RuO_2 with graphene sheets using this technique forms a porous particle-sheet structure that eases ionic transportation (Wu et al. 2010c). Vanadium oxide (V_2O_5) with porous nanosize honey comb structure also can be produced *via* sol-gel method (Reddy and Reddy 2006). Although the structure is advantageous for charge storage purpose, the poor stability of V_2O_5 still persists, where the specific capacitance loss is very significant during the first 100 cycles upon cycling. Such problem can be improved by integrating the material with graphene. Graphene–V_2O_5 composite can retain 73% of its initial specific capacitance after 1000 cycles (Fu et al. 2016). By combining graphene–V_2O_5 with MWCNTs, the cycle stability is further enhanced. There is only around 17% of initial specific capacitance loss after 32,500 cycles (Bi et al. 2017). This good stability can be attributed to the island-chain structure of the composite material. On the other hand, integration with two carbon materials has promoted a better charge storage condition. As the raw V_2O_5 and V_2O_5 aerogel only have specific capacitances of 169 F g^{-1} and 251 F g^{-1} respectively, the ternary composite has obtained 540 F g^{-1}. It is 219% higher than the specific capacitance value of raw V_2O_5. This proves the significance of synergistic effect between these compositions. In order to fabricate a desired nanostructure, template can be applied along with the sol-gel synthesis. For instance, a sol-gel prepared manganese oxide may exhibit a particle-like structure in microscale while a nanowire structure can be formed by using AAO template during the synthesis process (Reddy and Reddy 2003, Wang et al. 2005). The length and thickness of nanowires is determined by the pore diameter and thickness of the template employed.

3. Electrochemical Performance of Nanomaterial for Single Electrode

In 1995, a high specific capacitance of 768 F g^{-1} obtained by a hydrous ruthenium oxide ($RuO_2 \cdot xH_2O$) had been reported (Zheng et al. 1995). Compared to the highest specific capacitance (350 F g^{-1}) achieved by anhydrous ruthenium oxide (RuO_2) at that time, the possible application of amorphous $RuO_2 \cdot xH_2O$ in electrochemical capacitor had been unlocked (Raistrick 1992). Further investigation revealed that Ru ions in the bulk participated in the charge storage processes. In an amorphous $RuO_2 \cdot xH_2O$, H^+ ions can diffuse into the bulk and interact with the active sites (Zheng and Jow 1995). It turns out that a higher specific capacitance should be achieved by a thicker film which has higher mass loading. Nevertheless, a higher mass loading results in higher resistance for ionic transport and electron diffusion. Furthermore, the cycle stability will be reduced due to the structural degradation caused by the repeated ion penetration through the thick film. As the solid state diffusion of ions is relatively slower in a thick film, the deterioration upon cycling for a pseudocapacitive material is significant. As a result, pseudocapacitive material has shorter cyclic charge-discharge life compared to EDLC (Chen 2013). Thus, fabricating the electrode material in nanostructure is the very first step for achieving a better electrochemical performance. In general, the nanoscale architectures are beneficial in terms of: (1) shortening the ion transport lengths, (2) reducing electron diffusion time, (3) increasing the surface area, and (4) improving the surface-to-volume ratio which enhances the effective contact between the electrolyte and electrode surface.

3.1 Ruthenium oxide and its composite

The proposed charge storage mechanism of RuO_2 based on electrochemical protonation is shown as in Reaction 3.1 (Trasatti and Buzzanca 1971). On the other hand, the reactions for two-terminal pseudocapacitor based on RuO_2 electrodes are described as in Reactions 3.2–3.4 (Zheng et al. 1997).

$$RuO_2 + xe^- + xH^+ \rightarrow RuO_{2-x}(OH)_x \quad (3.1)$$

Positive electrode :

$$HRuO_2 \underset{\text{Discharge}}{\overset{\text{Charge}}{\Leftrightarrow}} H_{1-\delta}RuO_2 + \delta H^+ + \delta e^- \quad (3.2)$$

Negative electrode :

$$HRuO_2 + \delta H^+ + \delta e^- \underset{\text{Discharge}}{\overset{\text{Charge}}{\Leftrightarrow}} H_{1+\delta}RuO_2 \quad (3.3)$$

Overall reaction :

$$HRuO_2 + HRuO_2 \underset{\text{Discharge}}{\overset{\text{Charge}}{\Leftrightarrow}} H_{1-\delta}RuO_2 + H_{1+\delta}RuO_2 \quad (3.4)$$

where $0 < \delta < 1$ and H^+ represents the proton. As shown, proton participates in the charge storage process along with the electron. One can notice that the double layer capacitance is co-existing with the pseudocapacitance. Nevertheless, the pseudocapacitance is predominant as its value can be ten times larger than the double layer capacitance for the same surface area.

The specific capacitance estimated from the Reaction 3.1 above is in the range of 1400–2000 F g^{-1}. Hydrous RuO_2 prepared by using sol-gel method and annealed at proper temperature can retain water content effectively and thus, it contributes to the charge storage process (Zheng et al. 1995). To apply as electrode material for electrochemical capacitor, RuO_2 needs to anchor on a substrate. At the same time, the ionic and electronic transportation should be balanced in order to maximize the utilization of RuO_2 (Rolison et al. 2009). For this reason, a conductive support is required. When $RuO_x \cdot nH_2O$ of 3.01 nm nanodots is incorporated with conductive activated carbon (AC) on graphite electrode, an astonishing specific capacitance is obtained. The value was 1340 F g^{-1} as measured from cyclic voltammetry curve at 25 mV s^{-1} (Hu et al. 2004). Impregnation with other material is not the only way to achieve high specific capacitance. It is known that the degree of crystallinity can determine the pseudocapacitance achievable for hydrous RuO_2 as the proton can intercalate easily into the bulk when RuO_2 is in amorphous form (Zheng et al. 1995). The transformation to crystalline structure from amorphous can lead to the decrease of specific capacitance. A RuO_2–carbon composite electrode had achieved the highest specific capacitance of 407 F g^{-1} when RuO_2 was heated at 100°C (Kim and Popov 2002). Another study also showed the enhancement of specific capacitance from 740 F g^{-1} to 1300 F g^{-1} after being annealed at 200°C for 2 hours (Hu et al. 2006). The role of RuO_2 nanoparticles is not only limited as redox active sites, but they also can act as spacer. A RuO_2–graphene composite electrode with particle-sheet structure possesses 570 F g^{-1} of total capacitance (Wu et al. 2010c). The agglomeration of RuO_2 nanoparticles and re-stacking of graphene sheets are reduced effectively. Further modification on conductive backbone, such as incorporating graphene with carbon nanotubes (CNT), can promote the rate capability and cycle stability. Although the specific capacitance obtained is only around 503 F g^{-1}, the nanocomposite made up of RuO_2 nanoparticles with graphene-CNT hybrid backbone supported on nickel foam can achieve excellent cycle stability (Wang et al. 2014). Without significant degradation, the composite electrode was shown to be self strengthening upon cycling. The electrochemical performance of RuO_2 nanoparticles is dependent on fabrication technique. As mentioned above, the specific capacitance varies within the range of 407 to 1340 F g^{-1}. The highest value is very close to the minimum theoretical specific capacitance value. Even though the electrochemical characteristic of RuO_2 is attractive, the commercial application is impeded by the material cost and its toxicity.

3.2 Manganese oxide and its composite

Like $RuO_2{\cdot}xH_2O$, the charge storage mechanism of a manganese oxide involves proton (Toupin et al. 2004). However, the low diffusion coefficient of protons into bulk manganese oxide has limited the utilization for redox charge transfer (Zhao et al. 2011). Hence, heat treatment is important in determining the structure, degree of hydration, and electrochemical performance of manganese oxide.

Electrodeposition technique, which includes anodic deposition, cathodic deposition, and electrophoretic deposition, is a common preparation technique for manganese oxide (Babakhani and Ivey 2010, Cheong and Zhitomirsky 2009, Hu and Tsou 2002, Nagarajan et al. 2006). As-deposited manganese oxide film is amorphous in nature. The amorphousness starts to transform to crystalline form at around 500°C (Chang et al. 2004). Besides, the structure changed from nanofibrous to spherical, and then rod-like when the annealing temperature increased from 100°C to 600°C. Therefore, excellent electrochemical performance is usually displayed by the deposited manganese oxide film which has been annealed at relatively lower temperature such as around 200°C. Nevertheless, manganese oxide is well-known for its poor conductivity. In this case, a carbon material is usually employed as its conductive backbone. Carbon nanotubes (CNT) have high electrical conductivity and high chemical stability. However, it is not easy to effectively deposit manganese oxide nanoparticles onto carbon nanotubes as the nanoparticles may not penetrate thoroughly. The specific capacitance value that can be achieved lies in the range between 100 and 300 F g^{-1} (Chen et al. 2010b, Ma et al. 2008, Xie and Gao 2007, Zhang et al. 2008a). The same situation is encountered by multi-walled carbon nanotubes

(MWCNTs) (Guimin et al. 2008). The connection between manganese oxide nanoparticles with MWCNTs can be improved by PEDOT and PSS (Sharma and Zhai 2009). PSS is dispersed on MWCNTs for stabilization purpose while PEDOT offers a bridge for manganese oxide nanoparticles to connect with MWCNTs. When the electrostatic interaction takes place between PEDOT and PSS, the link between nanoparticles and MWCNTs is completed.

The composite electrode that consists of graphene and MnO_2 nanocomposite usually exhibits 300–400 F g^{-1} of specific capacitance (Cheng et al. 2011, Sawangphruk et al. 2013, Yu et al. 2011, Zhang et al. 2011). As the electrochemical performance is morphology dependent, it can also be affected by the electrode design. A simple electrode that involved only deposition of MnO_2 on graphene film shows 328 F g^{-1}. When this MnO_2 anchored-graphene design is placed on a conductive substrate, the specific capacitance achieved will be altered slightly. For example, the employment of porous textile fibers and flexible carbon fiber paper as substrates have gained specific capacitances of 315 F g^{-1} and 393 F g^{-1} respectively (Sawangphruk et al. 2013, Yu et al. 2011). Regardless of the conductive substrate, the design purpose is to support the nanocomposite and improve ionic and electronic transportation. When MnO_2 nanosheets are combined with poly(diallydimethylammonium chloride)-functionalized reduced graphene oxide (RGO) nanosheets, the intimate interaction offers short diffusion pathway and facilitates an efficient charge transfer (Zhang et al. 2011). Other than anchoring the nanoparticles onto the conductive support, the conductive material can also be used to decorate the manganese oxide nanoparticles. By coating a conductive carbon on the surface of nanoparticles, the ionic transport at the interface between electrolyte and electrode is facilitated. The improvement is presented as the enhancement in specific capacitance around 70% (Jiang et al. 2011). This design also greatly reduces the dissolution of oxide upon cycling.

3.3 *Carbon materials and their composites*

Among the carbon materials that have been studied for electrochemical capacitor are activated carbon, graphene, carbon nanotubes, and carbon nanofibers. As the charge storage of carbon materials is governed by double layer capacitance, the surface area and porosity of the material are important. Activated carbon has a wide range of pore sizes: micropores (< 2 nm), mesopores (2–50 nm), and macropores (> 50 nm). However, most of the pores are in micropores range which are not accessible by ions resulting in a low electrolyte accessibility (Frackowiak and Béguin 2001). On the other hand, graphene and carbon nanotubes have better electrolyte accessibility which makes them a good candidate for electrochemical capacitor application. Carbon materials also readily combine with metal oxide or conducting polymer to serve as conductive backbone and thus improve the charge storage.

Conducting polymer is a redox material. Its pseudocapacitance is contributed by the reversible redox reactions which takes place on the conjugated double bonds inside the polymer, as shown by Reactions 3.5 and 3.6 (Vol'fkovich and Serdyuk 2002).

$$P_m - xe + xA^- \Leftrightarrow P_m^{x+} A_x^- \tag{3.5}$$

$$P_m + ye + xM^+ \Leftrightarrow P_m^{y-} M_y^+ \tag{3.6}$$

where P_m represents a polymer with conjugated double bonds network, m indicates the degree of polymerization while A^- and M^+ stand for anions and cations, respectively. The charge storage process is displayed in Fig. 10.

Conducting polymer can be prepared *via* chemical polymerization or electrochemical polymerization. The former technique is employed for mass production because it is cost effective. On the contrary, electrochemical polymerization can directly produce a conducting polymer film which makes this technique suitable for electronic application (Toshima and Hara 1995). In general, the composite which consists of carbon material and conducting polymer, is usually prepared *via* two steps: (1) carbon material is dispersed into solution to form a suspension and (2) polymerization takes place on the carbon material. The conducting polymer formed is then coated on the surface of carbon material.

In a graphene–conducting polymer composite, graphene as a support material is not only enhancing the electronic transportation but also providing more redox active sites. A high specific capacitance of 1046 F g^{-1} was obtained with graphene nanosheets–PANI (Yan et al. 2010). The

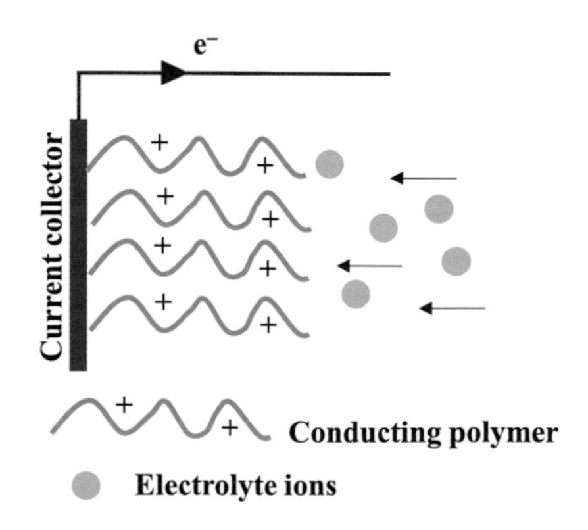

Figure 10. Schematic illustration of charge storage in a conducting polymer.

successful and complete coating of nanoscale PANI on graphene nanosheets has shortened the ion diffusion length into the material. The ionic and electronic transports are balanced which enable the high utilization of PANI. An intimate interaction between these two materials offers an enhancement of specific capacitance around 50% compared to the one obtained with a single material of PANI (Chini and Chatterjee 2017, Xu et al. 2010). The high capacitance is usually contributed by the PANI with the aid from graphene as a conductive pathway. An appropriate amount of graphene in the graphene–PANI composite can boost the conductivity about ten times higher compared to PANI film (Wu et al. 2010a). The improved cycle stability also shows that the shrinkage/swelling problem of conducting polymer is reduced. For example, 92% of initial capacitance was retained for graphene/PANI composite after 2000 cycles. Compared to the cycle stability of PANI, which retained 74% of its initial capacitance, the improvement is significant (Xu et al. 2010). Other than graphene, PANI can also be integrated with CNT. CNT has good mechanical strength and electrical properties. The addition of PANI with CNT can raise 289% of the specific capacitance compared to CNT alone (Deng et al. 2005). When graphene, PANI, and CNT are combined, a good specific capacitance of 569 F g^{-1} can be obtained (Lu et al. 2011b). Besides, an excellent cycle stability is achieved: 96% of initial capacitance is retained after 5000 cycles. This ternary composite, with graphene and CNT as the outer and inner current collectors respectively, forms a rigid network for PANI to undergo volumetric change during the repeated cycling processes. Nanocomposite made up of graphene, polypyrrole (PPy), and CNT also exhibits outstanding cycle stability where it lost only 5% of its capacitance after 5000 cycles (Lu et al. 2012). As a similar method is employed for graphene–PANI–CNT and graphene–PPy–CNT fabrications, the morphologies formed are alike which is layer-by-layer structure that ease the accessibility of the electrolyte ions.

Other than carbon–conducting polymer, nanocomposite of carbon material and metal oxide has appeared to be a promising electrode material as well. A nickel oxide (NiO)–CNT nanocomposite prepared by simple chemical precipitation method exhibits 160 F g^{-1} at 10 mA g^{-1} (Lee et al. 2005). It is around 31% higher than the specific capacitance achieved by bare NiO. The crystallite size of NiO nanoparticles coated on the CNT is smaller than bare NiO. The smaller particle size indicates a higher surface area that can offer more active sites for the occurrence of redox reaction. Intertwined CNT–vanadium oxide (V_2O_5) nanocomposite also possesses hierarchically porous structure that consists of small pores which can further increase the surface area effectively (Chen et al. 2011). An excellent cycle stability has been achieved by a composite of CNT arrays coated with manganese oxide nanoflower. It experiences only 3% of capacity loss after 20,000 cycles (Zhang et al. 2008a). The morphology study revealed that the nanoflower was evolved from nanosheets. Besides, the manganese oxide nanoflower was found to form favorably at the junction of CNT (Fig. 11). However, the growth mechanism is unclear. Thus, the main contributor for cycle stability is yet to be confirmed. Compared to the ternary composite that made up of carbon material and conducting polymer such as graphene–PANI–CNT and graphene–PPy–CNT, CNT–RGO–MnO_2 can retain 70% of initial capacitance after 5000 cycles (Lu et al. 2016). The good stability could be attributed to the 3D structure formed by RGO and CNT with uniform distribution of MnO_2 nanoparticles. At some extent, the dissolution of MnO_2 which will greatly degrade the stability performance is restrained. RuO_2–CNT–RGO ternary composite also has good cycle stability. The final specific capacitance is enhanced by 6% after 8100 cycles (Wang et al. 2014). This composite has a carbon hybrid backbone (CNT–RGO) coated on nickel foam, which assists in electrochemical stability of the substrate in electrolyte, bridges the RuO_2 nanoparticles and substrate effectively, and provides a good conductive pathway (Fig. 12).

Electrode of RuO_2 nanodots coated-graphene exhibits good cycle stability as well. It retains 92% of its initial specific capacitance after 3000 cycles (Chen et al. 2012). Different structure of transition metal oxide on

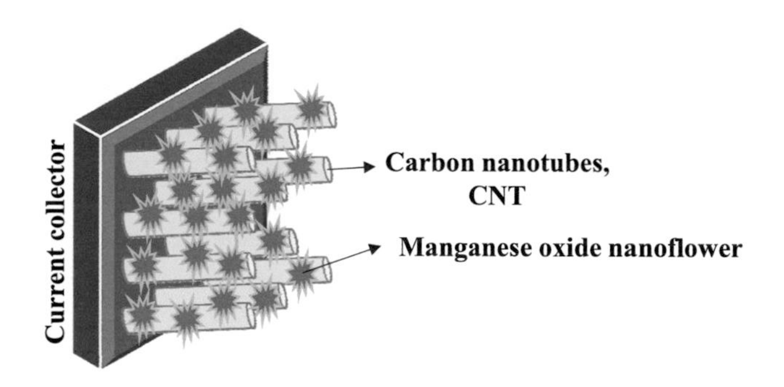

Figure 11. CNT arrays coated with manganese oxide nanoflower.

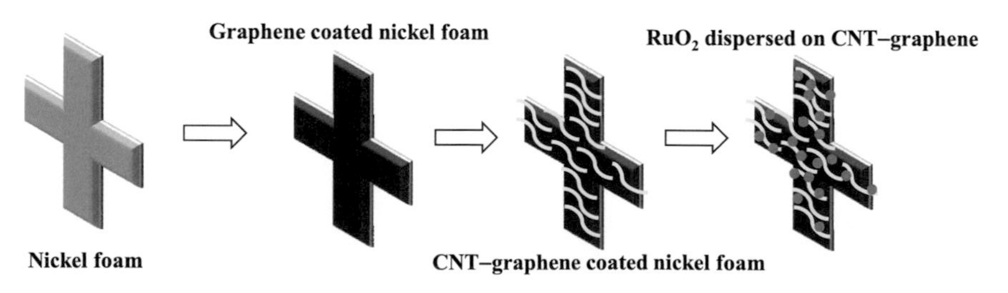

Figure 12. Schematic illustration of the preparation of RuO_2–CNT–RGO ternary composite.

graphene sheet leads to different level of cycle stability. The irregular shape of Mn_3O_4 nanoparticles contribute to around 92% of capacitance retention after 3000 cycles (Lee et al. 2015). Mn_3O_4 with nanoparticle size range of 37–50 nm has improved specific capacitance by 43% after 1000 cycles while it loses 5% of initial specific capacitance during the first 500 cycles with the particle size range from 3–5 nm (Wu et al. 2013, Zhang et al. 2012). Both composites are prepared using solvothermal method but employs different heating temperature. The bigger particle size is obtained when the heating condition applied is 180°C for 11 hours while smaller particle size grows at slightly lower temperature at 120°C for 24 hours. Hence, graphene–Mn_3O_4 with larger particle size also has higher crystallinity. Additionally, the solvent used for former electrode is a mixture of ethanol and water while the latter uses dimethyl sulfoxide (DMSO). From here we can notice the importance of optimized preparation conditions as it will lead to a significant difference in electrochemical performance. On the other hand, MnO_2 nanowires coated graphene oxide sheets retains around 93% of initial specific capacitance after 1000 cycles (Dai et al. 2014).

Other than RuO_2 and MnO_x, graphene–zinc oxide (ZnO) has also been studied as potential electrode for electrochemical capacitor application. As a semiconductor material, ZnO is well-known for its electrical and optical properties. The graphene–ZnO with similar structure presents a wide range of specific capacitance. For instance, irregular shaped ZnO nanoparticles may exhibits 11 F g^{-1} to 62 F g^{-1} while nanoparticle size of 5–10 nm achieves 146 F g^{-1} (Lu et al. 2011a, Lu et al. 2010, Zhang et al. 2009). The variation in specific capacitance can be caused by the different crystallinity introduced by different fabrication methods. As the nanoparticle structure contributes low specific capacitance, nanorod structure of ZnO on graphene sheets achieves total specific capacitance of 400 F g^{-1} (Dong et al. 2012). When the ZnO nanorods are deposited on both sides of the reduced graphene oxide (RGO) sheets, a sandwich structure is formed. This ZnO–RGO–ZnO nanocomposite electrode obtains maximum specific capacitance of 275 F g^{-1} (Li et al. 2014). Such nanocomposite electrode performs good cycle stability that its specific capacitance drops only 2% after 2000 cycles. In addition, graphene–ZnO nanocomposite has been reported to have slightly increased specific capacitance upon cycling (Haldorai et al. 2014). The specific capacitance is enhanced by 10% during the first 100 cycles and maintained until 1000 cycles. This good cycle stability can be attributed to the stable connectivity between the ZnO and graphene.

4. Electrochemical Capacitor

In order to ensure the applicability of any electrode in an electrochemical capacitor, the electrode should be examined by assembling into a two-terminal device. The two-terminal device can be divided into two groups: symmetric and asymmetric. A symmetry electrochemical capacitor possesses two identical electrodes separated by electrolyte while asymmetry type is made up of distinct electrodes. Due to the different potential change of two electrodes, asymmetry electrochemical capacitor usually has a higher energy and power density. In addition, a hybrid electrochemical capacitor refers to a relatively simpler capacitor made up of capacitive and pseudocapacitive electrodes.

As a composite electrode of graphene–manganese oxide that can exhibit specific capacitance of 478 F g^{-1}, the corresponding symmetric electrochemical capacitor can only achieve around half of the specific capacitance obtained by the single electrode, i.e., 240 F g^{-1} (Bello et al. 2013). This electrochemical capacitor has good cycle stability where there is no capacitance loss after 1000 cycles. Recently, an onion-like carbon (OLC) has been found that can be integrated well with Mn_3O_4. Its performances as a composite electrode and electrochemical capacitor have been examined and compared with other Mn_3O_4-based composite electrode such as activated carbon (AC)–Mn_3O_4, graphene oxide (GO)–Mn_3O_4, and carbon nanotubes (CNT)–Mn_3O_4 (Makgopa et al. 2017). OLC–Mn_3O_4 based electrochemical capacitor achieves the highest specific capacitance of 195 F g^{-1}. CNT–Mn_3O_4 based device exhibits slightly lower specific capacitance value, i.e., 180 F g^{-1} while AC–Mn_3O_4-based shows the lowest value (124 F g^{-1}). Additionally, the optimum energy and power density estimated are around 4 W h kg^{-1} and 52 kW kg^{-1}, respectively. The PANI–MnO_x electrode based symmetric electrochemical capacitor retains one fourth of the specific capacitance achieved by the single electrode, which is 112 F g^{-1} (Sun et al. 2008).

A symmetric electrochemical capacitor based on manganese oxide is limited by the irreversible redox reaction of Mn^{4+}/Mn^{2+} at negative electrode and Mn^{4+}/Mn^{7+} at positive electrode. In order to overcome this limitation, alternative material can be employed as negative electrode. An asymmetric electrochemical capacitor, composed of graphite oxide and MnO_2 as positive electrode and graphite oxide as negative electrode, has been reported to retain around 81% of capacitance after 5000 cycles (Zhao et al. 2012). The maximum energy and power density achieved are 24 W h kg^{-1} and 32.3 kW kg^{-1}. The graphite oxide employed in this electrochemical capacitor has high electrochemical stability where it lost only 3% of its specific capacitance after 10,000 cycles. It is exfoliated with the assistance of microwave before chemically activated by KOH. Thus, it has high specific surface area and an ultraporous structure with pore size range from 0.6 to 5 nm.

The electrodes can be a hybrid composite electrode in order to boost the electrochemical performance. For example, positive electrode is made up of MnO_2–SWNT while negative electrode consists of In_2O_3–SWNT

Table 1. Electrochemical performances of various electrochemical capacitors.

Type	Electrode material	Electrolyte	Potential window, V	Specific capacitance, F g^{-1}	Energy density, W h kg^{-1}	Power density, kw kg^{-1}	Reference
Symmetric	graphene–MnO_2	1 M Na_2SO_4	1.0	240	8.3	20	(Bello et al. 2013)
	AC–Mn_3O_4		0.8	124	2.8	18	(Makgopa et al. 2017)
	GO–Mn_3O_4			160	3.6	24	
	CNT–Mn_3O_4			180	3.9	33	
	Onion-like carbon–Mn_3O_4			195	4.3	52	
Asymmetric	Graphite oxide–MnO_2 ǀ graphite oxide	1 M Na_2SO_4	2.0	175	24	32.3	(Zhao et al. 2012)
	MnO_2–SWNT ǀ In_2O_3–SWNT			184	25.5	50.3	(Chen et al. 2010a)
	Graphene ǀ graphene–MnO_2			-	30.4	5	(Wu et al. 2010b)
Hybrid	Activated mesocarbon microbeads ǀ MnO_2	1 M Et_4NBF_4 in acetronile	3.0	228	128.0	-	(Wang et al. 2010b)
	Activated carbon ǀ MnO_2	0.1 M K_2SO_4	2.0	20	11.7	-	(Brousse et al. 2007)
		0.65 M K_2SO_4	2.2	29	10.0	3.6	(Brousse et al. 2004)
		1 M LiOH	0.7	60	4.0	-	(Yuan and Zhang 2006)
	Carbon ǀ MnO_2	Nafion 115 membrane	1.6	48	17.0	-	(Staiti and Lufrano 2010)
	Anthraquinone-modified carbon fabric ǀ ruthenium oxide	1 M H_2SO_4	1.3	109	26.7	17.3	(Algharaibeh et al. 2009)

*The values that are not given is due to non availability from original articles.

(Chen et al. 2010a). MnO_2 and In_2O_3 are nanowire structured and they contribute to charge storage effectively, compared with MnO_2 nanoparticles. A hybrid electrochemical capacitor made up of activated mesocarbon microbeads and MnO_2 shows a good specific capacitance of 228 F g^{-1} and it retains 96% of its initial specific capacitance after 1200 cycles (Wang et al. 2010b). The electrolyte employed was 1 M Et_4NBF_4 in acetronile which has a working potential window of 3.0 V. Hybrid electrochemical capacitor made up of activated carbon and MnO_2 electrodes are quite common. Its performance can be determined by the electrolyte used. For example, K_2SO_4 and LiOH can lead to different specific capacitance value of 29 and 60 F g^{-1} respectively (Brousse et al. 2004, Yuan and Chang 2006). Besides, the usage of KOH can cause the formation of irreversible Mn_3O_4 due to air oxidation (Im and Manthiram 2003). The combination of activated carbon, MnO_2, and K_2SO_4 has been found to possess excellent cycle stability. The capacitance loss is limited to 12.5% for 195,000 cycles. Meanwhile, a hybrid capacitor of ruthenium oxide and anthraquinone-modified carbon fabric obtains the highest energy density of 26.7 W h kg^{-1} and power density of 17.3 F g^{-1} (Algharaibeh et al. 2009). Compared to a symmetric capacitor with ruthenium oxide electrode, hybrid capacitor utilizes less Ru content but performs satisfactorily, achieving 109 F g^{-1} of specific capacitance. As a comparison, the symmetric capacitor obtains 193 F g^{-1}. As the carbon fabric offers double layer capacitance, an appropriate treatment or modification can enhance the capacitance. The importance of modification is revealed when the unmodified carbon fabric is assembled with ruthenium oxide in order to design another hybrid capacitor. It only achieves 95 F g^{-1} of specific capacitance. Furthermore, the modified carbon fabric–ruthenium oxide based hybrid capacitor has slightly higher energy density than its symmetric counterpart. This privilege can be due to the wider operating potential window, i.e., 1.3 V as ruthenium oxide based symmetric capacitor operates at 1 V.

5. Future Prospects of Nanomaterial

The expansion of global network has led to the rapid growth of worldwide economy. This globalization has encouraged the advancement of telecommunication and transportation fields. Instead of agriculture, many developing countries are emphasizing on the importance of information industry and informational activities which can offer higher productivity and efficiency. This transformation of economy requires a proper design of structures in industry and financial sectors, along with appropriate policies in a developing country. However, globalization also results in some environmental issues such as global warming, pollution, and overfishing. More importantly, it can bring about a rapid depletion of natural resources. In order to utilize the energy effectively, energy storage device is a necessity. The diversity in the types of energy storage system reflects a wide range of roles in various field and applications. The energy storage system is classified based on the energy storage method, e.g., electrical, mechanical, chemical, electrochemical, thermal, and biological. In order to synthesize an effective energy storage device, various materials have been studied intensely. Regardless of the material type, particle size has been recognized as playing an important role in determining the performance of energy storage device. The technology that is related to nanomaterial is named as nanotechnology. Nowadays, nanotechnology has entered into industries related to food, cosmetic, and healthcare. The energy devices and components are foreseen to be lighter, smaller, and more efficient. Thus, there is no doubt that the field of nanomaterials will continue to grow and expand.

References

Algharaibeh, Z., X. Liu and P.G. Pickup. 2009. An asymmetric anthraquinone-modified carbon/ruthenium oxide supercapacitor. J. Power Sources 187(2): 640. doi: http://doi.org/10.1016/j.jpowsour.2008.11.012.

Ardizzone, S., G. Fregonara and S. Trasatti. 1990. "Inner" and "outer" active surface of RuO_2 electrodes. Electrochim. Acta 35(1): 263.

Arikado, T., C. Iwakura and H. Tamura. 1977. Electrochemical behavior of the ruthenium oxide electrode prepared by the thermal decomposition method. Electrochim. Acta 22(5): 513.

Augustyn, V., P. Simon and B. Dunn. 2014. Pseudocapacitive oxide materials for high-rate electrochemical energy storage. Energy Environ. Sci. 7(5): 1597.

Babakhani, B. and D.G. Ivey. 2010. Anodic deposition of manganese oxide electrodes with rod-like structures for application as electrochemical capacitors. J. Power Sources 195(7): 2110.

Bard, A.J. and L.R. Faulkner. 2001. Electrochemical Methods: Fundamentals and Applications. New York: John Wiley and Sons.

Becker, H.I. 1957. United States Patent No. 2800616. U. S. P. Office.

Bello, A., O.O. Fashedemi, J.N. Lekitima, M. Fabiane, D. Dodoo-Arhin, K.I. Ozoemena et al. 2013. High-performance symmetric electrochemical capacitor based on graphene foam and nanostructured manganese oxide. AIP Adv. 3(8): 082118.

Bi, W., G. Gao, Y. Wu, H. Yang, J. Wang, Y. Zhang et al. 2017. Novel three-dimensional island-chain structured V_2O_5/graphene/MWCNT hybrid aerogels for supercapacitors with ultralong cycle life. RSC Adv. 7(12): 7179.

Boos, D.L. 1970. United States Patent No. 3536963. U. S. P. Office.

Brinker, C.J. and G.W. Scherer. 2013. Sol-gel science: the physics and chemistry of sol-gel processing: Academic press.

Brousse, T., P.-L. Taberna, O. Crosnier, R. Dugas, P. Guillemet, Y. Scudeller et al. 2007. Long-term cycling behavior of asymmetric activated carbon/MnO_2 aqueous electrochemical supercapacitor. J. Power Sources 173(1): 633.

Brousse, T., D. Bélanger and J.W. Long. 2015. To be or not to be Pseudocapacitive? J. Electrochem. Soc. 162(5): A5185.

Brousse, T., M. Toupin and D. Bélanger. 2004. A hybrid activated carbon-manganese dioxide capacitor using a mild aqueous electrolyte. J. Electrochem. Soc. 151(4): A614.

Brousse, T., M. Toupin, R. Dugas, L. Athouël, O. Crosnier and D. Bélanger. 2006. Crystalline MnO_2 as possible alternatives to amorphous compounds in electrochemical supercapacitors. J. Electrochem. Soc. 153(12): A2171.

Cao, Y. and T.E. Mallouk. 2008. Morphology of template-grown polyaniline nanowires and its effect on the electrochemical capacitance of nanowire arrays. Chem. Mater. 20(16): 5260.

Chang, J.-K., Y.-L. Chen and W.-T. Tsai. 2004. Effect of heat treatment on material characteristics and pseudo-capacitive properties of manganese oxide prepared by anodic deposition. J. Power Sources 135(1–2): 344.

Chang, K.-H. and C.-C. Hu. 2004. Hydrothermal synthesis of hydrous crystalline RuO_2 nanoparticles for supercapacitors. Electrochem. Solid-State Lett. 7(12): A466.

Chang, K.-H., C.-C. Hu and C.-Y. Chou. 2007. Textural and capacitive characteristics of hydrothermally derived $RuO_2 \cdot xH_2O$ nanocrystallites: independent control of crystal size and water content. Chem. Mater. 19(8): 2112.

Chapman, D.L. 1913. Philosophical Magazine, 25.

Chen, G.Z. 2013. Understanding supercapacitors based on nano-hybrid materials with interfacial conjugation. Prog. Nat. Sci.: Materials International 23(3): 245.

Chen, P.-C., G. Shen, Y. Shi, H. Chen and C. Zhou. 2010. Preparation and characterization of flexible asymmetric supercapacitors based on transition-metal-oxide nanowire/single-walled carbon nanotube hybrid thin-film electrodes. ACS Nano 4(8): 4403.

Chen, W., Z. Fan, L. Gu, X. Bao and C. Wang. 2010. Enhanced capacitance of manganese oxide via confinement inside carbon nanotubes. Chem. Commun. 46(22): 3905.

Chen, Y., X. Zhang, D. Zhang and Y. Ma. 2012. One-pot hydrothermal synthesis of ruthenium oxide nanodots on reduced graphene oxide sheets for supercapacitors. J. Alloys Compd. 511(1): 251.

Chen, Z., V. Augustyn, J. Wen, Y. Zhang, M. Shen, B. Dunn et al. 2011. High-performance supercapacitors based on intertwined CNT/V_2O_5 nanowire nanocomposites. Adv. Mater. 23(6): 791.

Cheng, Q., J. Tang, J. Ma, H. Zhang, N. Shinya and L.-C. Qin. 2011. Graphene and nanostructured MnO_2 composite electrodes for supercapacitors. Carbon 49(9): 2917.

Cheong, M. and I. Zhitomirsky. 2009. Electrophoretic deposition of manganese oxide films. Surf. Eng. 25(5): 346.

Chini, M.K. and S. Chatterjee. 2017. Hydrothermally reduced nanoporous graphene–polyaniline nanofiber composites for supercapacitor. FlatChem. 1: 1.

Chuang, C.-M., C.-W. Huang, H. Teng and J.-M. Ting. 2012. Hydrothermally synthesized RuO_2/carbon nanofibers composites for use in high-rate supercapacitor electrodes. Compos. Sci. Technol. 72(13): 1524.

Conway, B.E., V. Birss and J. Wojtowicz. 1997. The role and utilization of pseudocapacitance for energy storage by supercapacitors. J. Power Sources 66(1–2): 1.

Conway, B.E. 1999. Electrochemical Supercapacitors: Scientific Fundamentals and Technological Applications: Springer US.

Cook, J.B., H.S. Kim, Y. Yan, J.S. Ko, S. Robbennolt, B. Dunn et al. 2016. Mesoporous MoS_2 as a transition metal dichalcogenide exhibiting pseudocapacitive Li and Na–ion charge storage. Adv. Energy Mater.

Dai, K., L. Lu, C. Liang, J. Dai, Q. Liu, Y. Zhang et al. 2014. *In situ* assembly of MnO_2 nanowires/graphene oxide nanosheets composite with high specific capacitance. Electrochim. Acta 116: 111.

Decher, G., J.D. Hong and J. Schmitt. 1992. Buildup of ultrathin multilayer films by a self-assembly process: III. Consecutively alternating adsorption of anionic and cationic polyelectrolytes on charged surfaces. Thin Solid Films 210: 831.

Deng, M., B. Yang and Y. Hu. 2005. Polyaniline deposition to enhance the specific capacitance of carbon nanotubes for supercapacitors. J. Mater. Sci. 40(18): 5021.

Devarayan, K., D. Lei, H.-Y. Kim and B.-S. Kim. 2015. Flexible transparent electrode based on PANi nanowire/nylon nanofiber reinforced cellulose acetate thin film as supercapacitor. Chem. Eng. J. 273: 603.

Dong, X., Y. Cao, J. Wang, M.B. Chan-Park, L. Wang, W. Huang et al. 2012. Hybrid structure of zinc oxide nanorods and three dimensional graphene foam for supercapacitor and electrochemical sensor applications. RSC Adv. 2(10): 4364.

Frackowiak, E. and F. Béguin. 2001. Carbon materials for the electrochemical storage of energy in capacitors. Carbon 39(6): 937.

Fu, X., Y. Chen, Y. Zhu and S.C. Jana. 2016. Synergetic hybrid aerogels of vanadia and graphene as electrode materials of supercapacitor C 2(3).

Galizzioli, D., F. Tantardini and S. Trasatti. 1974. Ruthenium dioxide: a new electrode material. I. Behavior in acid solutions of inert electrolytes. J. Appl. Electrochem. 4(1): 57.

Gouy, L.G. 1910. C. R. Chim., 149.

Grahame, D.C. 1947. The electrical double layer and the theory of electrocapillarity. Chem. Rev. 41: 441.

Guimin, A., Y. Ping, X. Meijun, L. Zhimin, M. Zhenjiang, D. Kunlun et al. 2008. Low-temperature synthesis of Mn_3O_4 nanoparticles loaded on multi-walled carbon nanotubes and their application in electrochemical capacitors. Nanotechnology 19(27): 275709.

Haldorai, Y., W. Voit and J.-J. Shim. 2014. Nano ZnO@reduced graphene oxide composite for high performance supercapacitor: Green synthesis in supercritical fluid. Electrochim. Acta 120: 65.

Helmholtz, H. 1853. Ueber einige Gesetze der Vertheilung elektrischer Ströme in körperlichen Leitern mit Anwendung auf die thierisch-elektrischen Versuche. Ann. Phys. 165(6).

Hou, Y., Y. Cheng, T. Hobson and J. Liu. 2010. Design and synthesis of hierarchical MnO_2 nanospheres/carbon nanotubes/conducting polymer ternary composite for high performance electrochemical electrodes. Nano Lett. 10(7): 2727.

Hu, C.-C. and T.-W. Tsou. 2002. Ideal capacitive behavior of hydrous manganese oxide prepared by anodic deposition. Electrochem. Commun. 4(2): 105.

Hu, C.-C., W.-C. Chen and K.-H. Chang. 2004. How to achieve maximum utilization of hydrous ruthenium oxide for supercapacitors. J. Electrochem. Soc. 151(2): A281.

Hu, C.-C., K.-H. Chang, M.-C. Lin and Y.-T. Wu. 2006. Design and tailoring of the nanotubular arrayed architecture of hydrous RuO_2 for next generation supercapacitors. Nano Lett. 6(12): 2690.

Hwang, J.Y., M.F. El-Kady, Y. Wang, L. Wang, Y. Shao, K. Marsh et al. 2015. Direct preparation and processing of graphene/RuO_2 nanocomposite electrodes for high-performance capacitive energy storage. Nano Energy 18: 57.

Im, D. and A. Manthiram. 2003. Role of bismuth and factors influencing the formation of Mn_3O_4 in rechargeable alkaline batteries based on bismuth-containing manganese oxides. J. Electrochem. Soc. 150(1): A68.

Innocenzi, P., L. Malfatti, B. Lasio, A. Pinna, D. Loche, M.F. Casula et al. 2014. Sol-gel chemistry for graphene-silica nanocomposite films. New J. Chem. 38(8): 3777.

Jiang, H., L. Yang, C. Li, C. Yan, P.S. Lee and J. Ma. 2011. High-rate electrochemical capacitors from highly graphitic carbon-tipped manganese oxide/mesoporous carbon/manganese oxide hybrid nanowires. Energy Environ. Sci. 4(5): 1813.

Kim, H.-S., J.B. Cook, S.H. Tolbert and B. Dunn. 2015. The development of pseudocapacitive properties in nanosized-MoO_2. J. Electrochem. Soc. 162(5): A5083.

Kim, H. and B.N. Popov. 2002. Characterization of hydrous ruthenium oxide/carbon nanocomposite supercapacitors prepared by a colloidal method. Journal of Power Sources 104(1): 52.

Lee, H.-M., G.H. Jeong, D.W. Kang, S.-W. Kim and C.-K. Kim. 2015. Direct and environmentally benign synthesis of manganese oxide/graphene composites from graphite for electrochemical capacitors. J. Power Sources 281: 44.

Lee, H.Y. and J.B. Goodenough. 1999. Supercapacitor behavior with KCl electrolyte. J. Solid State Chem. 144(1): 220.

Lee, H.Y., V. Manivannan and J.B. Goodenough. 1999. Electrochemical capacitors with KCl electrolyte. C. R. Acad. Sci.—Series IIC—Chemistry 2(11): 565.

Lee, J.W., A.S. Hall, J.-D. Kim and T.E. Mallouk. 2012. A facile and template-free hydrothermal synthesis of Mn_3O_4nanorods on graphene sheets for supercapacitor electrodes with long cycle stability. Chem. Mater. 24(6): 1158.

Lee, J.Y., K. Liang, K.H. An and Y.H. Lee. 2005. Nickel oxide/carbon nanotubes nanocomposite for electrochemical capacitance. Synth. Met. 150(2): 153.

Li, Z., J. Wang, X. Liu, S. Liu, J. Ou and S. Yang. 2011. Electrostatic layer-by-layer self-assembly multilayer films based on graphene and manganese dioxide sheets as novel electrode materials for supercapacitors. J. Mater. Chem. 21(10): 3397.

Li, Z., P. Liu, G. Yun, K. Shi, X. Lv, K. Li et al. 2014. 3D (Three-dimensional) sandwich-structured of ZnO (zinc oxide)/rGO (reduced graphene oxide)/ZnO for high performance supercapacitors. Energy 69: 266.

Liu, Y., G. Yuan, Z. Jiang and Z. Yao. 2014. Solvothermal synthesis of Mn_3O_4 nanoparticle/graphene sheet composites and their supercapacitive properties. J. Nanomater. 151.

Lu, L., S. Xu, J. An and S. Yan. 2016. Electrochemical performance of CNTs/RGO/MnO_2 composite material for supercapacitor. Nanomater. Nanotechno. 6.

Lu, T., Y. Zhang, H. Li, L. Pan, Y. Li and Z. Sun. 2010. Electrochemical behaviors of graphene–ZnO and graphene–SnO2 composite films for supercapacitors. Electrochim. Acta 55(13): 4170.

Lu, T., L. Pan, H. Li, G. Zhu, T. Lv, X. Liu et al. 2011a. Microwave-assisted synthesis of graphene–ZnO nanocomposite for electrochemical supercapacitors. J. Alloys Compd. 509(18): 5488.

Lu, X., H. Dou, S. Yang, L. Hao, L. Zhang, L. Shen et al. 2011b. Fabrication and electrochemical capacitance of hierarchical graphene/polyaniline/carbon nanotube ternary composite film. Eletrochim. Acta 56(25): 9224.

Lu, X., D. Zheng, T. Zhai, Z. Liu, Y. Huang, S. Xie et al. 2011c. Facile synthesis of large-area manganese oxide nanorod arrays as a high-performance electrochemical supercapacitor. Energy Environ. Sci. 4(8): 2915.

Lu, X., H. Dou, C. Yuan, S. Yang, L. Hao, F. Zhang et al. 2012. Polypyrrole/carbon nanotube nanocomposite enhanced the electrochemical capacitance of flexible graphene film for supercapacitors. J. Power Sources 197: 319.

Ma, S.-B., K.-W. Nam, W.-S. Yoon, X.-Q. Yang, K.-Y. Ahn, K.-H. Oh et al. 2008. Electrochemical properties of manganese oxide coated onto carbon nanotubes for energy-storage applications. J. Power Sources 178(1): 483.

Makgopa, K., K. Raju, P.M. Ejikeme and K.I. Ozoemena. 2017. High-performance Mn_3O_4/onion-like carbon (OLC) nanohybrid pseudocapacitor: Unravelling the intrinsic properties of OLC against other carbon supports. Carbon 117: 20.

Nagarajan, N., H. Humadi and I. Zhitomirsky. 2006. Cathodic electrodeposition of MnOx films for electrochemical supercapacitors. Electrochim. Acta 51(15): 3039.

Pan, C., Y. Lv, H. Gong, Q. Jiang, S. Miao and J. Liu. 2016. Synthesis of Ag/PANI@MnO_2 core-shell nanowires and their capacitance behavior. RSC Adv. 6(21): 17415.

Paunovic, M., M. Schlesinger and D.D. Snyder. 2010. Fundamental considerations. *In*: Schlesinger, M. and M. Paunovic (eds.). Modern Electroplating: John Wiley and Sons.

Raistrick, I.D. 1992. Electrochemical capacitors. pp. 297. *In*: McHardy, J. and F. Luduig (eds.). Electrochemistry of Semiconductors and Electronics: Processes and Devices. Park Ridge, New Jersey, USA: Noyes Publications.

Reddy, R.N. and R.G. Reddy. 2003. Sol–gel MnO_2 as an electrode material for electrochemical capacitors. J. Power Sources 124(1): 330.

Reddy, R.N. and R.G. Reddy. 2006. Porous structured vanadium oxide electrode material for electrochemical capacitors. J. Power Sources 156(2): 700.

Rightmire, R.A. 1966. United States Patent No. 3288641. U. S. P. Office.
Rolison, D.R., J.W. Long, J.C. Lytle, A.E. Fischer, C.P. Rhodes, T.M. McEvoy et al. 2009. Multifunctional 3D nanoarchitectures for energy storage and conversion. Chem. Soc. Rev. 38(1): 226.
Santhanagopalan, S., A. Balram and D.D. Meng. 2013. Scalable high-power redox capacitors with aligned nanoforests of crystalline MnO_2nanorods by high voltage electrophoretic deposition. ACS Nano 7(3): 2114.
Sato, J., H. Kato, M. Kimura, K. Fukuda and W. Sugimoto. 2010. Conductivity of ruthenate nanosheets prepared via electrostatic self-assembly: characterization of isolated single nanosheet crystallite to mono- and multilayer electrodes. Langmuir 26(23): 18049.
Sawangphruk, M., P. Srimuk, P. Chiochan, A. Krittayavathananon, S. Luanwuthi and J. Limtrakul. 2013. High-performance supercapacitor of manganese oxide/reduced graphene oxide nanocomposite coated on flexible carbon fiber paper. Carbon 60(0): 109.
Schwartz, M. 1994. Deposition from aqueous solutions: an overview. Handbook of Deposition Technologies for Films and Coatings—Science, Technology and Applications, 506.
Sharma, R.K. and L. Zhai. 2009. Multiwall carbon nanotube supported poly(3,4-ethylenedioxythiophene)/manganese oxide nano-composite electrode for super-capacitors. Electrochim. Acta 54(27): 7148.
Staiti, P. and F. Lufrano. 2010. Investigation of polymer electrolyte hybrid supercapacitor based on manganese oxide–carbon electrodes. Electrochim. Acta 55(25): 7436.
Stern, O. 1924. Z. Elektrochem., 30.
Sun, L.-J., X.-X. Liu, K.K.-T. Lau, L. Chen and W.-M. Gu. 2008. Electrodeposited hybrid films of polyaniline and manganese oxide in nanofibrous structures for electrochemical supercapacitor. Electrochim. Acta 53(7): 3036.
Toshima, N. and S. Hara. 1995. Direct synthesis of conducting polymers from simple monomers. Prog. Polym. Sci. 20(1): 155.
Toupin, M., T. Brousse and D. Bélanger. 2004. Charge storage mechanism of MnO_2 electrode Used in Aqueous Electrochemical Capacitor. Chem. Mater. 16(16): 3184.
Trasatti, S. and G. Buzzanca. 1971. Ruthenium dioxide: A new interesting electrode material. Solid state structure and electrochemical behavior. J. Electroanal. Chem. Interfacial Electrochem. 29(2): A1.
Vol'fkovich, Y.M. and T.M. Serdyuk. 2002. Electrochemical Capacitors. Russ. J. Electrochem. 38(9): 935.
Wang, K., J. Huang and Z. Wei. 2010a. Conducting polyaniline nanowire arrays for high performance supercapacitors. J. Phys. Chem. C 114(17): 8062.
Wang, H.-Q., Z.-S. Li, Y.-G. Huang, Q.-Y. Li and X.-Y. Wang. 2010b. A novel hybrid supercapacitor based on spherical activated carbon and spherical MnO_2 in a non-aqueous electrolyte. J. Mater. Chem. 20(19): 3883.
Wang, W., S. Guo, I. Lee, K. Ahmed, J. Zhong, Z. Favors et al. 2014. Hydrous ruthenium oxide nanoparticles anchored to graphene and carbon nanotube hybrid foam for supercapacitors. Sci. Rep. 4: 4452.
Wang, X., X. Wang, W. Huang, P.J. Sebastian and S. Gamboa. 2005. Sol–gel template synthesis of highly ordered MnO_2 nanowire arrays. J. Power Sources 140(1): 211.
Wu, M.-S. and P.-C. Julia Chiang. 2004. Fabrication of nanostructured manganese oxide electrodes for electrochemical capacitors. Electrochem. Solid State Lett. 7(6): A123.
Wu, Q., Y. Xu, Z. Yao, A. Liu and G. Shi. 2010a. Supercapacitors based on flexible graphene/polyaniline nanofiber composite films. ACS Nano 4(4): 1963.
Wu, Y., S. Liu, H. Wang, X. Wang, X. Zhang and G. Jin. 2013. A novel solvothermal synthesis of Mn_3O_4/graphene composites for supercapacitors. Electrochim. Acta 90: 210.
Wu, Z.-S., W. Ren, D.-W. Wang, F. Li, B. Liu and H.-M. Cheng. 2010b. High-energy MnO_2 nanowire/graphene and graphene asymmetric electrochemical capacitors. ACS Nano 4(10): 5835.

Wu, Z.-S., D.-W. Wang, W. Ren, J. Zhao, G. Zhou, F. Li et al. 2010c. Anchoring hydrous RuO_2 on graphene sheets for high-performance electrochemical capacitors. Adv. Funct. Mater. 20(20): 3595.

Xie, X. and L. Gao. 2007. Characterization of a manganese dioxide/carbon nanotube composite fabricated using an *in situ* coating method. Carbon 45(12): 2365.

Xu, J., K. Wang, S.-Z. Zu, B.-H. Han and Z. Wei. 2010. Hierarchical nanocomposites of polyaniline nanowire arrays on graphene oxide sheets with synergistic effect for energy storage. ACS Nano 4(9): 5019.

Yan, J., T. Wei, B. Shao, Z. Fan, W. Qian, M. Zhang et al. 2010. Preparation of a graphene nanosheet/polyaniline composite with high specific capacitance. Carbon 48(2): 487.

Yu, D. and L. Dai. 2010. Self-assembled graphene/carbon nanotube hybrid films for supercapacitors. J. Phys. Chem. Lett. 1(2): 467.

Yu, G., L. Hu, M. Vosgueritchian, H. Wang, X. Xie, J.R. McDonough et al. 2011. Solution-processed graphene/MnO_2nanostructured textiles for high-performance electrochemical capacitors. Nano Lett. 11(7): 2905.

Yuan, A. and Q. Zhang. 2006. A novel hybrid manganese dioxide/activated carbon supercapacitor using lithium hydroxide electrolyte. Electrochem. Commun. 8(7): 1173.

Zhang, H., G. Cao, Z. Wang, Y. Yang, Z. Shi and Z. Gu. 2008a. Growth of manganese oxide nanoflowers on vertically-aligned carbon nanotube arrays for high-rate electrochemical capacitive energy storage. Nano Lett. 8(9): 2664.

Zhang, J., J. Jiang and X.S. Zhao. 2011. Synthesis and capacitive properties of manganese oxide nanosheets dispersed on functionalized graphene sheets. J. Phys. Chem. C 115(14): 6448.

Zhang, X., H. Chen and H. Zhang. 2007. Layer-by-layer assembly: from conventional to unconventional methods. Chem. Commun. (14): 1395.

Zhang, X., W. Yang and D.G. Evans. 2008b. Layer-by-layer self-assembly of manganese oxide nanosheets/polyethylenimine multilayer films as electrodes for supercapacitors. J. Power Sources 184(2): 695.

Zhang, X., X. Sun, Y. Chen, D. Zhang and Y. Ma. 2012. One-step solvothermal synthesis of graphene/Mn_3O_4 nanocomposites and their electrochemical properties for supercapacitors. Mat. Lett. 68: 336.

Zhang, Y., H. Li, L. Pan, T. Lu and Z. Sun. 2009. Capacitive behavior of graphene–ZnO composite film for supercapacitors. J. Electroanal. Chem. 634(1): 68.

Zhao, G.-Y. and H.-L. Li. 2008. Preparation of polyaniline nanowire arrayed electrodes for electrochemical supercapacitors. Micropor. Mesopor. Mat. 110(2–3): 590.

Zhao, X., B.M. Sanchez, P.J. Dobson and P.S. Grant. 2011. The role of nanomaterials in redox-based supercapacitors for next generation energy storage devices. Nanoscale 3(3): 839.

Zhao, X., L. Zhang, S. Murali, M.D. Stoller, Q. Zhang, Y. Zhu et al. 2012. Incorporation of manganese dioxide within ultraporous activated graphene for high-performance electrochemical capacitors. ACS Nano 6(6): 5404.

Zheng, J.P., P.J. Cygan and T.R. Jow. 1995. Hydrous ruthenium oxide as an electrode material for electrochemical capacitors. J. Electrochem. Soc. 142(8): 2699.

Zheng, J.P. and T.R. Jow. 1995. A new charge storage mechanism for electrochemical capacitors. J. Electrochem. Soc. 142(1): L6.

Zheng, J.P., J. Huang and T.R. Jow. 1997. The limitations of energy density for electrochemical capacitors. J. Electrochem. Soc. 144(6): 2026.

Index